Hans Lengerer is an acknowledged authority on the Imperial Japanese Navy. A entitled *Imperial Japanese Warships Illustrated* was published between 2016 a *The Aircraft Carriers of the Imperial Japanese Navy and Army* in January 2019, of the Imperial Japanese Navy 1868–1945 in 2021. He is coauthor, with Lars A *Ships of the Imperial Japanese Navy 1868–1945*, of which Vols I and III have been published.

Peter Marland is a former RN Weapon Engineer Officer with sea jobs in HM Ships *Blake*, *Bristol* and *Euryalus*, and postings ashore in research and in procurement. He subsequently worked as an Operational Analyst, and has contributed a number of articles to *Warship* on postwar Royal Navy weapons and electronics. He is a Chartered Engineer, with post-graduate qualifications in Project Management and in teaching.

Stephen McLaughlin retired in 2017 after working for 35 years as a librarian at the San Francisco Public Library. In addition to contributing regularly to *Warship*, he is the author of *Russian and Soviet Battleships* (US Naval Institute Press, 2003) and is co-editor of an annotated version of the controversial *Naval Staff Appreciation of Jutland* (Seaforth Publishing, 2016).

Kathrin Milanovich has been researching the history of the Imperial Japanese Navy and has contributed a number of recent articles to *Warship*.

Dirk Nottelmann is a marine engineer by profession, and is currently working for the German shipping administration. He has contributed to *Warship*, *Warship International* and various German magazines, as well as being author of *Die Brandenburg-Klasse* (Mittler, 2002) and co-author of *Halbmond und Kaiseradler* (Mittler, 1999). His book *The Kaiser's Cruisers* (with Aidan Dodson) was published by Seaforth Publishing in late 2021.

Jan Radziemski is a lawyer who for many years held managerial positions in large Polish and multinational corporations in the telecommunication sector. A naval enthusiast for over fifty years, he has one of the largest private libraries in Poland on naval subjects and is a regular contributor to leading Polish naval and maritime magazines. His specialism is the history of the Soviet Navy, and he has written some 150 articles on the subject.

Conrad Waters is the author of numerous articles on modern naval matters and editor of *World Naval Review* (Seaforth Publishing). He also edited Seaforth's *Navies in the 21st Century*, shortlisted for the 2017 Mountbatten Award. His history of the Royal Navy's Second World War 'Town' class cruisers was published in 2019 and he is currently writing a sequel on the *Fiji* class and their successors.

Michael Whitby is Senior Naval Historian at the Directorate of History and Heritage, National Defence Headquarters, Ottawa. He has published widely on Second World War and Cold War naval history, including co-authoring the official histories of the Royal Canadian Navy in the Second World War, and editing *Commanding Canadians: The Second World War Diaries of Commander AFC Layard, RN* (UBC Press, 2007) and *50 North: An Atlantic Battleground* (Lewin of Greenwich Organisation, 2021).

WARSHIP 2022

WARSHIP 2022

Editor: **John Jordan**

Assistant Editor: **Stephen Dent**

OSPREY
PUBLISHING

Title pages: The war-built frigate HMS *Grenville* is seen here in 1973, sporting an early Type 1030 antenna fitted for trials purposes. Peter Marland's feature article on postwar radar developments in the Royal Navy is published on pages 98–114 of this year's annual. (C & S Taylor)

OSPREY PUBLISHING
Bloomsbury Publishing Plc
Kemp House, Chawley Park, Cumnor Hill, Oxford OX2 9PH, UK
29 Earlsfort Terrace, Dublin 2, Ireland
1385 Broadway, 5th Floor, New York, NY 10018, USA
E-mail: info@ospreypublishing.com
www.ospreypublishing.com

OSPREY is a trademark of Osprey Publishing Ltd

First published in Great Britain in 2022

© Osprey Publishing Ltd, 2022

All rights reserved. No part of this publication may be reproduced or transmitted in any form or by any means, electronic or mechanical, including photocopying, recording, or any information storage or retrieval system, without prior permission in writing from the publishers.

A catalogue record for this book is available from the British Library.

ISBN: HB 9781472847812; eBook 9781472847829; ePDF 9781472847836; XML 9781472847805

22 23 24 25 26   10 9 8 7 6 5 4 3 2 1

Cover design by Stewart Larking
Page layout by Stephen Dent
Printed and bound in India by Replika Press Private Ltd.

Osprey Publishing supports the Woodland Trust, the UK's leading woodland conservation charity.

To find out more about our authors and books visit www.ospreypublishing.com. Here you will find extracts, author interviews, details of forthcoming events and the option to sign up for our newsletter.

CONTENTS

Editorial	6

Feature Articles

THE BEGINNINGS OF SOVIET NAVAL POWER: THE 1927 FLOTILLA LEADERS — 8
Przemysław Budzbon and Jan Radziemski describe the large, fast *Leningrad* class, built by the Soviet Union to counter modern destroyers then under construction for the navies of Poland, Romania and Turkey.

THE CHALLENGE OF OPERATION 'TUNNEL', SEPTEMBER 1943 – APRIL 1944 — 29
Michael Whitby analyses the series of Channel 'sweeps' that the Royal Navy undertook during 1943–44.

THE IJN CARRIERS *SŌRYŪ* AND *HIRYŪ* — 47
Kathrin Milanovich looks at the design of these medium fleet carriers in the context of the London Treaty.

THE DEVELOPMENT OF THE SMALL CRUISER IN THE IMPERIAL GERMAN NAVY PART III: THE GUNBOATS — 63
Dirk Nottelmann continues the story of the development of the German *Kleiner Kreuzer* by looking at the development of German gunboats, the tale of which has rarely been told.

THE BATTLESHIP *JAURÉGUIBERRY* — 80
Philippe Caresse looks at the origins and service career of this unusual vessel, generally regarded as the most successful of the early battleships of the 'Fleet of Samples' (*Flotte d'échantillons*).

POSTWAR RADAR DEVELOPMENT IN THE ROYAL NAVY — 98
Peter Marland gives an account of the evolution of radar in the Royal Navy during the postwar era.

AFTER THE *SOVETSKII SOIUZ*: SOVIET BATTLESHIP DESIGNS 1939–1941 — 115
Stephen McLaughlin describes the course of the follow-up design studies that followed the commencement of construction of the *Sovetskii Soiuz* class but which ceased with the German invasion of 22 June 1941.

THE GENESIS OF YOKOSUKA NAVY YARD — 132
Hans Lengerer puts the development of the major IJN dockyard at Yokosuka into its historical context.

***ESPLORATORI* OF THE *REGIA MARINA*, 1906–1939** — 147
Enrico Cernuschi looks at the development of the Italian 'scout' and its influence abroad.

MODERN EUROPEAN FRIGATES — 161
Conrad Waters assesses how recent design trends are reflected in the latest generation of European frigates.

THE AUSTRALIAN *BATHURST*-CLASS MINESWEEPER CORVETTE — 178
Mark Briggs studies the the largest group of warships ever built in Australia and the role they played in the revival and expansion of Australian shipbuilding.

C 65 *ACONIT*: FRANCE'S PROTOTYPE OCEAN ESCORT — 191
John Jordan looks at the origins and key systems of a ship inspired by the US Navy's 'ocean escorts'.

Warship Notes	198
Reviews	205
Warship Gallery	219

Aidan Dodson presents a selection of photographs documenting the scrapping of three iconic British warships.

EDITORIAL

The lead article in this year's *Warship* is an account of the development, construction, trials and service of the Soviet 'flotilla leaders' of the *Leningrad* class by Przemysław Budzbon and Jan Radziemski, whose article on the early Soviet submarines we published in 2020. It is a cautionary tale that demonstrates the importance of continuity in naval construction. The Russian revolution swept aside not only a political system but established naval force structures and procedures. By purging the upper and middle classes of counter-revolutionary influences, the Soviets lost at a stroke an accumulation of knowledge and expertise in ship design and construction that then became difficult, if not impossible to recover. When in the mid-1920s, faced with a Europe hostile to Russian communism, they saw the need to recreate powerful, sophisticated military forces incorporating the latest technology for self-defence, they ran up against huge infrastructure problems resulting from the virtual collapse of the military industries. Constructors with limited experience of ship design were confronted with the need to deliver what had been decided by committees that often had unrealistic expectations, in a climate of fear and paranoia. It was therefore unsurprising that priorities became distorted, and that meeting one particularly demanding requirement – in this case high speed – came at the expense of other essential qualities. Soviet ambitions for the construction of a fleet of new, technically advanced ships were in many ways admirable, but this 'brute force' approach to achieving them was always doomed to failure. Nevertheless, as the authors of the article point out, the Soviets learned from their mistakes, and the flawed designs that resulted from these early programmes formed the basis for the more successful ship types that followed.

Our second article by Canadian historian Michael Whitby also focuses on painful lessons learned, but in a totally different context. Operation 'Tunnel' was the codename for a series of Channel sweeps by the British Royal Navy during the Second World War targeting German maritime traffic off the French coast. Conducted by a mixed force generally comprising a light cruiser, Fleet and 'Hunt'-class escort destroyers and opposed by a uniform group of fast, well-armed torpedo boats referrred to by the Admiralty as 'Elbings', the sweeps resulted in numerous failures and the occasional disaster, the most prominent of which was the loss of the light cruiser *Charybdis* in October 1943. For the author, the essential weakness of the British operations was the *ad hoc* nature of the force compositions and the lack of continuity of command. Once the Admiralty reinforced Plymouth with the types of warship considered necessary for a permanent strike force and intensive training took place in which formations and suitable tactics were evolved, the German torpedo boat force was eliminated, virtually ending the coastal traffic off the northern coast of Brittany.

This year's coverage of the Imperial Japanese Navy features two contrasting articles, one technical, the other historical. Kathrin Milanovich looks at the design of the IJN medium fleet carriers *Sōryū* and *Hiryū*, conceived in the shadow of the London Naval Armaments Treaty of 1930, while Hans Lengerer begins a two-part account of the development of Yokosuka Navy Yard. The latter article highlights the influence of the European powers, and in particular France, in these early years, when Japan was on the cusp of modernisation both in terms of her political organisation and military/industrial infrastructure. *Sōryū* and *Hiryū*, on the other hand, were distinctively Japanese in conception, and broke with western carrier design in a number of important respects; they proved remarkably successful in service and, in addition to providing the basis for the later *Shōkaku* and *Zuikaku*, arguably the best carriers of the early years of the Pacific War, spawned a new generation of war-built ships (the *Amagi* class).

Four of this year's feature articles are a continuation of established series on historical warships. Dirk Nottelmann continues his series on German 'Small Cruisers' (*Kleiner Kreuzer*) with a study of ships initially classified as 'gunboats' but which eventually merged with the light cruiser category. In the process he examines the history (and the validity) of the 'gunboat' classification. From a slightly later period, Philippe Caresse continues his series on the French *Flotte d'échantillons* with a focus on *Jauréguiberry*, the most successful of the early battleships of the 1890 programme. *Jauréguiberry* marked a break from earlier French practice in a number of respects. Designed by the innovative naval architect Amable Lagane, she was the first French battleship to be built in a private shipyard, the first major unit to have electrical, as opposed to hydraulic training for her main guns, and the first to have her secondary battery of 138.6mm guns mounted in twin turrets.

Peter Marland continues his study of Royal Navy postwar weapons and electronics with an account of radar development from 1945 to the present day. It is a story punctuated by major successes followed by failures and dead ends, and Peter's comprehensive analysis investigates the reasons behind this halting progress. Finally, Stephen McLaughlin continues his series on the Soviet battleship designs of the prewar and early war period with a study of the in-house projects intended to follow the four 'super-battleships' of the *Sovetskii Soiuz* class. Similar in size and overall capabilities to *Sovetskii Soiuz*, Projects 23*bis* and 24 would have ironed out some of the 'rough edges' of the latter ships by simplifying the protection system (and the manufacture of the armour plates – a major issue given the USSR's embryonic military/industrial infrastructure). The Navy also wanted an increase in

EDITORIAL

The Armstrong-built second class cruiser *Takasago* anchoring off the south coast of the UK (possibly Portsmouth) in 1898, shortly before she made the passage to Japan. *Takasago* and her two US-built half-sisters *Kasagi* and *Chitose* are the subject of an article by Kathrin Milanovich to be published in next year's *Warship*. (Hans Lengerer collection)

speed from 28 to 30 knots (which proved problematic) and a doubling of the HA battery from eight to sixteen guns (which was achieved, albeit by sleight of hand).

Enrico Cernuschi gives us an illuminating account of the lightly-protected 'scout' concept pioneered by the Italian Navy as the *esploratore* and its influence on other navies, in particular the French, during the interwar era. It was a concept that made sense during the battleship era, when a battle fleet had only a vague idea of the location and strength of the enemy main body, but which was rendered obsolete by the rise in air power during the 1930s. The need to operate in the face of potential enemy air (and missile) attack is likewise a key characteristic of the latest European surface combatants that form the subject of a detailed study by Conrad Waters, who looks at the latest developments in propulsion, weaponry and electronics and the trend towards increased size, modularity and mission flexibility, while Mark Briggs' article on the 'minesweeper corvettes' of the *Bathurst* class demonstrates how, under the pressure of war, Australia was able to expand its shipbuilding capacity and produce vessels capable of a range of local missions in impressive numbers in a variety of locations.

Our feature section this year concludes with a short article by the Editor on the French anti-submarine 'corvette' *Aconit*. Designed as a sophisticated modern 'ocean escort' on the pattern of the US Navy's *Bronstein* and *Garcia* classes, *Aconit* introduced a range of weaponry of French design and manufacture that would become standard on French warships of the 1970s and 1980s. This article is the first in a series on major French postwar ship designs that will begin with a study of the aircraft carriers *Clemenceau* and *Foch* which is scheduled to be published in *Warship 2023*.

Warship Notes this year features an unusual contribution by regular contributor Enrico Cernuschi on the medallions, in gold, silver or cheaper metals that were traditionally minted to commemorate Italian warships and acquired by those serving in them or presented to officers' wives, who often wore them on formal occasions, while Kenneth Fraser continues his series on warship names by looking at 'national' names in the Royal Navy. We continue to review major naval books published during the past year, and this year's Gallery has a selection of photographs documenting the breaking up of three iconic British warships, HM Ships *Agincourt*, *New Zealand* and *Princess Royal* at Rosyth Naval Dockyard in 1923–1925.

Besides the article on the French postwar carriers, next year's *Warship* will include features on the German Flak ship conversions of the Second World War (Aidan Dodson and Dirk Nottelmann), the Italian 26,500-ton battleship designs of the early 1930s (Michele Cosentino), the IJN 8in-gun protected cruisers of the *Takasago* class (Kathrin Milanovich – see accompanying photo), the fate of the 12in guns of the Russian WWI battleship *General Alekseev* (Sergei Vinogradov), the 1936 Ansaldo UP.41 battleship design for the Soviet Union (Stephen McLaughlin), and postwar electronic warfare in the Royal Navy (Peter Marland). Philippe Caresse will continue his series on the French battleships of the *Flotte d'échantillons* with a study of *Masséna*, and Hans Lengerer will conclude his two-part article on the development of Yokosuka Navy Yard.

John Jordan
March 2022

THE BEGINNINGS OF SOVIET NAVAL POWER:
THE 1927 FLOTILLA LEADERS

During the period 1926–29 a number of hypothetical security threats emerged on the western and southern borders of the Soviet Union, leading to a major new military programme that included a class of large, fast flotilla leaders intended to counter the modern destroyers being built in French, British and Italian shipyards for Poland, Romania and Turkey. The resulting ships of the *Leningrad* class, which are the subject of a detailed study by **Przemysław Budzbon** and **Jan Radziemski**, encountered major technical problems due to the embryonic state of development of Soviet military infrastructure.

The backbone of the Soviet destroyer force after the Civil War was constituted by the *Novik* series.[1] Based on a German-built prototype, they proved to be excellent ships in 1916 but after the Great War were rapidly outclassed by newer foreign destroyers. It was considered that a ship with greater firepower at the head a flotilla of *Novik*-type destroyers was required to give them support when breaking through the defensive line – a concept similar to that of the British flotilla leader. This was combined with a fascination with the tactical value of speed that was very much typical of the era.

When the Central Maritime Commission for Determining the Priority of the Restoration of the Navy was established in October 1921, these views were reflected in a proposal for an 'Improved *Novik*' capable of 40 knots with an armament of six 130mm (5.1in) guns and twelve torpedo tubes. The baton was taken up in November 1923 by the Naval Scientific-Technical Committee (NTKM), assembled from the prominent shipbuilding experts who remained in the country after

The destroyer *Petrovskii*, one of the later destroyers of the *Novik* series (*Kerch* class), in the 1930s. Laid down in 1915 at Nikolaev as *Korfu*, she was launched in 1924 and commissioned with the Naval Forces of the Black Sea in 1925. Following the disgrace of her patron she was renamed *Zhelezniakov* in 1939 and served successfully during the Second World War. *Petrovskii* was armed with four single 4in/45 guns and nine 450mm torpedo tubes in triple mountings. (Przemysław Budzbon collection)

THE BEGINNINGS OF SOVIET NAVAL POWER

Balance of Destroyer Forces in the Principal Naval Theatres (status on 31 December), As Seen by the Soviet Naval Planners

Baltic	1922	1923	1924	1925	1926	1927	1928	1929	1930
Naval Forces of the RKKA	7	4	3	3	5	7	10	11	12
Most likely enemies in active service + planned or building	2	2+4[1]	2	2	2+2[2]	2+2[2]	2+2[2]	2+2[2]	3+1[3]
Estonian Navy	2	2	2	2	2	2	2	2	2
Latvian Navy		0+4[1]							
Polish Navy					0+2[2]	0+2[2]	0+2[2]	0+2[2]	1+1[3]

Black Sea	1922	1923	1924	1925	1926	1927	1928	1929	1930
Naval Forces of the RKKA	0	0	1	1	3	3	4	4	5
Most likely enemies in active service + planned or building	2	2	2	0+2[4]	2	2+2[5]	2+2[5]	2+4[7]	6+4[8]
Romanian Navy	2	2	2	0+2[4]	2	2+2[5]	2+2[5]	2+2[5]	4
Turkish Navy								0+2[6]	0+4[8]

Notes:
1. Four 1,000-tonne destroyers planned by the Latvian General Staff in 1923 but not supported by the government during a vote in the Saeima (Latvian Parliament) in Dec 1923.
2. Two *Wicher* class under construction in France.
3. *Burza*, the second ship of the *Wicher* class under construction.
4. Two *Mărăști* class under refit.
5. Two *Regele Ferdinand* class under construction in Italy.
6. Two *Kocatepe* class under construction in Italy.
7. Two Romanian and two Turkish under construction.
8. Two *Tinaztepe* class and two *Kocatepe* class under construction in Italy.

1917 and headed by Pyotr Leskov, a former Rear Admiral in the Imperial Navy. Taking into account the conclusions of the Commission, the preliminary requirements for the so-called 'Type 1922 Fleet Torpedo Boat' were drafted: 4,000 tonnes normal displacement with a power plant comprising twelve boilers and four sets of geared turbines, a speed of 40 knots, and an armament of eight 130mm (in three twin and two single mountings), anti-aircraft guns and twelve torpedo tubes. A similar approach was adopted by the Operational Directorate of the Naval Staff, which in March 1925 issued requirements for a 'Large Torpedo Boat Destroyer' with the following characteristics: 4,000 tonnes displacement, a maximum speed of 40 knots, four 183mm (7.2in) guns plus two/four 4in or 5in HA guns, six 21in or 23in torpedo tubes, 100 mines, 20 depth charges, and a catapult capable of launching a floatplane fighter.

The current political climate was not conducive to the expansion of the Navy, which after the Kronshtadt mutiny of 1921 suffered a gradual curtailing of its independence. Following a number of setbacks in the bureaucratic struggle taking place in Moscow, the Navy failed to win acceptance for its warship construction programmes and in 1926 lost its status as a separate fighting service. It was deprived of all missions except direct support of the ground forces and lost its 'fleet' status, being renamed the Naval Forces of the RKKA (Red Army). On 26 November 1926 the Naval Construction Programme of the RKKA for 1926–32 was formally authorised by the Council of Labour and Defence (the central economic planning authority in Soviet Russia until the late 1920s, acronym STO). New construction included submarines, MTBs and patrol ships, but there was no mention of destroyers.

The 1927 War Scare

A new political crisis emerged in Poland when former Army Marshall Józef Piłsudski initiated a *coup d'état* against the new democratic government on 12 May 1926. Soviet-Polish relations rapidly deteriorated, while Poland strengthened its ties with Romania. In the summer of the same year the Soviet Union also attracted the hostility of the British government due to the interference of the Soviet-controlled Profintern (Red International of Trade Unions) in the British coalminers' strike. In December 1926 the *Manchester Guardian* broke a story about Soviet collaboration with the *Reichswehr*. These potential threats to the USSR were exacerbated by a series of further events in the following spring that were skilfully exploited by Stalin in his public pronouncements. These were:

- a Chinese government raid on the Soviet embassy in Beijing, and a purge of Communists in Shanghai and Canton by Chiang Kai-shek in April
- a raid by the British authorities on the Soviet trading company ARCOS and the severance of diplomatic relations by Prime Minister Stanley Baldwin in May
- the assassination of the Soviet ambassador to Poland in June 1927.

9

Later in June, the Politburo ordered the publication of a warning on the threat of war, and a special 'Defence Week' was declared. Propaganda that had already been initiated by the press ignited a wave of popular alarm. With the horrors and famine of the Civil War still fresh in their minds, the urban population rushed to buy consumer goods while the peasants – still the core of the Red Army – reacted unfavourably to calls to prepare for hostilities.

The war scare is nowadays perceived as a manipulated panic created to discredit Trotsky and to open the way for Stalin's seizure of power. With the country facing this fictitious external threat, it was easy to transform the legitimate indignation of the 'Left opposition' and the 'Right deviation' into a treacherous agitation. Freed from the customary restraints, Stalin pushed the already unattainable targets of the First Five-Year Plan (1928–1932) even higher. This was described as 'acceleration … of the sectors having defence significance'. It was in this new context that destroyers re-emerged as key elements of the naval programme.

1929 Addendum

By the mid-1920s the East European nations had begun to recover economically from the Great War, and most of their surplus resources were being invested in the renewal of their defence capabilities. These relatively small sums were augmented by French military loans to Poland and Romania that made possible modernisation and naval orders.

At the very beginning it was the qualitative advantage of new construction in the Baltic that stimulated planning. Despite this, the balance of naval power in the Baltic remained favourable to the Soviet Union – Germany was not considered a likely enemy at that time. However, the situation was very different in the Black Sea. The attention of the RKKA naval planners therefore became focused on the latter theatre, where the relative parity that currently prevailed was to be disturbed in the years to come. Romania ordered two modern destroyers in 1927, while in January 1928 the Turkish naval command was energised by the formation of the Undersecretariat of the Sea, and orders for new destroyers were placed in Italy.[2]

These events had an effect on the planning studies for new construction. The Director of Naval Operations issued requirements for a 'Black Sea Destroyer', which following an upgrade in the calibre of the main guns to 130mm (to counter the new Romanian boats, whose characteristics had become known) formed the basis for the May 1928 staff requirements of the RKKA. These were revised and finally approved on 1 November 1928 by the Director of the Naval Forces (*Namorsi*), Romuald Muklevich.

On 4 February 1929 the Council of Labour and Defence amended the 1926 Naval Programme, approving the construction of three destroyers for the Black Sea.

A Bumpy Road

The preliminary design was prepared by the NTKM under the leadership of Iulian Shimanskiy (since 1916 head of the team preparing workshop documentation for the *Novik* class at the Putilov Works) and was presented on 11 March 1929. Following submissions by the Soviet of the Naval Forces of the Black Sea, the specification was modified to include a speed of 42 knots, a range of 250nm at full speed, depth charge throwers and a catapult, on a displacement of 2,100 tonnes.

In June 1929 the destroyer programme was jeopardised by the decision of the Politburo to increase the size of the fully-mobilised RKKA from 2.6 million to 3 million men, 2,000 aircraft and 1,500 tanks. The only branch of the service to suffer would be the Navy, its budget being cut by more than 26 per cent. Fortunately for the project, the

Evolution of the Staff Requirements

	DNO Black Sea Destroyer	RKKA Staff Requirements	DNF Destroyer Leader
Date	Apr 1928	May 1928	Nov 1928
Displacement (normal)	1,750 tonnes	1,750 tonnes	2,000 tonnes
130mm guns	–	4–5	5
4in guns	8 (2xII DP, 2xII)	–	2
AA guns	12 – 37 mm (3xIV)	–	–
21in TTs	6 (2xIII)	6–9 (2xIII *or* 3xIII)	6 (2xIII)
DCs	15–20	–	–
Mines	60–80	60–80	–
Seaplane[1]	1 Ju-20	1 Ju-20	1 Ju-20
Speed	40 knots	40 knots	40 knots
Range at 40 kts	700nm[2]	700nm[2]	240nm

Notes:
[1] The Junkers Ju-20 seaplane was to compensate for the weakness of the Soviet cruiser force.
[2] Range at full speed was to allow strikes on the Bosporus and Constanta.

Stalinist armaments programme was expanding all the time, and funds were restored six months later.

For the second time the project avoided the threat of cancellation by a hair's breadth when a wave of arrests swept through the NTKM Technical Department in April 1930 in response to the failure of the design of the *Uragan* class. The head of the department, Nikolai Vlas'ev, and the commanders of the key sections became victims. There is no evidence relating the timing of these repressive measures to the completion of the preliminary design of the new destroyers; nevertheless, it is puzzling. Muklevich eagerly approved the project within two weeks and placed the formal order for the technical design with the *Soyuzverf* (All-Union Shipbuilding Industry Association), with a deadline fixed for 1 November 1930. Four ships were to be completed within eighteen months following approval of the design.

On 13 June 1930 the *Revvoyensoviet*, the Revolutionary War Soviet and the highest Soviet military authority, which was headed by Kliment Voroshilov (People's Commissar of Military and Naval Affairs), added three destroyers for the Baltic and three for the Black Sea to the naval construction programme, and formally approved the project while putting the 42-knot speed proposal to one side.

The execution of the technical design was commissioned from the team with the greatest experience in the design of combat ships in the Soviet Union at that time. It had carried out projects for the renovation and modernisation of the *Novik*-type destroyers, and from November 1926 was responsible for the design and construction of the first Soviet-built warships, the 'patrol ships' of the *Uragan* class. The team was assembled for the Special Ships (a euphemism for 'warships' in Soviet practice) Design Bureau (acronym: BSPS), established a few months earlier, using as its basis the Technical Bureau of the Severnaia Shipbuilding Yard (formerly the Putilov Works). The bureau was headed by Vladimir Nikitin; Pavel Trakhtenberg was to have overall responsibility for the design of the new destroyers, with Anatoliy Maslov supervising the hull and Aleksandr Sperianski the machinery compartments.

The 1929 NTKM Project

Specification	1929 NTKM design
Date approved	July 1930
Displacement (normal)	2,100 tonnes
130mm guns	5
37mm AA guns	4
0.5in MGs	4
21in TTs	6 (2×III)
Seaplane	1 Ju-20
Catapult	1
DCs	20
Mines	80
Speed (6h sustainable)	40 knots

Voroshilov's approval and the fact that the order for the technical design had already been placed did not prevent further tinkering with the requirements. Two months later the Staff of the RKKA again raised the issue of a 42-knot speed. With displacement already fixed at 2,100 tonnes, weight reductions were proposed for the hull and fittings with a view to allocating the savings to the machinery. Mindful of the tight timeline for the design work, Muklevich proposed restricting the maximum speed to 40 knots for the first series of ships, and to return to the question after experience had been gained.

The project documentation for the destroyer, specifying a normal displacement of 2,250 tonnes and a speed of 40.5 knots, was delivered on time by the BSPS and by the *Revvoyensoviet* on 7 December 1930. The first three ships (Series I) were to be built at the Black Sea shipyards within 22 months (*ie* by October 1932) – a decision quickly amended by Voroshilov, who reallocated one of the ships to the Severnaia Shipbuilding Yard in the Baltic. This yard was to be the lead yard for the project and was to be responsible for completion of the prototype.

The proposed construction schedule was completely unrealistic given the sophistication of the new destroyers

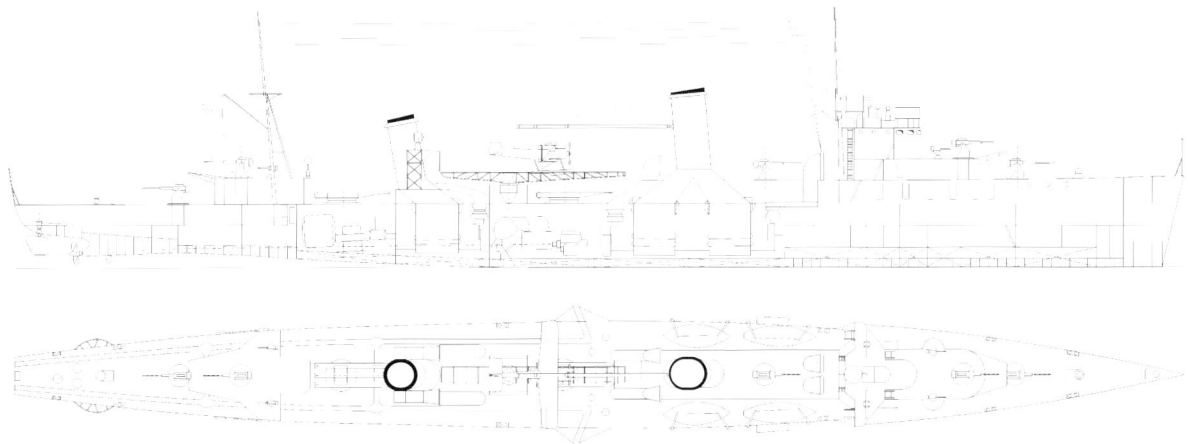

General Arrangement of the 1929 NTKM Black Sea Fleet destroyer. (Drawn by Jarosław Dzierżawski)

„УРАГАН" (проект 2) ГУСП-ЦКБС №1 ЧЕРТ. № АС-770

The project number of drawing No AS-700. A close inspection reveals that there is a figure '1' inside the '2' (see text).

and the precarious state of industrial development in the Soviet Union at that time. However, according to Stalinist doctrine ambition in planning was no vice, while moderation was not a virtue. Therefore, despite the chronic scarcity of resources, unfeasibly tight deadlines were imposed on all the projects. Ignorant for the most part of the technical constraints, the upper tier of 'Red' management had only a limited selection of instruments in its toolbox, so reorganisation seemed the only remedial measure available. In order to secure timely delivery of the destroyers, the BSPS was reorganised to include engineers taken from the technical bureaux of the Baltic and Andre Marti Yards, plus the 'survivors' from the Severnaia Yard. The new organisation was established on 18 January 1931 and named the Central Construction Bureau of Special Shipbuilding (acronym TsKBS), with Nikitin as the Chief Engineer. Its most significant and urgent task was the preparation of the detailed plans for the Series I destroyers.

Proekt 1

The destroyer design was allocated the code *Proekt 1* (Project 1).[3] The authors have often wondered why it was the destroyer project that was assigned number 1 in 1931, while the *Uragan* project begun in 1926 by the same team became *Proekt 2*. The solution to this puzzle turned out to be 'political'. The design of the *Uragan* class – the first warships be designed and built in the Soviet Union – had the code *Proekt 1* allocated in the Technical Bureau of the Severnaia Shipbuilding Yard in 1926, and when the project was reassigned in the summer of 1930 to the newly created BSPS it retained this number. However, sea trials of the lead ship of the class that took place in the late autumn of that year revealed very significant design and workmanship flaws, the most serious of which was the failure to meet the contractual speed of 29 knots by three knots. An error that in normal circumstances could mean the loss of a job, in Soviet conditions could literally mean the loss of a head. The team of designers of the Series I submarines (D class) had been confined to the dungeons of OGPU (the secret police, later known as the NKVD, then the KGB) for five months – see the authors' article in *Warship 2020*.

Nikitin and his team happily avoided imprisonment. It is possible that the lack of speed in these relatively minor vessels was of less consequence in the eyes of OGPU than the suspected lack of stability in the submarines. Another mitigating factor was that the arrest of the project team responsible for the *Uragan* class would have resulted in a serious delay to the destroyer project; this would then be deemed a 'wrecking' measure for which OGPU would be held responsible.

It was in this paranoid atmosphere that the BSPS was reorganised as TsKBS. Nikitin lost his job as head of the team and had to settle for the position of chief engineer. The new management team of TsKBS had no intention of launching their careers with a project generally regarded as a failure, so the decision was made to allocate the *Proekt 1* designation to the new destroyers, while the *Uragan* class project became *Proekt 2*. However, there was someone in the design team who was determined that this change should be recorded for posterity, given that it did not reflect the reality of the situation. At that time the drawings were made on technical tracing paper, so any change involved scratching a thin layer of tracing paper to remove the previous content and applying a new layer in such a way that the change was invisible. However, in the drawing of the longitudinal section in the General Arrangement plan of the *Uragan* class (No AS-700), the draughtsman left the original number '1' visible, surrounding it with the outline of the new designation, the number '2'. (The authors wish to thank their Russian friends for pointing out this detail.)

However, the success of the new destroyer design had become a serious matter. Speed, which might not have been crucial for the escorts, was the most important tactical element of the destroyer for the military who, having reluctantly relinquished the excessive requirement of 43 knots, insisted on a speed of 40.5 knots. Thus, on the one hand Trakhtenberg pressed Maslov to fine the lines of the hull to minimise drag at high speeds, while at the same time Sperianski worked on the power plant to achieve the desired output and to secure a margin that might prove to be a lifesaver for all the team members in the event of errors in the calculations.

The Race for Speed

The hull lines were strictly subordinated to minimising drag, regardless of the impact on manoeuvrability or technological difficulties. The slender hull (L/B ratio: 10.9) had a spindle-shaped underwater form with sharply-tapered ends, and the wing shafts were within streamlined skegs so as to minimise turbulence.

Happily, the testing tank at Leningrad was still in operation, and the knowledge of flow hydrodynamics had not been lost. The former head of the tank Aleksei Krylov, the internationally-renowned mathematician and theoretician of shipbuilding, survived the atrocities of the Civil War and was still active. In total 45 different hull models were tested. This was the reason Maslov was strongly opposed to taking into account the suggestion of the Soviet naval mission, which was able to gain insight into Italian construction. The Italian scout *Antonio da Noli* of the *Navigatori* class, which the mission was able to inspect in September 1930, had the classic rounded

THE BEGINNINGS OF SOVIET NAVAL POWER

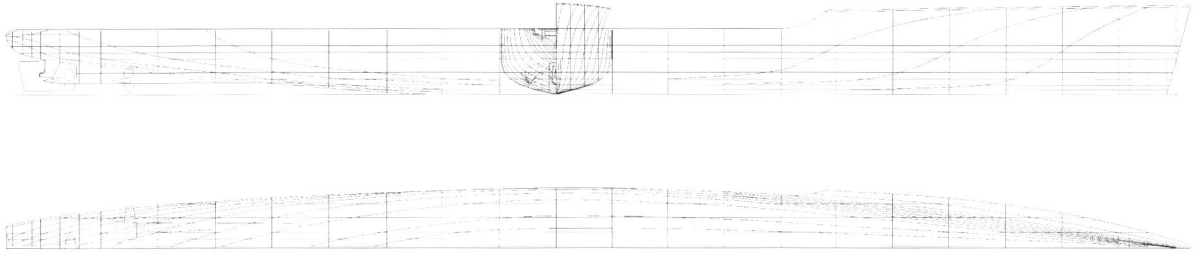

The body plans for No AS-38 were signed by Trakhtenberg and Maslov. One can only imagine how much courage it took to sign this drawing. (Drawn by Jarosław Dzierżawski)

stern and shafts supported by 'A' brackets. The adoption of a similar solution for the new Soviet destroyers would have had the effect of simplifying construction of the stern, while at the same time reducing technological complexity and construction costs. However, the increased costs were irrelevant in the context of the Soviet economy; moreover, this aspect was outside Maslov's area of responsibility.

The power plant, which was in theory capable of delivering 66,000shp, comprised three boilers and three sets of turbines driving three shafts in a unit arrangement. The two forward boiler rooms, which were in line in separate compartments, supplied steam for the geared turbines driving the wing shafts, which were located side by side in a midship engine room in the broadest part of the hull, while the third boiler room supplied steam for the after set of turbines driving the centre shaft (see the General Arrangement plan).

The boilers were rated at 21.5kg/cm^2 (310psi) with a maximum steam temperature of 335°C, and the unit weight of the machinery as planned was 8.8kg/hp. These figures are comparable with those associated with Italian or French vessels of the period, but Russian lack of experience in the design of turbines and high-performance marine boilers was to prove a major obstacle. An additional factor was the lack of suitable materials and technology for the production of modern components, which would hamper efforts to achieve the designed performance. Unlike hull construction, in which the Russians already had experience, the power plant for the new destroyers posed a significant design and construction risk. The only way to reduce this risk was to build in as high a power margin as possible, possibly 20 per cent or more. From an economic point of view such an approach was wasteful but, as already stated, cost was irrelevant given a centralised economy, and this aspect was outside Sperianski's area of responsibility.

The game was played on many levels: on the one hand to provide ships with the required characteristics, on the other to reduce the stringent performance requirements. In March 1931, the new *Namorsi* Vladimir Orlov informed Voroshilov that, with 60 per cent of the hull documentation and 90 per cent of the machine documentation now complete, it could be concluded that achieving the required speed of 40.5 knots was unrealistic; any changes would involve a significant increase in

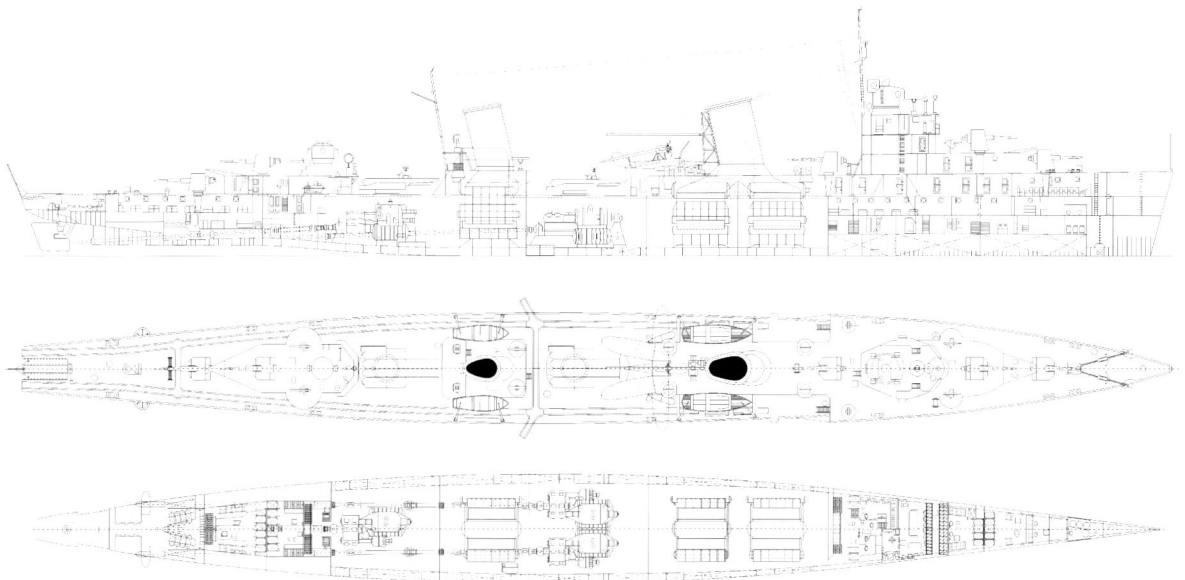

The General Arrangement plan of *Proekt 1* in June 1932. A seaplane still featured at this phase, but without the catapult, which was suppressed. The lower drawing shows the arrangement of the machinery. (Drawn by Jarosław Dzierżawski)

the machinery output and thereby extend the timetable for the design process. He proposed reducing the speed figure to 40 knots; Voroshilov gave his consent.

Six Years to Come

At the peak of the design process in January 1932, a reorganisation swept through Soviet industry which, to the surprise of the Politburo, proved to be incapable of delivering unrealistic results. To Stalin strict centralisation appeared to be the best solution. The Supreme Soviet of the National Economy (acronym VSNKh) was dissolved as well as its branches in the republics, together with the Trusts that governed the various enterprises. In their place there were to be three ministries (designated 'People's Commissariats' in Soviet parlance; acronym *Narkomat*), to be responsible for the following: heavy industry, light industry, and forestry (the sale of timber was a key source of foreign exchange). The shipbuilding industry became subordinated to the *Narkomat* of heavy industry, headed by Stalin's personal friend, Georgian Sergo Ordzhonikidze. Two more shipbuilding design teams were formed, so TsKBS was renamed TsKBS-1.

The *Proekt 1* workshop documentation was ready in June 1932 and approved by the Technical Directorate of the Naval Forces. The ships were to be named after the largest Soviet cities, the capitals of the republics of the union, and were initially designated the *Moskva* class. They were to be classified as 'destroyer leaders'.[4]

Khar'kov and *Moskva* were laid down on 19 and 29 October 1932 respectively at the Andre Marti Works at Nikolaev. Greater ceremony was accorded to the keel laying of *Leningrad*, which as the prototype was to become the name-ship for the whole class. This took place at the Severnaia Shipbuilding Yard on 5 November 1932, on the eve of the 15th anniversary of the Bolshevik revolution, with Politburo member Sergei Kirov driving in the first rivet. The hull was launched a year later on 17 November 1933, and the fact that the launch did not take place ten days earlier (the date of the following anniversary) suggests that the construction of the ship was not seen at that time as a success. Although the hull was now in the water, the delivery of the propulsion machinery, armament and other systems was drastically delayed. Even the word 'delay' is euphemistic, as most of the afore-mentioned systems existed only as a range of prototypes, and construction of the hulls of *Leningrad*'s two sisters had been suspended for some time.

Delivery of the turbines was two years late. The first problem was a 90 per cent rejection rate of casings supplied by the Baltic Works; then the turbine blades proved to be defective because the workshop's measuring gauge was never properly calibrated before use and proved to be inaccurate. Some key components had been ordered abroad, and when delivered on time were stored in the open air, so that when the time came for them to be fitted they suffered from oxidisation. Even when finally complete, the turbine sets had to wait for the test equipment, which was acquired too late.

The construction of the boilers encountered equally serious setbacks. Manufacturers of the thin-walled forged steam collectors did not exist in the Soviet Union, so it was necessary to order them from Germany. However, the time spent looking for a local manufacturer and then undergoing bureaucratic import procedures was lost forever. Boiler testing began only in 1933 and lasted for almost a year, during which time the designers worked hard to stabilise the water flow in the feed system. Thus, at a time when the boilers had become the most serious cause of delay, the decision was made to put them into production in their current, precarious state of development.

Initially it was planned to import steel for construction of the hulls. Finally, local production of a manganese alloy steel was used, but when worked into plates and frames it tended to crack. The duralumin used for the deckhouses and some less critical parts of the hulls led to *Leningrad* being nicknamed the 'silver ship'. The aluminium smelters had supplied sheets that had already been rejected by the aircraft industry, and the shipyard had no experience in storing and handling such components, so corrosion developed. A similar story was repeated with almost every component or device.

The weapons that the ships were due to carry did not exist at that time. Since 1930 the Bolshevik Works (former Obukhov) had been working on the development of a new version of the 130mm/55 Model 1913 gun, shortened to 45 calibres. However, in May 1932 it was decided to increase barrel length to 50 calibres and to introduce substantial changes. Unsurprisingly, the B-13 prototype failed its trials two years later. Incomplete guns passed manufacturer's trials only in April 1935 despite the barrel life not exceeding 150 rounds. Time was now pressing to such an extent that, without waiting for further tests, twelve Model B-13 Series I guns were ordered by the Directorate of the Naval Forces. Even so, these were delivered too late for installation in *Leningrad*, which on completion was armed with five B-7 type 130mm/55 guns, a modification of the 1913 model.

The 3in model 34-K HA mountings were not ready, while the 37mm 11-K guns (planned as a re-barrelled version of the Vickers 2pdr Mk II delivered to Russia during the Great War) did not enter series production at all, nor did the 20mm 2-K light AA guns. Thus, the anti-aircraft armament initially comprised two elderly 76mm HA 'Lender' guns[5] plus two of the only modern weapon available on time: the 45mm 21-K AA gun (an enlarged copy of the Rheinmetall 37mm anti-tank gun purchased from the *Reichswehr* in the early 1930s). Four 0.5in DK machine guns completed the firepower of the ship; in place of torpedo tubes, mock-ups were mounted.

As there was no opportunity to design and manufacture a fire control system by Elektropribor (a company established from the remnants of the Geisler and Ericsson teams), delay was avoided by placing an order in Italy for the fire control computers, torpedo control gear and FC directors, the contract being signed in 1931. Four sets were delivered by the Italian company Officine

Building Dates

Name	Project No	Yard No	Builder	Laid down	Launched	Accepted[1]
Series I						
Leningrad	1	S-450	No 190	5 Nov 1932	17 Nov 1933	5 Dec 1936[2]
Moskva	1	S-223	No 198	29 Oct 1932	30 Oct 1934	–
			No 201	–	–	10 Nov 1938[3/4]
Khar'kov	1	S-224	No 198	29 Oct 1932	9 Sept 1934	–
			No 201	–	–	10 Nov 1938[3]
Series II						
Minsk	38	S-471	No 190	5 Oct 1934	6 Nov 1935	10 Nov 1938[5]
Kiev[6]	38*bis*	S-267	No 190	15 Jan 1935[7]	–	–
Ordzhonikidze[6]			No 199	10 Mar 1936[8]	25 Jul 1938	–
Sergo Ordzhonikidze[6]			No 202	–	–	27 Dec 1939[8]
Tiflis	38*bis*	S-268	No 190	15 Jan 1935[7]	–	–
			No 199	10 Aug 1936[8]	24 Jul 1939	–
Tbilisi[9]			No 202	–	–	11 Dec 1940[10]

Notes:
1. Date of formal signature of acceptance and official commissioning; exceptions noted.
2. Formal acceptance; left shipyard in July 1938.
3. Following fitting-out, towed to the delivery workshop at Sevmorzavod Yard in Sevastopol for completion.
4. Commissioning date according to S S Berezhoi, *Korabli i suda VMF SSSR 1928–1945* (Moscow 1988), was 10 Aug 1938, which may indicate that the ship was conditionally accepted and commissioned in August, while the final acceptance took place in November.
5. Commissioned 15 Feb 1939.
6. *Kiev* renamed *Ordzhonikidze* 25 Jul 1938, then *Sergo Ordzhonikidze* 27 Dec 1939, then *Baku* 25 Sept 1940.
7. Laying down date at the Andre Marti Yard in Nikolaev for the purpose of pre-assembly.
8. Keel laying at the Leninskii Komsomol Yard in Komsomolsk-on Amur.
9. *Tiflis* renamed *Tbilisi* 25 Sept 1940.
10. Completion at the Dal'zavod in Vladivostok.

Shipyard numbers were assigned on 30 December 1936 as follows:

Zavod No 190	Severnaia Sudostroitel'naia Verf' im A A Zhdanova, Leningrad; formerly the Putilov Works, renamed Severnaia in 1918, dedicated to Andrei Zhdanov in 1935.
Zavod No 198	Nikolaevskii Sudostroitel'nyi Zavod im Andre Marti, Nikolaev; formerly the Naval Yard, renamed Chernomorskii Sudostroitel'nyi Zavod in 1918, dedicated to André Marti in early 1930s.
Zavod No 199	Sudostriotel'nyi Zavod im Leninskogo komsomola, Komsomolsk-on-Amur; newly built shipyard.
Zavod No 201	Sevmorzavod im Sergo Ordzhonikidze, Sevastopol; formerly the Sevastopol Admiralty Yd, renamed Sevmorzavod in 1921, dedicated to Sergo Ordzhonikidze in 1936.
Zavod No 202	Dal'zavod im K E Voroshilova, Vladivostok; formerly the Mechanical Plant of the Vladivostok Naval Base, renamed Dal'zavod in 1919, dedicated to Kliment Voroshilov in 1931.

Galileo in 1933 and were to be the subject of thorough evaluation studies.

The idea of operating a seaplane was postponed, despite the Soviet-manufactured SPL-type flying boat being ready and accepted for service in 1935.[6]

In September 1934 the prototype *Leningrad* was ready to begin static trials. With her completion in prospect, construction of the hulls of *Moskva* and *Khar'kov* had been resumed and was sufficiently advanced to launch them within a few weeks. However, almost two years were to pass before *Leningrad* first went to sea. Some of this time was spent modifying the steering gear. On 25 August 1936, at a displacement of 2,411 tonnes, *Leningrad* easily attained her contractual speed of 40 knots with 61,500shp, while at the rated output of 66,000shp she achieved 41 knots in calm water. The fly in the ointment was the adverse effect on trim at these higher speeds; the bow lifted out of the water and the stern trimmed by 1.5 metres, increasing to 2 metres at 42 knots. Admitting 35 tonnes of water into the bow ballast tanks reduced the trim and speed increased to 42.5 knots.

The tendency of the bow to lift at high speed was a serious obstacle to firing the forward guns; nevertheless, the day was a success for Nikitin and his team. Speed was a fetish for the military, and one for which they were willing to make considerable sacrifices. During the acceptance trials on 5 November 1936 on the 60-metre deep Gogland Mile (off Suursaari Island) in Sea State 1–2, at a displacement of 2,225 tonnes and with 67,250shp, *Leningrad* achieved 43 knots without having exhausted the power output margin of 10–15 per cent. However, at 43 knots the quarterdeck and the depth charge chutes were completely submerged and the ship trimmed by 2.8 metres at the stern. The euphoric mood

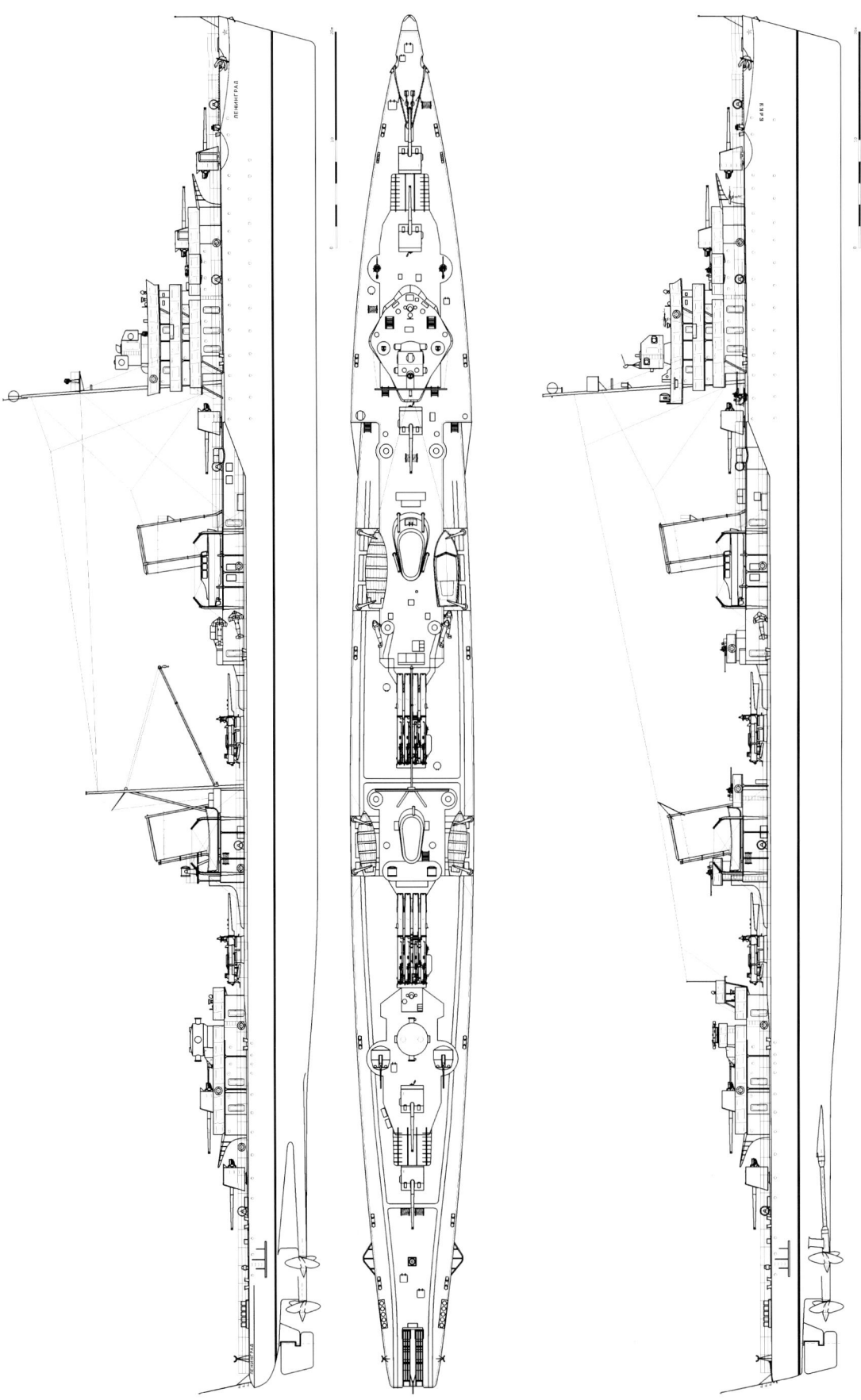

Proekt 1 *Leningrad* in 1938 (top) and Proekt 38-bis *Baku* (ex-*Sergo Ordzhonikidze*, ex-*Ordzhonikidze*, ex-*Kiev*) in 1944 (bottom). (Drawn by Tomasz Grotnik)

was not even punctured by the failure of the turbine gearing, which cracked because it had been welded to save weight.

There were too many other design and construction flaws to list here, although in mitigation it should be noted that *Leningrad* was virtually the first combat ship completely designed and built in Russia after a 20-year hiatus, and her designers and builders were therefore unable to benefit from the experience of older colleagues or existing documentation, nor from established standards of design or construction. The major issues were as follows:

- poor seaworthiness (shipping water, tendency to yaw in a beam sea)
- poor manoeuvrability at the full range of speeds
- considerable vibration amidships and aft above 35 knots
- hull too fragile to withstand sagging in heavy seas or the shock when all the guns were fired at once
- rapid corrosion of boiler tubes and turbine blades
- a turbulent waterflow in the boilers when all the 28 nozzles were on line
- severe pitting of the propellers due to cavitation.

For all this, the high-speed spell clearly enchanted the acceptance commission, which commended the vessel as 'a modern ship with good running and seakeeping qualities ... The achievement of 43 knots speed instead of the 40 knots stipulated in the contract should be specially noted.' These favourable remarks were attributed to a ship with poor manoeuvrability, outdated guns, mocked-up torpedo tubes and a fragile hull. However, the ability to focus on the positive (or the negative) to the exclusion of all else is a characteristic of authoritarian regimes. The ship was officially handed over to the Navy on 5 December 1936, the day Stalin's much-heralded 'constitution' for the Soviet Union was announced. Stalin received a worthy present, and the officials and the builders responsible for *Leningrad* received awards and bonuses. The reality was different.

The turn of the year was also marked by another stage in the progressive militarisation of the Soviet Union and the consolidation of the defence sector. The *Narkomat* of the Defence Industry (acronym NOP) was created, bringing together all companies producing for the needs of the RKKA. As part of this third consecutive reorganisation, TsKBS-1 was redesignated TsKB-17. The shipyards were assigned numbers instead of names, so the Severnaia Shipbuilding Yard became Shipbuilding Works No 190. And it was in the docks of Yard No 190 that *Leningrad* was to spend the next two years, during which time her defects were eliminated (or at least smoothed out) and the designed weapon systems installed. She finally left the shipyard in July 1938, six years after the completion of her workshop documentation. She was assigned to the Baltic Fleet – from 1937 the Navy regained its independent status from the Red Army, so the term Naval Forces was no longer in use. Within a few months her two sisters joined the Black Sea Fleet. Of these *Moskva* broke the speed record established by the prototype *Leningrad*, attaining 43.57 knots during trials. Reliable torpedoes were delivered only in 1939.

By this time most of the key individuals associated with

Leningrad in 1938 steaming at full speed; at higher speeds she had a marked trim by the stern. Note the prominent stern wake, evidence of the high wave resistance of the spindle-shaped hull. *Leningrad* has a single band on the first funnel. (Witalii Kostrichenko collection, courtesy of Tomasz Grotnik)

Results of Speed Trials

Name	Date	Displacement	Horsepower	Speed
Series I				
Leningrad	5 Nov 1936	2,225 tonnes	67,250shp	43.00 knots
Moskva	6 Apr 1938	2,330 tonnes	77,725shp	43.57 knots
Series II				
Minsk	Spring 1938	–	68,000shp	40.50 knots
Ordzhonikidze	2 Nov 1939	–	70,710shp	41.60 knots
Tbilisi	Sep 1940	–	–	41.80 knots

Note:
Figures for *Khar'kov* are not available.

Recognition Markings

Funnel bands on completion; hull markings introduced in 1939.

Name	Fleet	Hull markings Cyrillic	Latin equiv	Funnel bands 1st	2nd
Series I					
Leningrad	Baltic	ЛН	LN	narrow	–
Moskva	Black Sea	МС	MS	narrow[1]	narrow[1]
Khar'kov	Black Sea	ХР	KhR	narrow	–
Series II					
Minsk	Baltic	МН	MN	–	narrow
Sergo Ordzhonikidze	Pacific	ОЖ[2]	OZh[2]	wide	–
Tbilisi	Pacific	ТБ	TB	wide	wide

Notes:
[1] From mid-1939 two narrow bands on the first funnel and none on the second.
[2] ВЛ (VL) was carried during trials for disinformation purposes to suggest the ship's name was *Vladivostok*.

Moskva following completion in 1938. Like *Leningrad*, she was armed with the B-13 Series I guns in rectangular shields, which had a prominent box-shaped extension for the trainer on the right side – a feature rarely seen on A- and B-type mountings. Note the bands on both funnels. (Przemysław Budzbon collection)

THE BEGINNINGS OF SOVIET NAVAL POWER

Leningrad in 1939, with a prominent hull marking. Although a poor-quality newspaper print, the photo is nevertheless impressive. (Przemysław Budzbon collection)

Sergo Ordzhonikidze (ex-*Ordzhonikidze*, ex-*Kiev*) following completion in December 1939. She was armed with the B-13 Series II guns without the box extension for the trainer on the right side of the shields – see 'P' mounting. (Witalii Kostrichenko collection, courtesy of Tomasz Grotnik)

the project were dead: they had either committed suicide (Vlas'ev), been executed by firing squad, were missing without trace, or had been assassinated (Kirov and Ordzhonikidze). Apart from Stalin and Voroshilov, who were the directors of this terrible spectacle, only Nikitin and Maslov survived.

Proekt 38 and *Proekt 38bis*

The first Five-Year Plan ended in failure. When in 1932 the time came for a review, its authors claimed the exact opposite and trumpeted its success. By skilfully juggling the data it was concluded that the most important goals of the plan had been achieved; where it fell short 'wreckers' were blamed. The Central Statistical Board was accused of under-selling the data and was purged. Its successors provided a different set of figures and nobody dared to contest them. In a popular Moscow joke from the time, a zoo guide states, indicating a recently-acquired crocodile: 'He is five metres long from the tail to the head, and six metres from the head to the tail'. 'Why the difference?', asks one of the visitors. 'Go check', the guide replies.

This putative success served as an excuse for a six-fold increase in military expenditures in the second Five-Year

Leningrad in heavy ice; she is following icebreakers and is being assisted by a tug. The photo was taken during the Winter War with Finland, in late December 1939 or early January 1940. The box extension for the trainer is seen on the shields of 'X' and 'Y' mountings. Note the unusual configuration of the stern, the lower part of which was fined for speed. (Przemysław Budzbon collection)

Plan. The Red Navy duly received its share, not as generous as for the Army or Air Force, but sufficient to announce a 1933–38 expansion programme involving the construction of more than 1,500 ships and boats, including ten – the number was later reduced to eight – destroyer leaders. The programme, which was presented by *Namorsi* Orlov, was formally approved by the Soviet of Labour and Defence on 11 July 1933. In the absence of other projects, it was decided to repeat the existing design and at the same time seek foreign assistance for further construction.

The decision to embark on a repeat series was delayed until dock tests of *Leningrad* could begin and it was possible to confirm that the design was viable. Modifications were carried out on the basis of the experience gained during the construction of the prototype. The first was to the shape of the stern. The information obtained by the Soviet naval mission in Italy suggested that a fuller hull-form aft was required, and that the skegs be replaced by conventional shaft brackets. This view was reinforced by technological difficulties experienced during the construction of the hull of *Leningrad* caused by confined spaces in the whole stern area. Project modifications were carried out by TsKBS-1 early in 1934. The other changes were introduced in line with the progress in the fitting-out of *Leningrad*.

To avoid confusion in the workshop documentation, in 1935 Series II was assigned the official number *Proekt 38*. This change, which aimed to bring order to the flow of workshop documentation to the shipyard – both the Series I and the Series II prototypes were in hand at the time – was much appreciated by the TsBKS-1 team.[7]

Externally, in addition to the stern structure, the Series II ships differed from Series I in the following respects:

- small changes in the bridge layout (shape of screens and the companionway)
- curved shape of the blast screens on the bow and stern deckhouses

Minsk during a review in the Neva River in 1939 or 1940. Note the Series II gunshields without the box extension for the trainer. (Boris Lemachko collection)

- third 3in gun abaft 'X' mount (actually to be installed on *Tbilisi* only)
- funnel rails parallel to the upper edge of the funnel top.

The allocation of orders for the eight ships of Series II was planned to be as follows:

- Severnaia Shipbuilding Yard, Leningrad: four, including the prototype *Minsk* (laid down 5 October 1934)
- Andre Marti Yard, Leningrad: two ships.
- Andre Marti Yard, Nikolaev: two ships.

However, given the slow progress in the construction of *Leningrad* and the possibility of ordering modern ships from Italy, it was decided to limit the construction of Series II. Following the start of work on the two Black Sea leaders (*Kiev* and *Tiflis*) at the Andre Marti Yard in Nikolaev, it was decided to allocate them to the Far East, as since the Manchurian Crisis of 1932 Soviet forces there had undergone a major expansion. Construction of four further units was cancelled.

Minsk in 1940. (Przemysław Budzbon collection)

THE BEGINNINGS OF SOVIET NAVAL POWER

Tbilisi after completion. Views from the searchlight platform looking forward and aft. In the bow view, note the 130mm B-13 Series II gun mountings with streamlined shields. (Przemysław Budzbon collection)

Khar'kov at Poti during the late summer of 1942. Note the 130mm B-13 Series II gun mountings with streamlined shields. She was assigned to the 1st Destroyer Flotilla and is wearing the tactical number 12. (Witalii Kostrichenko collection, courtesy of Tomasz Grotnik)

Characteristics

The initial figures are for *Leningrad* as completed; the characteristics of other ships of the class are beneath.

Displacement (designed)	2,030 tons standard, 2,265 tonnes normal, 2,675 tonnes full load
Displacement (as built)	2,032 tons standard, 2,282 tonnes normal, 2,693 tonnes full load
	Moskva & *Khar'kov* 2,030 tons, 2,675 tonnes, 3,080 tonnes
	Minsk 1,952 tons, 2,237 tonnes, 2,597 tonnes
	Sergo Ordzhonikidze 2,029 tons, 2,350 tonnes, 2,680 tonnes
	Tbilisi 2,063 tons, 2,298 tonnes, 2,708 tonnes
Length oa	127.5m
Breadth max	11.7m
Depth to the main deck	7.0m
Draught max (full load)	4.18m
Immersion	9.1t/cm
Metacentric height	80cm at normal load; 104cm at full load
	Moskva, Khar'kov, Ordzhonikidze 91cm; 117cm
	Minsk 76cm; 96cm
	Tbilisi 80.5cm; 100.7cm
Rolling cycle	8.3sec
	Ordzhonikidze, Tbilisi 10sec
Turning circle with 25° helm	4.0 cables at 28.5kts, 3.5 cables at 18kts
Boilers:	three three-drum small watertube
working pressure	21.5 kg/cm^2
steam temperature	335°C
heating surface	1,582m^2
capacity	135t/h
feed water:	78 tonnes
Main machinery:	three sets of impulse/reaction geared turbines
rated output	66,000shp total
propellers	three of 2.5m diameter
fuel	Marine Oil Fuel
capacity	210t normal, 600t full load
consumption	32t/h at full speed; 5.0t/h at economical speed
	Moskva, Khar'kov 29t/h; 5.6t/h
	Minsk 28.5t/h; 4.9t/h
	Tbilisi 27.4t/h; 4.65t/h
range	*Leningrad* 873nm at 40kts; 2,100nm at 20kts
	Moskva, Khar'kov 615nm at 33kts; 1,540nm at 16kts
	Minsk 835nm at full speed; 2,100nm at 20kts
	Ordzhonikidze 752nm at 40kts; 2,000nm at 20.45kts
	Tbilisi 930nm at 40.54kts; 2,780nm at 20.67kts

Kiev and *Tiflis* were to be assembled at the new shipyard erected in 1932 in the area of the village of Permskoe on the left bank of the River Amur. It was one of the 'great constructions of socialism', officially carried out under the auspices of the Soviet youth organisation Komsomol but in fact built using the forced labour of prisoners. Its purpose was to strengthen the military and economic presence of the Soviet Union in the Far East. The village changed its name to Komsomolsk-on-Amur, and the same name was allocated to the shipyard.

Using the imported Vickers bearing/range calculator (called AKUR by the Russians), the Sperry data transmitters and the Italian Galileo computer, Elektropribor was able to develop in 1937 its own computer TsAS-2, which became the core element of the *Mina* destroyer fire control system. Some internal arrangements of the ships and cabling had to be modified accordingly and the workshop documentation adapted. The design documentation that included all these changes was designated *Proekt 38bis*.

Construction was planned in two main phases: prefabrication at Nikolayev and assembly at Komsomolsk. The hull and ship systems sections were manufactured at the Andre Marti Yard, where they were subject to preliminary assembly and adjusting. They were then broken down into modules not exceeding 10 tonnes and transported on flat railway cars to Moscow, where they were rolled onto the tracks of the Trans-Siberian Railway. After covering a total of 9,700 kilometres to Khabarovsk, the platforms with the blocks were rolled onto a long pier on the Amur shore and loaded onto barges using a floating crane. The last part of journey was some 350km along the Amur, and the cargo was unloaded at the docks near building shed 'A' where the branch of the Andre Marti Yard had been established. Once the concrete of

Electricity:	
turbo-generators	two PST30, 100kV/h total
diesel generators	two PN, 60kV/h total
voltage	115 V
Navigation equipment	Kurs-1 gyrocompass, two 5in magnetic compasses, GO-II log, ZMI sounding lead
Main guns	*Leningrad* & *Moskva*: 5 – 130mm/50 B-13 in single mounts Series I with rectangular shields; ammunition capacity 750 rounds normal, 1,000 max, 25 ready-use
	Minsk & *Sergo Ordzhonikidze* B-13 Series II guns with rectangular shields
	Khar'kov and *Tbilisi* B-13 Series II guns with rounded shields
Main fire control	*Proekt 31* & *Proekt 38*: DCT (forward) Officine Galileo Duplex with two OG-3 rangefinders and APG director sight; CT (aft) Officine Galileo with one OG-3 rangefinder; 2 Galileo DM-1.5m rangefinders (bridge); fire control calculators (main and auxiliary) Officine Galileo Centrale
	Proekt 38bis: DCT (forward) B-12-4 with two DM-4 rangefinders and VMTs-2 director sight; 1 DM-3 rangefinder (aft), 2 ZD rangefinders (bridge); fire control calculator MinaL
Medium AA guns	2 (*Tbilisi* 3) – 3in/55 34-K in single mounts; ammunition capacity 600 rounds normal, 738 max, 48 ready-use
Light AA guns	2 – 45mm/46 21-K in single mounts; ammunition capacity 2,630 rounds normal, 3,975 rounds max
AA fire controls	none
Machine guns	4 – 0.5in DK on single mounts
	Moskva & *Khar'kov* 12 – 0.5in Browning on twin mounts
	Minsk & *Ordzhonikidze* 6 – 0.5in DK on single mounts
	Tbilisi 6 – 0.5in DShK on single mounts)
Searchlights	2 Galileo 90cm
	Proekt 38bis 60cm MPE-E6
Torpedo tubes	8 – 21in N-7 in quadruple mounts; capacity 16 (8 spare)
Torpedo fire control	*Leningrad*: Officine Galileo Centrale
	Moskva & *Khar'kov*: Mobiletto
	Minsk: Mina
	Proekt 38bis: Mech
Anti-submarine	2 (*Ordzhonikidze* 1) DC chutes; 10 B-1 charges; 20 BM-1 charges
A/S detection	none
Sonic station	Arktur
Mines	maximum load (both TTs and 'X' gun inactive) 76 Model 1926 or 68 Model 1931, 84 MZ-26 sweep obstructers
Paravanes	2 K-1
Complement	225 (19 officers, 11 petty officers, 195 ratings)
Seaplane (optional)	1 SPL type; boom, avgas tank of 1,160 litres

building dock No 3 had dried and hardened, the keel of *Kiev* was laid on 10 March 1936, while five months later, following the completion of construction of building dock No 2, the keel of *Tiflis* was laid. The latter dock, originally intended for the assembly of submarines, was too short to accommodate the hull. The solution was to build the hull with the bow slung above the lower part of the dock through the gate.

Both ships were to be fitted out by the Dal'zavod works at Vladivostok, as the necessary workshops were as yet not available at Komsomolsk. Following her launch on 25 June 1938 *Kiev* was renamed *Ordzhonikidze* and towed via the Amur River and Tatar Strait to the ship repair yard at Sovetskaya Gavan (former Imperatorskaya Gavan). There she was fitted for open sea towage; she then left via the Sea of Japan for Vladivostok, arriving at the Dal'zavod works on 25 October 1938. She was completed there within the year. In the meantime, her name was again changed – this time to *Sergo Ordzhonikidze*. The last of the Series II ships to be completed was *Tiflis* (renamed *Tbilisi* in 1940 after the town changed its name). Her completion on 11 December 1940 effectively completed the long and tortuous story of the design and construction of this class.

The fitting out of *Minsk* took three years after launch. That of her sisters proceeded more smoothly, each taking approximately 18 months to complete due to the installation of equipment and machinery during the construction of the hull. This was not possible in the shipyards of the western part of the Russia, where the capacity limitations of traditional slipways and the influx of new orders imposed the shortest possible hull construction cycles, with virtually empty shells launched as a result.

The *Proekt 38* ships proved to be no better than their

Proekt 1 counterparts. With their rounded sterns, they did not have the excessive trim at high speeds, but they were generally slower. They inherited other shortcomings of the earlier design:

- poor seaworthiness
- poor manoeuvrability
- fragile hulls
- rapid pitting and corrosion of the propellers due to cavitation.

Conclusions

The first Soviet destroyers were the children of their time. The concept of the super-destroyer came to prominence at a moment of great instability in Soviet war doctrine, particularly with regard to the role of the Navy. It was shaped by a generation of revolutionary military commanders (seamen, corporals or at best lieutenants of the former Imperial armed forces) whose theoretical military knowledge did not extend beyond barrack drill principles. The designers and former junior constructors who survived the 1917 Revolution were faced with the task of implementing hugely unrealistic staff requirements. The shipbuilding industry had to be rebuilt from scratch from 1927, after ten years in which no new experience of ship design and construction had been accumulated. In addition, the designers were working in an atmosphere of growing terror, at a time when taking a few ears of rye lying on the ground from a field was a serious crime. All design decisions had to be made under this threatening cloud. The current fetish for speed eclipsed all other design considerations, such as seaworthiness and reliability, and the power plant incorporated in these ships was oversized and over-powered simply to keep the designers out of prison. In consequence the ships' range was reduced, and draconian savings had to be made in the construction of the hull.

The ships were designed in a chaotic environment characterised by constant policy and organisational changes. This was the period during which Stalin assumed absolute power over the Soviet Union; it was a period marked by 'the terror', the progressive militarisation of society, forced industrialisation and rearmament, to which the entire economy was subordinated. All this was done in a highly uncoordinated manner, building tensions and imbalances in all areas. Added to these phenomena were technological backwardness, low cultural capital and a work ethic that had been completely destroyed. Ambitious production plans drawn up by the Politburo decision makers ended up 40 per cent short of their targets, and of the 60 per cent realised only half met the standards originally envisaged.

The difficulties encountered by the shipyards is best illustrated by the steam valve box of *Minsk*, which had to be cast no fewer than 46 times before a satisfactory result was obtained. No wonder that construction and fitting-out were prolonged, while the quality of the final product remained poor. Initially the delivery time for the *Proekt 1* ships was set at 18 months following approval of the preliminary design; at the time of approval this was revised to 22 months. After two years the planned delivery time was extended first to 37 months, then to 50 under the Second Five-Year plan. The prototype was provisionally accepted by the Soviet Navy 73 months after the approval date, and *Leningrad* finally entered service 93 months after the approval date.

Taking all of the above into account, one should not be surprised that Russian authors consider these ships to be a success. Knowing all too well under what conditions their design and construction took place, they appreciate the fact that these ships were completed at all. And this was arguably the greatest achievement of the *Leningrad*-class leaders: that they prepared the ground for future Soviet ship design and construction. All the Soviet warships of the first generation served this purpose: the *Uragan* class patrol ships (see *Warship* No 22/1982), the MTBs of the *Sh-4* type (*Warship* No 8/1978), the *Dekabrist* class submarines (*Warship 2020*) and the minesweepers of the *Fugas* class (*Warship 2016*). The next generation of warships would benefit from greater maturity in design and construction, thanks largely to access to German and Italian military technology.

Activity before 22 June 1941

17–25 November 1938
Moskva visited Istanbul with a Soviet delegation for the funeral of the Turkish President Kemal Atatürk.
14–17 June 1939
Great Training Cruise of the Baltic Fleet around Gotland, with *Minsk* as flagship of the Light Force Detachment.
31 July–late August 1939
Leningrad immobilised in the dockyard; 732 tubes replaced in boiler No 2.
30 August–1 September 1939
Minsk was with the destroyer flotilla on a training exercise in Koporskaya Gulf.
1 September 1939
German invasion of Poland. All Soviet warships recalled from exercises; began loading live ammunition and supplies.
6 September 1939
Secret mobilisation of the Baltic Fleet.
17 September 1939
Red Army entered Poland. The Polish submarine *Orzeł* escaped from internment at Tallinn. The highest state (No 1) of combat readiness was ordered for the three Soviet Fleets (Northern, Baltic and Black Sea).
18 September 1939
The hunt for *Orzeł* by Soviet warships under Deputy *Narkom* Ivan Isakov with his flag in *Minsk* began in the Gulf of Finland. *Leningrad* participated as flagship of the 2nd Destroyer Flotilla of the Light Force Detachment.
21 September 1939
Following the release of two kidnapped Estonian guards on the coast of Gotland, it became apparent that *Orzeł* had headed westward.

THE BEGINNINGS OF SOVIET NAVAL POWER

Close-up views of *Khar'kov* at Batum following the German air raid of 18 January 1943. The burning tanker *Peredovik* in the background. (Przemysław Budzbon collection)

22 September 1939
Minsk, with the destroyers *Gordyi* and *Smetlivyi* in company, escorted the Soviet freighter *Sibir'*.

22 September 1939
The hunt for *Orzeł* was called off, but Stalin decided to continue and to explore means of putting military pressure on the Baltic States.

24–26 September 1939
Leningrad and the destroyer *Stremitel'nyi*, which were engaged in a feigned hunt for 'unidentified foreign submarines', entered the territorial waters of Estonia in Hara Bay, and fired on what was claimed to be a 'secret Polish submarine base'. Estonian motor launches and a plane were machine-gunned.

28 September 1939: Estonia
5 October 1939: Latvia
10 October 1939: Lithuania
Forced signing of mutual assistance pacts between the Soviet Union and the Baltic States. In exchange for dubious trade concessions, the Soviet Union offered 'protection' in the form of establishing military bases in these countries; this opened up the possibility of stationing ships of the Baltic Fleet in their harbours.

25

Leningrad photographed in June 1944 by a US Naval Attaché at Leningrad. Note the 130mm B-13 Series II gun mountings with streamlined shields that replaced the Series I guns with the rectangular shields; the blast screens were modified accordingly. The twin 3in 81-K HA mounting is seen abaft 'X' gun. (US Naval History & Heritage Command, 80-G-176369)

10 October 1939
Leningrad and the 1st Destroyer Flotilla were assigned to Liepaja.

11 October 1939
Minsk, accompanied by the destroyers *Gordyi* and *Smetlivyi*, arrived in the roadstead of Tallinn and announced the intention of entering the harbour the following day.

20–23 October 1939
Moskva, flying the flag of commander of the Destroyer Brigade *Kapitan 2 ranga* (Captain) Sergei Gorshkov, and the destroyer *Besposhchadnyi* visited Istanbul.

12–15 November 1939
Leningrad, flying the flag of the C-in-C Baltic Fleet *Flagman 2 ranga* (Rear Admiral) Vladimir Tributs, visited Tallinn.

14 November 1939
Minsk moved her base from Tallinn to Kronshtadt to prepare for the war with Finland.

30 November 1939
The Soviet Union invaded Finland.

Tbilisi, with the destroyer *R'ianyi* to starboard, on VJ day. Note the third 3in 34-K gun mounting aft. (Przemysław Budzbon collection)

Tbilisi in 1945 with the full outfit of Lend-Lease-supplied radar: US SF (in radome atop foremast) and British Types 291 (air search) and 284 (gun fire control). Note the 130mm B-13 Series II gun mountings with streamlined shields. (Witalii Kostrichenko collection, courtesy of Tomasz Grotnik)

Leningrad in 1947, photographed on the Neva River close to the University Wharf. The cutaway stern is prominent. (Przemysław Budzbon collection)

Leningrad in her last years of combat service, photographed from the English Wharf in the Neva River in 1956. (Przemysław Budzbon collection)

Leningrad shelled Seiskaari Island (72 rounds of 130mm).

6–7 December 1939
Minsk (16 rounds of 130mm), with the older destroyers *Karl Marx* and *Volodarskiy*, conducted a reconnaissance sortie of Kilpisaari Island; the bombardment produced no results due to poor visibility in a snowstorm. Two attempts to provoke return fire were unsuccessful. On the third day the bombardment was abandoned. *Minsk* stopped a German merchantman *Oliva* for inspection off Naissaari Island.

8 December 1939
Minsk, with destroyers of the 3rd Flotilla, acted as a cover force for minesweepers clearing approaches to the 10in battery on Koivisto Island off Saarenpää. When she was fired on by the battery *Minsk* sighted the Finish submarine *Saukko*, but allowed her to proceed.

9–10 December 1939
Minsk, *Leningrad* and the 3rd Destroyer Flotilla screened the battleship *Oktyabrskaya Revolutsiya* during a bombardment of the Finnish batteries on the Tiurinsaari and Koivisto Islands.

13 December 1939
Minsk (59 rounds of 130mm) and *Leningrad* (78 rounds of 130mm), with the destroyer *Steregushchiy* and two patrol ships, shelled the Finnish battery on Koivisto Island. *Leningrad* fired on a Finnish aircraft (2 rounds of 3in, 4 rounds of 45mm).

14 December 1939
Minsk (2 rounds of 130mm) and *Leningrad* formed part of the cover force for the battleship *Oktyabrskaya Revolutsiya* during an unsuccessful attempt to shell the Finnish batteries on Saarenpää Island.

15 December 1939
Minsk (37 rounds of 130mm), in company with the destroyer *Steregushchiy* and other ships, engaged the Finnish 8in battery on Pitkäpaasi Island during a reconnaissance sortie.

18–19 December 1939
Minsk (143 rounds of 130mm), with the destroyers *Steregushchiy* and *Lenin*, screened the battleships *Oktyabrskaya Revolutsiya* and *Marat* during an engagement with the 10in battery on Koivisto Island.

30 December 1939
Minsk and *Leningrad* were engaged in an unsuccessful attempt to reach Koivisto and Tiurinsaari for a planned bombardment. After two days of struggling through heavy ice the operation was cancelled and *Minsk* sailed to Liepaja, where she was to be based; *Leningrad* had her hull damaged over a length of 30 metres, and arrived in Liepaja on 4 January 1940 for repairs.

4 January 1940
A planned bombardment of Russarö Island by *Minsk* and *Leningrad* was cancelled.

17 January 1940
Leningrad stopped and inspected the German steamer *Geir* off Hiiuma.

20–22 January 1940
Minsk searched for two days for the Soviet submarine *S-2*, which sank on a Finnish minefield.

13 June 1940
The decision was taken by the Politburo to seize the Romanian provinces of Bessarabia and Bukovina.

15 June 1940
The highest state (No 1) of combat readiness was ordered for the Black Sea Fleet.

26 June 1940
An ultimatum was delivered to Romania requesting the 'return' of Bessarabia and Bukovina to the Soviet Union. The ultimatum was accepted by the Romanian Government.

18–25 July 1940
Minsk was dispatched to Riga as a demonstration of power during a vote in the Latvian Parliament on a proposal to join the Soviet Union.

26 July–23 August 1940
Minsk was under repair.

September 1940
When steaming at 30 knots in a gale with wind speeds of Force 8–9, *Minsk* sustained severe damage to her forecastle deck plating.

October 1940–17 June 1941
Minsk was again under repair.

31 May–20 June 1941
Leningrad completed repairs, but subsequently had to be

Leningrad was reclassified as a destroyer in 1949. She is seen here following her 1951–54 modifications. (Przemysław Budzbon collection)

docked no fewer than nine times because of continuing leaks.

14–19 June 1941

Moskva was part of the cover force during a landing exercise of the 150th Infantry Division in the Crimea.

22 June 1941: Order of Battle

Leningrad, Minsk: 3rd Destroyer Flotilla, Squadron, Baltic Fleet.

Khar'kov, Moskva: 3rd Destroyer Flotilla, Light Force Detachment, Black Sea Fleet.

Baku: 1st Destroyer Flotilla, Destroyer Brigade, Pacific Fleet.

Tbilisi: 2nd Destroyer Flotilla, Destroyer Brigade, Pacific Fleet.

Endnotes:

[1] No fewer than 53 destroyers based on the prototype *Novik* were laid down for the Imperial Russian Navy between 1910 and 1915. Thirty were commissioned before the Revolution, but only nine remained in service after the Civil War. The Soviets subsequently completed six of the unfinished hulls and repaired a further two. The ships in the *Novik* series had a normal displacement of around 1,300 tonnes and mounted three/four 4in guns and eight/nine 450mm torpedoes in four twin or three triple mountings.

[2] As part of the 1927 Naval Programme Romania ordered two powerful destroyers from the Pattison Shipyard in Naples. Based on a British flotilla leader design (*Shakespeare* class), they carried five 120mm (4.7in) guns and six 533mm (21in) torpedo tubes on a displacement of 1,400 tonnes. The two 1,250-tonne destroyers of the Turkish *Kocatepe* class were ordered from Ansaldo, and were armed with four 120mm guns and six 533mm torpedo tubes. They would be followed by two similar ships of the *Tinaztepe* class.

[3] For anyone interested in the history of the Soviet ship-building, there is a well-known centralised system of assigning subsequent numbers to projects. The first was the destroyer project described here, while today the recognised list of numbers extends as far as 2550.

[4] In Russian to the present day the word for 'destroyer' is *esminets*, a contraction of *eskadrennyi minonosets* ('fleet torpedo boat'). The word *minonosets* literally means 'mine carrier', as torpedoes were originally designated 'mobile mines' in the 19th century.

[5] In 1908 the development of an anti-aircraft gun was suggested by teachers at the Mikhailovsky Artillery Academy. Officers of the school M V Dobrovolsky, E K Smyslovsky and P N Nikitin drew up the technical requirements, while the author of the paper V V Tarnovsky proposed installing the gun on a truck chassis. In the summer of 1913, Tarnovsky completed the design of the gun and the project was approved by the GAU in 1914. Due to political infighting, Tarnovsky was forced to sell his design to the Putilov plant in St Petersburg and the design was modified by F F Lender. The Pattern 1914/15 guns were thereafter known as 'Lender Guns'.

[6] The Chyetverikov SPL was a small flying boat with folding wings that was designed to be launched from a submarine. It had a twin-boom tail and a single over-wing forward-facing engine. The fuselage was 7.4m long and the deployed wingspan 9.5m. It was designed to fit into a 2.5m-diameter cylinder with wings folded. Development began in 1931.

[7] The author, who worked with the Soviet military in the ship-yard during the time of the communist regime in Poland, experienced similar changes on a number of occasions. They were always associated with the allocation of almost a full-scale budget to the project work, which in fact consisted mostly of alterations of existing documentation.

"FOOLIN' AROUND THE FRENCH COAST":
THE CHALLENGE OF OPERATION 'TUNNEL', SEPTEMBER 1943 – APRIL 1944

Michael Whitby's in-depth study looks at the series of operations in the Channel that the Royal Navy undertook in 1943–44 under the code-name 'Operation Tunnel', and analyses the successes and failures.

In the early evening of Friday 22 October 1943, the light cruiser HMS *Charybdis* led six destroyers out of Plymouth and headed south across the English Channel. They were on Operation 'Tunnel', a generic series of offensive missions initiated by Plymouth Command the previous month that had destroyers, sometimes led by cruisers, sweeping enemy shipping routes off northwest Brittany. Nine 'Tunnels' had been carried out thus far, with only one spirited brush with the enemy. That would change this night when torpedoes from five German torpedo boats ripped into the British formation, sinking *Charybdis* and so damaging the 'Hunt'-class destroyer HMS *Limbourne* that she had to be dispatched by friendly forces; the five British destroyers that survived returned home without firing a shot or even sighting the enemy. Some 500 men lost their lives. It was a disaster by any measure, and official *post mortem*s identified flaws with the operations. Yet, despite reassessment that resulted in the establishment of a permanent strike force, which underwent a comprehensive training regime, the second phase of 'Tunnels' that ran into April 1944 continued to be plagued with problems, with the margin between success and failure remaining slim. If one accepts that such operations, a method of asserting maritime presence, are necessary when forces go on the offensive, as Plymouth Command did in the autumn of 1943, then focus can sharpen on the challenges of meeting those commitments in wartime – when demands on ships and personnel are severe – as well as the consequences when that is not done effectively.

The Setting

The target of Operation 'Tunnel' was enemy shipping running under the cover of darkness between ports in the western Channel and the Bay of Biscay. The most accessible section of the shipping route from Plymouth was a section of the coast of northwest Brittany between Lézardrieux and Ile Vierge. It lay 90–100nm from Plymouth, or about five hours steaming at 20 knots, which meant 'Tunnels' typically lasted about twelve hours. Since the *Luftwaffe* remained active over the Channel during daylight – the devastating success of Fritz X and Hs-293 glider bombs against surface forces was particularly alarming – forces had to be away from the French coast before dawn, limiting the window for interception to only a few hours. In consequence, 'Tunnels' followed a predictable pattern, leaving Plymouth in the evening and steaming south until they were a few miles off the coast. At that point, they could implement their only real variation by sweeping either east or west down the shipping lane. However, that too could be problematic: too far east and they would come within range of the enemy's coastal artillery in the Channel Islands; too far west and there lurked the formidable defences surrounding Brest.

Theatre of Operations
Canadian Tribals in the Channel

Moreover, certain directions at certain times could expose them to poor conditions of light. Alternative routes and timings were, thus, quite limited – somewhat reminiscent of a tunnel one might say. Against that, 'Ultra' intercepts often provided notice of enemy plans.

Enemy defences were another determinant. Coastal radar stations at Ushant, Ploumanach or Paimpol invariably detected 'Tunnel' units as they approached the French coast. Although radar detectors in the British ships made them aware they were being tracked, they nonetheless lost surprise, and the forewarned enemy could duck into one of the many small harbours along Brittany's rugged coast. As for the Kriegsmarine's surface units, it was destroyers that concerned Plymouth Command. Although they principally filled a defensive role, they were a potential offensive menace and, with the invasion on the horizon, planners wanted to reduce the threat. Marine Gruppe West usually kept their few powerful Type 36A 'Narvik' class destroyers away from Allied bombers at bases in the southern Bay of Biscay, leaving the smaller Type 39 Flottentorpedoboot (lit 'Fleet Torpedo Boats') of the 4.Torpedoboot flotille as the workhorses in the western Channel. Small destroyers that displaced 1,780 tonnes, Type 39s featured three 10.5cm (4.1in) guns and six 53.3cm (21in) torpedo tubes, and were fast at 33 knots. Dubbed 'Elbings' by the Allies, they began moving into the western Channel during the summer of 1943, and proved to be a skilled, well-trained adversary. Both destroyers and merchant shipping were 'Tunnel' objectives; however, paring down opposition destroyer strength was the overarching goal; as a senior Admiralty official acknowledged: 'It has always been hoped by all, including the forces concerned, that enemy destroyers w[ou]ld be the prize.'[1]

An action on 10 July 1943 demonstrated the Type 39's capability and sparked a change in Plymouth's considerations. Engaging three 'Hunt'-class destroyers attacking minesweepers off Ile de Batz, T-24 and T-25 damaged all three, one seriously, in an intense gunnery duel while suffering only splinter damage themselves. Plymouth Commmand had previously relied upon 'Hunts' to carry out its anti-shipping sweeps but the 10 July action demonstrated that the small British destroyers, with their armament of four 4in guns and two 21in torpedo tubes, could be overmatched by the Type 39; the latter's six/eight-knot speed advantage was particularly troubling. Plymouth required more capable warships, but they were hard to come by. There were no cruisers or Fleet destroyers permanently attached to the command, and the only such forces available were on loan from the Home Fleet for operations in the Bay of Biscay.

In an effort to reconcile these challenges, in the first week of September 1943, Vice Admiral Sir Ralph Leatham launched Operation 'Tunnel'. Just one week into his appointment as C-in-C Plymouth and under instructions from the Admiralty to increase pressure on the Kriegsmarine, Leatham mounted the sweeps under the generic name 'Tunnel'. They therefore resembled Plymouth's ongoing 'Hostile' series of minelaying

Vice Admiral Sir Ralph Leatham as Flag Officer Malta in 1943. Under direction to take the offensive, Leatham initiated Operation 'Tunnel' in September 1943. Although he had previously run anti-shipping sweeps from Malta, 'Tunnels' introduced new challenges. (IWM A-7229)

missions in that the targets of individual missions and the forces carrying them out varied, but they retained the same general character. Governed by a standing operation order dubbed 'OPTU', the generic aspect of 'Tunnels' streamlined the process of slotting ships rotating through Plymouth into the missions and minimised the need for signalling at designated navigation waypoints. The first 'Tunnel' ran on the night of 5/6 September when, in what the Admiralty war diary vaguely described as 'a small mission', the light cruiser Phoebe led the Fleet destroyers Grenville and Ulster on an eastward sweep up the Breton coast. They repeated the operation three times over the next five nights; although they never encountered the enemy, they at least achieved a degree of familiarity, both with themselves and the task in hand.

First Action

The next 'Tunnel', run after a three-week pause on 3/4 October, emphasised the challenges confronting Leatham in mounting the operations. Although Grenville and Ulster remained available, Phoebe had been transferred to the Mediterranean, leaving the 'Hunt'-class destroyers Limbourne, Tanatside and Wensleydale to fill out 'Force X', as Plymouth dubbed the improvised

"FOOLIN' AROUND THE FRENCH COAST": THE CHALLENGE OF OPERATION 'TUNNEL', SEPTEMBER 1943 – APRIL 1944

The *Dido*-class light cruiser HMS *Phoebe* off the US east coast in June 1943 after her repair by the New York Navy Yard. Her forward 5.25in. gun turret, removed following damage received when she was torpedoed off Africa, was replaced after she returned to the U.K. *Phoebe* led the Fleet destroyers HMS *Grenville* and *Ulster* on the first four 'Tunnels'. Her subsequent departure to the Mediterranean disrupted the teamwork honed by the three warships. (US Navy History and Heritage Command, 80-G-411)

The Type 39 *Flottentorpedoboot T-35* on trials with the US Navy in September 1945. The workhorses of *Marine Gruppe West* in the western Channel, these small, well-trained destroyers proved to be formidable adversaries to the forces carrying out 'Tunnels'.
(US Naval History and Heritage Command, NH-73329)

31

group.[2] Although Leatham would have desired to strengthen the force with a cruiser or additional Fleet destroyers, the 'Hunts' were all he had. Part of the 15th Destroyer Flotilla (15 DF) permanently assigned to Plymouth, the small destroyers' primary task was defending coastal convoys against E-boat attack. They did have the advantage of familiarity, but only *Wensleydale* had any offensive experience. There were also performance discrepancies between the 'Hunts' and the Fleets. In particular, the three Type 3 'Hunts' were about 5-6 knots slower than the Fleets,[3] and their two 21in torpedo tubes paled in comparison to the twin quadruple mounts of *Ulster* and *Grenville*; against that, they had the superior Type 271 search radar while the Fleets relied upon the obsolescent Type 291. Despite their shortcomings, 'Hunts' sailed on all remaining 'Tunnels' in 1943.

Staff officers at Plymouth weighed the disparities between Force X's destroyers in their pre-mission planning. Plymouth did not have an assigned Staff Officer (Operations), and in this instance the planning team likely included: Commander Reginald Morice, the Cdr (D) Plymouth; Captain NC Moore, the Staff Officer (Intelligence); and Commander Byron Alers-Hankey, D15 and Senior Officer (SO) of Force X, as well as staff from Leatham's combined headquarters. Of these, Morice and Alers-Hankey were the most influential as the senior type commanders ashore and afloat. Their objective for the 3/4 October 'Tunnel' was a coastal convoy running eastward from Brest and its escort of torpedo boats. The planners decided Force X should sweep from east to west, the opposite direction to the *Phoebe* 'Tunnels', since the 'Hunts' would be unable to overhaul the faster torpedo boats if a stern chase resulted; however, that meant Force X would be sillouetted against a lighter northern horizon as it ran down the Brittany coast. And, to raise the comfort level amongst ships unused to operating together and lacking tactical uniformity, they steamed in line ahead. Thus, force limitations shaped tactical considerations.

Steaming southwest down the Breton coast, shortly after 0100 on 4 October Force X's radar detected enemy destroyers at five miles on the port bow. Their identity was confirmed by specialist German-speaking 'Headache' operators in the 'Hunts' who monitored opposition voice transmissions – a capability the two Home Fleet destroyers lacked. In response, Cdr Alers-Hankey ordered a 40-degree turn together to port to approach the enemy in an open formation with the forward guns and radar unobstructed.[4] Although the Fleets complied, the 'Hunts' followed *Limbourne* around in line ahead. *Ulster*'s CO, Cdr William Donald, recalled: 'This was a pity as it reduced our fire power, and resulted in two of our side being left at a disadvantage as regards range when the battle started.'[5] Forewarned of Force X's presence by coastal radar, four Type 39s led by *Korvettenkapitän* Franz Kohlauf had sighted the British force against the brighter northern horizon shortly after they made their turn. The torpedo boats wheeled to the southeast, firing torpedoes. All missed, and the engagement became a confused *mêlée* 'with destroyers milling about all over the place' – *Grenville* was sprayed by pompom fire from a 'Hunt'.[6] Force X pursued the torpedo boats, but when *Ulster* and *Grenville* outpaced

The Type 3 'Hunt' HMS *Melbreak* in October 1942. Although their capabilities were not ideally suited to offensive sweeps, the 'Hunts' of Plymouth Command's 15th Destroyer Flotilla carried much of the load in the first phase of the 'Tunnels'. (Author's collection)

"FOOLIN' AROUND THE FRENCH COAST": THE CHALLENGE OF OPERATION 'TUNNEL', SEPTEMBER 1943 – APRIL 1944

HMS *Charybdis* on 22 February 1943, eight months before her catastrophic loss. Due to a dearth of 5.25in gun mountings, the *Dido*-class cruiser was armed with four twin 4.5in. disposed symmetrically fore and aft. Although *Charybdis* had seen challenging service screening convoys in the Mediterranean and conducting 'Stonewall' patrols in the Bay of Biscay, her sailors had no experience in the 'Tunnel' into which they were thrown with little warning. (ADM 3040, courtesy Conrad Waters)

the 'Hunts' they attracted the enemy's accurate fire, with shell splashes drenching their upperworks. *Grenville* veered away after being hit aft, leaving *Ulster* to fight alone before taking two hits forward. After snapping off a salvo of torpedoes, she too broke off. The torpedo boats suffered splinter damage; *Grenville* went into dock for a week while *Ulster*'s repairs took longer.

One would have thought that the first 'Tunnel' to meet the enemy would spark timely analysis, but that did not occur. Vice Admiral Leatham only submitted Plymouth's report to the Admiralty on 26 October, while analysts there did not begin circulating their commentary until 9 November. Meanwhile, the missions continued with 'Tunnels' on consecutive nights between 13/14 and 17/18 October. These were laid on in the hope of intercepting the merchant ship *Münsterland*, which intelligence revealed was being sent up-Channel from Brest. The blockade runner was an important objective. Germany was suffering from a severe shortage of troop transports and had recalled all vessels of suitable capacity home from ports in the Low Countries and the Bay of Biscay; *Münsterland* was among the most valuable of these vessels.[7] With one exception, the four mid-October 'Tunnels' featured the same mix of 'Hunts' and Fleets, with HMS *Rocket* replacing *Ulster*, but they came up empty since *Münsterland* had yet to leave Brest.

With the increased frequency of 'Tunnels' absorbing slim resources, Leatham sought reinforcements. With no cruisers and only the Fleets *Grenville* and *Rocket* available, Leatham still had to rely on the 'Hunts' of 15 DF, and they sailed on five 'Tunnels' in a little over two weeks. That, combined with their other duties, took a toll, and on 19 October Leatham alerted the Admiralty he was hamstrung by a dearth of 'Hunts'. Noting that German coastal convoys 'were more strongly escorted than before', offensive operations were 'now required to be carried out in greater strength than the 15th Flotilla can furnish.' Leatham explained that the situation was serious, since one of the five 'Hunts' remaining was scheduled to enter refit while another was due for a boiler clean. It was, therefore, 'a matter of urgency' that consideration be given to increasing the strength of 15 DF to eight 'Hunts'.[8] Regrettably, shortages existed everywhere and no relief was available.

The *Charybdis* Tragedy

The stage was now set for the 22/23 October 'Tunnel' that saw the loss of *Charybdis* and *Limbourne*. The calamity has been dissected in memoirs and monographs, but aspects have been overlooked, particularly the haste with which it was mounted. Much of this focus revolves around *Charybdis*. Since commissioning in late 1941, the cruiser had mainly been deployed as an anti-aircraft cruiser in the Mediterranean, and throughout much of 1943 had been occupied on 'Stonewall' patrols, supporting the anti-submarine offensive in the Bay of Biscay. Neither *Charybdis* nor her CO, Captain George Voelcker, had participated in a 'Tunnel,' nor, it appears, an offensive sweep of any kind.[9] When that prospect arose, the cruiser was thrown into the operation with scant notice. *Charybdis* returned to Plymouth from a Biscay patrol on the afternoon of 18 October, and two days later received notice to depart on another Operation 'Stonewall' at 1530 on 22 October. Those plans were abruptly scrapped when intelligence revealed *Münsterland* was being sent up the Channel on the night of 22/23 October. In the evening of 21 October Plymouth Command cancelled the 'Stonewall', and informed Voelcker that he would instead be leading a 'Tunnel' the

33

next night with a force comprising the Fleet destroyers *Grenville* and *Rocket* and the 'Hunts' *Limbourne*, *Steventstone*, *Talybont* and *Wensleydale*. Thus, with only 24-hours notice, Voelcker was thrust into command of a mission in which he had no experience, involving ships he knew slightly, at best.

This situation would not have been as sobering had the leadership of the 15 DF remained intact. Commander Alers-Hankey, D15, had accrued valuable experience as the 'Hunt' flotilla commander and had led two of the October 'Tunnels', including the engagement on 3/4 October, so he was well-versed in these operations. Earlier that month, however, he had been appointed to the carrier HMS *Formidable*, and on the afternoon of 20 October he relinquished command of *Limbourne* and 15 DF to Cdr Walter Phipps. Moreover, Lt-Cdr Frank Brown, who had led three sweeps in *Tanatside*, was also unavailable since his ship was having a boiler clean. Phipps was a seasoned destroyer officer but he had never served in a 'Hunt', and had scant opportunity to become acquainted with his flotilla or the nature of the operation ahead. Given the demands of war, it is unreasonable to suggest Alers-Hankey or Brown should have been made available; nonetheless Voelcker would have benefited from their experience. Consequently, Force 28's two divisions – *Charybdis* and the two Fleets in one, the *Hunts* forming the other – were led by officers lacking 'Tunnel' experience and unfamiliar with the ships under their command. The situation was evidently not helped by a poor pre-mission conference that left aspects of the operation confused. Phipps missed most of the briefing and was dissatisfied with what he did hear, complaining to his diary that 'I hadn't the least idea of [Voelcker's] intentions and could not get anything out of him.'[10] Such are the foibles of mutual inexperience.

The flawed preparation caused unease among some of those about to embark on the 'Tunnel'. This was most vividly expressed by *Grenville*'s Lt-Cdr Roger Hill, a veteran of nine 'Tunnels' – more than any other commanding officer. In his 1975 memoirs, Hill recounted his bitter reaction to the pre-mission briefing and the perils that lay ahead:

'Well best of luck', a staff officer closed off the briefing'.
I sat, furious, exasperated and impotent.
'Don't let's go', I murmured.
The staff officer looked at me sharply. 'I don't think I quite heard …?'
'I said, "Don't let's go", sir. Not tonight. Let us practice together every night for a week and then we'll crack those damned destroyers wide open.'
He glanced at me. 'I prefer to overlook that last remark,' he said coldly, and the meeting broke up.[11]

Reconstructed dialogue three decades after an event can be a trap for historians; moreover, *Münsterland* was too valuable a target to cancel the mission. However, whether or not the exchange – which, after all, bordered on insub-

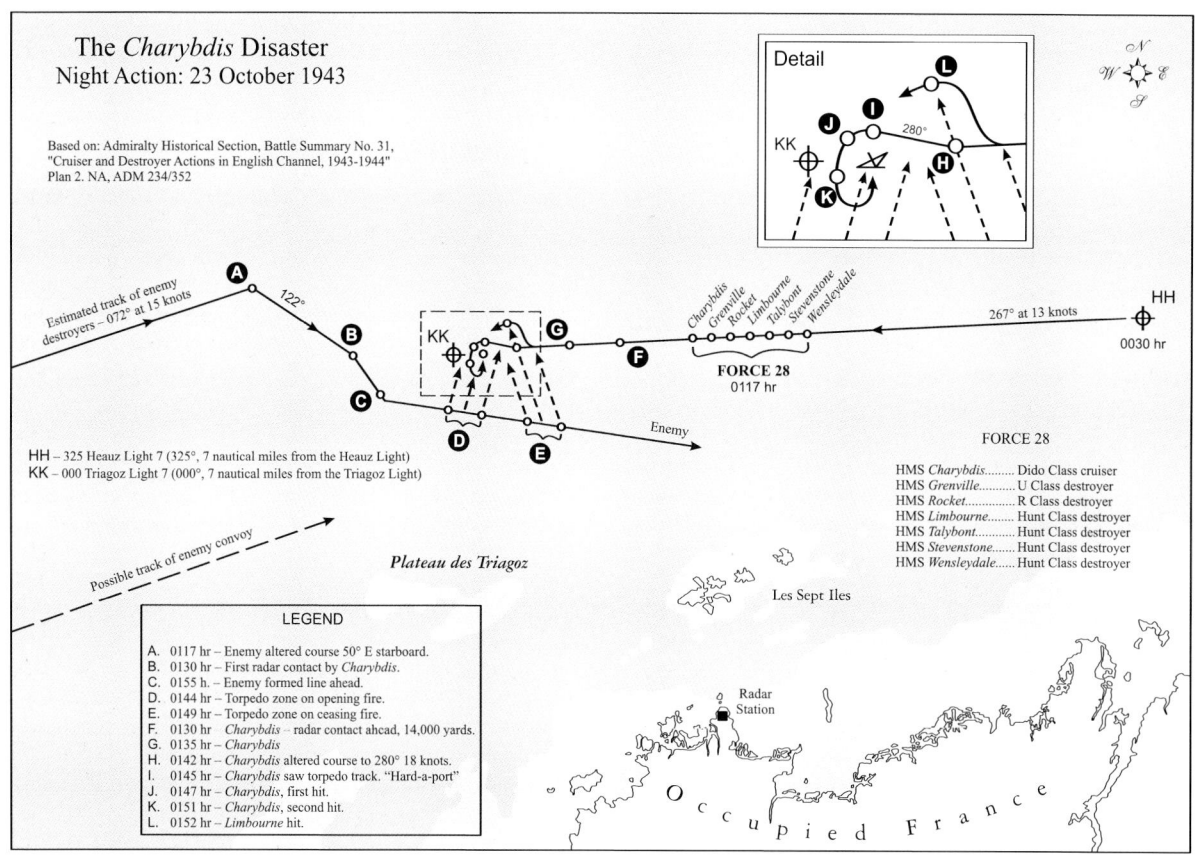

ordination – occurred as Hill recalled, that same day he did express similar sentiments on paper. Sometime before Force 28 departed Plymouth on 22 October, Hill submitted an analysis of the earlier 3/4 October 'Tunnel', including these recommendations:

8. General.
(a) The enemy's maximum speed was 33 knots. Our own forces were HM Ships *Grenville* and Ulster with a speed of 31.7 knots and the Hunts with a speed of 25 knots.
(b) This force had never been to sea together before.
(c) HMS *Grenville* had no Radar Type 271 fitted and had to keep radar silence until the enemy was sighted. Thus it was not possible to track or have any idea as to the number, disposition or course of the enemy forces until sighting took place. As ahead arcs are screened on Hunts 271 a zigzag of 20 degrees either side of the mean course had to be carried out.
(d) The ship was at sea seven days later having had HMCS *Athabaskan*'s 271Q fitted in lieu of the searchlight concurrently with repairs.
(e) No 'Headache' set was fitted in HMS *Grenville*.
(f) As this part of the French coast is one of the very few remaining places where enemy destroyers can be brought to action, it is respectfully submitted that a Division of fast destroyers, preferably 'Tribal' or 'M' class, (as the enemy always runs away) their Leader being fitted with a lattice mast, would very soon sink the lot, which might make the 'Narviks' come out.[12]

Hill's analysis cannot be confused with a normal after-action report since he had opportunity to submit it when *Grenville* was in dock for repairs after the 3/4 October action, plus he had sailed on four 'Tunnels' since. Rather, Hill was getting things off his chest, and whether he sat on the report for days or penned it directly in advance of the 22 October 'Tunnel', it confirms his disquiet. Sadly, it foreshadowed the disaster that befell Force 28.

As before, the inferior speed of the 'Hunts' affected tactical considerations. Plymouth Command deferred a decision over the direction the sweep should follow until the last minute. Given *Marine Gruppe West*'s practice of the destroyer screen preceding a convoy, Vice Admiral Leatham thought that if it could be established precisely when the *Münsterland* left Brest, a west-east sweep from astern 'might give the best chance of sinking the convoy.' However, 'if the time of sailing and position of the convoy is quite unknown, the low speed of the Hunts combined with the necessity for sailing from Plymouth at dusk in order to achieve surprise, reduces the chances of an overtaking contact.' As it was, inclement weather grounded the aerial reconnaissance, nor was there precise special intelligence; therefore, 'definite information about the convoy's sailing was not available.' Consequently, planners decided an east-west sweep 'would afford the best chance of interception.' Due to Force 28's lack of familiarity and tactical uniformity, line-ahead was again utilised; and in a perfect scenario, *Charybdis* and the

Lieutenant-Commander Roger Hill in 1942, looking every inch a destroyer officer. A veteran of the first nine 'Tunnels', including an action with enemy destroyers, Hill's warnings prior to the *Charybdis* 'Tunnel' proved prophetic. (CS Hill)

Fleets would engage the destroyer screen while the 'Hunts' tackled *Münsterland*. Even though a rising moon might silhouette Force 28, Leatham thought 'The weather conditions were expected to favour surprise, low cloud and showery conditions being forecast, and our forces would be expected to meet the enemy.'[13]

Instead, the enemy met them. At 0030, *Korvettenkapitän* Kohlauf, steaming ahead of the *Münsterland* convoy with five torpedo boats, received warning of Force 28's presence from coastal radar, which guided him to an intercept position after the British turned west. Force 28 had good tactical information of its own, but its utility was marred by miscalculations and flawed communications. *Limbourne* and *Talybont*'s 'Headache' operators monitored enemy manoeuvring signals as the two forces approached, but Voelcker failed to understand the significance of the information and the 'Hunts' neglected to pass on subsequent intercepts. At 0135, Voelcker informed his ships that *Charybdis* had a radar contact 14,000 yards ahead, but the subsequent signal for Force 28 to turn together to 280 degrees was received only by *Stevenstone*. As Leatham later described:

Thus at 0130 the situation in our force was that *Charybdis* knew that there was an enemy force 7 miles ahead and closing, but did not know its composition; while *Limbourne* and *Talybont* knew that there was a force of 5 ships (probably destroyers by the procedure) in the close vicinity, but did not know where.[14]

The range continued to drop, and at 0145 *Charybdis* fired starshell set at 4,000 yards to backlight the approaching contacts. It was too late. Two minutes earlier, Kohlauf's *T-23* sighted *Charybdis* against the brightening horizon bearing 350 degrees at an estimated 2,500 yards. He increased speed, led his force 120 degrees to starboard and began to launch torpedoes. *Charybdis* was staggered by the first of two torpedoes, with *Limbourne* hit moments later. Others narrowly missed *Grenville* and *Wensleydale* – personnel on the bridge of the latter ship actually heard one hiss past.

Force 28 lost all cohesion when the torpedoes struck. Lt-Cdr Hill recalled: 'for the next fifteen minutes the ship was manoeuvring to avoid collision.' A rain squall frustrated attempts to find Kohlauf who had moved off to the south. Forty minutes passed before Hill realised that the duties of senior officer had settled on him, but he hesitated to conduct rescue operations since he thought the attack had been carried out by E-boats that might be lurking in ambush by the two stricken ships. He sought instructions from Leatham who directed *Grenville* and *Rocket* to search to the west for *Münsterland*, leaving the 'Hunts' to rescue survivors. Attempts to tow *Limbourne* failed and when Hill returned to the scene after finding only empty seas, he ordered her to be torpedoed. The remnants of Force 28 left the scene at 0600; Kohlauf's torpedo boats made St Malo untouched, while *Münsterland* put into Lézardrieux.[15] George Voelcker and 500 others perished, while Walter Phipps, Force 28's No 2, was concussed after being catapulted off *Limbourne*'s bridge by the torpedo blast, leaving him with no recall of the action. Three months passed before Plymouth launched another 'Tunnel'.

The Reckoning

A defeat of this magnitude – one Admiralty staff officer described it as 'a very inglorious action' – sparked intense scrutiny. A Board of Inquiry considered only the direct cause of the losses and the fate of confidential books, but Plymouth Command and the Admiralty weighed the entire operation. Leatham praised Kohlauf's skill, but observed that Force 28's reaction to the initial enemy reports was flawed. He emphasised two errors: Force 28's neglect in exchanging radar and 'Headache' reports during the approach, and Voelcker's decision to close the enemy instead of gaining sea room to clarify the tactical situation. Plymouth's Commander (D), Cdr Morice, also thought Voelcker failed to appreciate the tactical possibilities:

> The enemy covering force was placed in circumstances in which it was forced to seek close action with a superior antagonist. Our own ships were sufficiently powerful to have enabled some of them to break through and engage the convoy and its close escort in addition to annihilating the covering enemy force.[16]

Admiralty analysts also criticised Voelcker's decision-making. Rear Admiral Patrick Brind, Director of the Operations Division (Home), summarised their viewpoint:

> The loss was due in large part to a lapse of judgement by an experienced and very well thought of officer – the captain of *Charybdis*. To continue on a steady course towards unknown vessels almost ahead was dangerous, particularly in view of the unfavourable conditions of light – a rising moon.[17]

These were telling points but, of course, Voelcker was unable to defend himself. Left unsaid was that an officer completely untried in such operations was given scant time – 24 hours – to plan the mission ahead, ponder appropriate tactics and weigh the capability of ships he barely knew. That, surely, was a factor in what unfolded.

Reset

Leatham closed his report with a detailed annex entitled 'Operations Against Enemy Coastal Convoys on the North Coast of France Between Cherbourg and Brest.' After reviewing the challenges posed by the sweeps and detailing the forces available to both sides, Leatham added recommendations to improve 'Tunnels'. These included deploying aerial strike forces and submarines, but these were ultimately rejected due to the inability of Coastal Command's strike wings to operate effectively at night and the challenges submarines would encounter along the tricky Breton coast. Three other proposals – increased aerial reconnaissance, sustained offensive mining and enhanced support from coastal forces – were adopted. Nonetheless, it was understood that surface forces would shoulder the burden. To that end, Leatham observed that a force similar to that deployed on 22/23 October should, in theory, 'be adequate to overwhelm the enemy force'; although he admitted 'the disadvantages of a mixed force are considerable':

> In the case of *Charybdis*, it was the cruiser's first trip to the North French coast, the Captain was unpractised in 'Headache' technique and had not worked previously with the Hunt destroyers, and the cruiser offers a larger target to radar as well as to gun and torpedo fire. The Fleet destroyers are a match for the enemy in speed though not in total gunpower, and the Hunts reduce the balance of gunpower, but have insufficient speed to maintain contact if the enemy retires. The Hunts have the best local knowledge of this type of warfare, but their opportunities for tactical training are few, as already mentioned. The Hunts have so many other duties that it is never certain how many and which, if any, will be available.

Force 28's composition was only part of the problem; equally important was a lack of familiarity with the task in hand. The right ships required the right experience. To achieve that, Leatham urged the establishment of a dedicated strike force formed of an 'adequate division of Fleet destroyers', preferably of the 'Tribal' class with their powerful gun armament. The Admiralty agreed, replying: 'In view of the importance attached to these operations, it is intended that in future you will have at your

"FOOLIN' AROUND THE FRENCH COAST": THE CHALLENGE OF OPERATION 'TUNNEL', SEPTEMBER 1943 – APRIL 1944

disposal, a more homogeneous and well-trained force.'[18]

But where to get the ships? There were simply too few cruisers and Fleet destroyers to meet demands. As a reminder of this, three days after the *Charybdis* operation, *Grenville* and *Rocket* were transferred to the Mediterranean, and any cruisers at Plymouth were required for 'Stonewall' patrols.[19] Even new construction could not satisfy needs. The 24 destroyers of the 'T', 'U' and 'V' classes emerging from shipyards had to be distributed amongst the Home, Mediterranean and Eastern fleets, with the home commands having lower priority; likewise with cruisers. The challenges in pulling warships together was revealed by preparations for the amphibious exercise 'Duck', scheduled to be run by Plymouth early in the New Year. Explaining that the 'Hunts' of 15 DF would be tied down by convoy escort duties, on 21 December Leatham informed the Admiralty: 'Prudence demands a covering force against the potential threat from enemy destroyers and T[orpedo].B[oats]. based in N.W. France. A force of 4 destroyers is suggested. It cannot be provided from Plymouth Command.'[20] On Christmas Day, the Admiralty issued a negative response, informing Leatham that no reinforcements were available and 'Duck' should be reduced accordingly.

The situation quickly changed. The destruction of *Scharnhorst* on 26 December, followed by victory over *Kriegsmarine* destroyers by the cruisers *Glasgow* and *Enterprise* in the Bay of Biscay two days later, enabled the Admiralty to concentrate a force 'of ultimate strength' at Plymouth. Within 48 hours of the Battle of North Cape, they notified the Home Fleet:

> It is intended to establish at Plymouth a force of two 6' and 2 5.25' cruisers and 8 heavily gunned Fleet destroyers with the intent of destroying enemy surface forces in the Channel and Bay areas, protection of convoys and craft engaged in amphibious exercises and to intercept blockade runners.

'As a first step', they directed the Home Fleet to 'sail 4 Tribals to Plymouth as soon as practicable', and to designate which destroyers would follow.[21] However, after *Glasgow* and *Enterprise* sank the 'Narvik'-class destroyer Z-27 and the torpedo boats T-25 and T-26 in a running daylight action on 28 December, the Admiralty scaled down the number of Fleet destroyers required at Plymouth from eight to five. Shorty thereafter, the 'Tribal'-class destroyer HMS *Ashanti*, and her Canadian sisters *Haida*, *Athabaskan* and *Iroquois* sailed south to form the 10th Destroyer Flotilla (10 DF). A fourth Canadian 'Tribal', HMCS *Huron*, was retained at Scapa Flow for an additional Russian convoy before she joined the flotilla. As for cruisers, the newly worked-up *Bellona* and *Black Prince* eventually joined Leatham's strike force, albeit on a rotating basis.

Leatham had long wanted to form Plymouth's strike force around 'Tribals'. As noted previously, he thought

HMCS *Haida* at Plymouth. The veteran of nine 'Tunnels' displays the powerful gun armament that made 'Tribals' Admiral Leatham's choice for the Plymouth strike force. Her radar-assisted gun director is atop the forward superstructure, and since she retains her original tripod foremast, the Type 271Q search radar is perched abaft the rear funnel where it was wooded by the forward superstructure. In contrast to their British sisters, the RCN 'Tribals' had the 2pdr pompom mounted high on the after canopy to improve firing arcs. (Department of National Defence)

HMS *Ashanti* sailed on eleven 'Tunnels', more than any other ship. In contrast to *Haida*, she was fitted with the latest fighting systems, including a lattice foremast robust enough to support a Type 272 aerial, thus enabling all-round search. Strangely, the censor appears to have erased her pendant number (G 51) but has left her various radar and HF/DF antenna. (Author's Collection)

In this image of HMS *Bellona* one can discern the reduced bridge superstructure and funnels that gave her a smaller silhouette than the earlier *Dido*s like *Phoebe* and *Charybdis*. Although *Bellona* did not see action on a 'Tunnel', she led the 10th Destroyer Flotilla through much of their training. (US Naval History and Heritage Command, NH-79175)

standard Fleet destroyers lacked the punch to overmatch the German torpedo boats, but the six 4.7-inch and twin 4-inch guns of the 'Tribals' satisfied that criterion. Some Admiralty officials argued for the 'J', 'K', or 'N'-class due to their greater torpedo armament and lower silhouette, an advantage in night warfare, but Leatham's view held – the fact there were only the five 'Tribals' at Scapa helped since there was no need to break up a larger, homogeneous flotilla. Besides their powerful gun armament, the big 'Tribals' featured a mix of modern fighting systems. *Ashanti*, *Athabaskan* and *Tartar*, which joined the 10 DF in February, had improved Type 272 or 276 search radar mounted on lattice foremasts with advanced Plan Position Indicator (PPI) displays; *Ashanti* and *Tartar* also had the new Action Information Organisation (AIO) to enhance tactical awareness. In contrast, *Haida*, *Huron* and *Iroquois* had the less capable Type 271Q mounted amidships with more rudimentary A-scan displays and plotting systems.[22] In terms of seasoning, the RCN 'Tribals' had sailed on several Russian convoys and their COs were all veteran destroyer captains, while *Ashanti* and *Tartar* boasted plenty of experience with the Mediterranean and Home fleets.

The choice of *Bellona* and *Black Prince* also made sense. Part of the five-ship *Bellona* sub-group of the *Dido* class, the small light cruisers were well-suited to 'Tunnels'. One role originally envisaged for the *Dido*s was as flagships for Rear-Admiral (D)s, leading destroyers on a range of missions, including offensive raiding and patrol operations. Their eight 5.25-inch guns and dual triple 21-inch torpedo tubes enabled them to handle destroyers, plus they were fast and relatively handy. Moreover, since they were not fitted with a fifth 5.25-inch mount in 'Q' position like the earlier *Dido*s, their forward superstructure had been lowered with their funnels proportionately reduced, presenting a lower silhouette. Since both cruisers were fresh from work-ups with the Home Fleet, their availability was as advantageous as their capabilities. The two 6in-gun cruisers the Admiralty originally envisaged allocating to the Plymouth strike force never materialised, perhaps because they were considered too large and unwieldy for the relatively close-quarters night action that characterised fighting in the Channel.

Launching the Second Phase

The inherent difficulty in running offensive sweeps off an enemy coast did not ease in the second stage of the 'Tunnel' programme, which ran from mid-January to the end of April 1944. Although the Plymouth strike force featured the desired capabilities and underwent a comprehensive training regime, circumstances continued to thwart successful outcomes to the majority of their sweeps.

Admiral Leatham – he was promoted on 9 December 1943 – dispatched four 'Tribals' on a 'Tunnel' with three 'Hunts' within days of their arrival at Plymouth in mid-January. Following a chaotic affair that saw the two divisions lose contact with one another off the French coast, Cdr Harry DeWolf, RCN, the veteran destroyer officer who commanded HMCS *Haida*, recommended comprehensive training in night operations. Perhaps Leatham had wanted to hammer home that very point, and he laid on an extensive sea training program to boost confidence and expose the ships to the demands of night operations, particularly the relatively new art of radar interception. DeWolf's report for March 1944 reveals the extent of the training:

> Night Encounter Exercises were carried out on the 3rd, 7th and 15th of March. The Force has also been exercised in manoeuvring, evading fighter attacks and radar tracking. Destroyers have carried out high and low angle shoots, and towing evolutions. Comprehensive communications and plotting exercises were carried out on the 4th, 8th and 16th of March.[23]

"FOOLIN' AROUND THE FRENCH COAST": THE CHALLENGE OF OPERATION 'TUNNEL', SEPTEMBER 1943 – APRIL 1944

Haida and the others also fulfilled operational commitments, participating in three defensive patrols, two invasion exercises, a minelaying sortie and a sweep into the Bay that was cut short. They also undertook four 'Tunnels'. The practice of mixing training and operations proved risky but, in the end, underscored the demands of night fighting.

Plymouth Command launched twelve 'Tunnels' from early February through to the end of April 1944, when the series terminated. Except for a sweep on the night of 4/5 February, all involved the 'Tribals' of 10 DF: that operation was initially assigned to them but they briefly returned to the Home Fleet for Operation 'Posthorn', a Fleet Air Arm strike off Norway. Instead, four 'Hunts' carried out the 'Tunnel' – the final one they would participate in – and they fought a sharp gun action against two minesweepers and the torpedo boat *T-29* off Ile Vierge. The 'Hunts' damaged *M-156*, but broke away when 'Headache' suggested E-boats were about to launch a torpedo attack. No such attack took place, nor were E-boats even present. Though successful – *M-156* became the first *Kriegsmarine* casualty associated with a 'Tunnel' when Typhoon fighter bombers finished her off the next day – the operation highlighted concerns over E-boats, and bogus reports of their presence marred the next two 'Tunnels'.

On the night of 25/26 February, *Bellona* led *Tartar*,

The 10th Destroyer Flotilla engaged in comprehensive training serials to help prepare the ships for Operation 'Tunnel'. Here, *Haida* in the foreground and a RN 'Tribal' at left, probably *Ashanti*, conduct high-speed manoeuvres under the guidance of *Bellona*. (Department of National Defence, DND R-1044)

Haida, *Athabaskan* and *Huron* south from Plymouth; *Tartar* had only recently joined 10 DF as flotilla leader. *Bellona*'s CO, Capt Charles Norris, was Senior Officer of Force 28. Norris had been with his first wartime command for only seven months, and although he had led the destroyers through many of their training serials,

Led by *Bellona*, four 'Tribals' manoeuvre from line ahead into sub-divisions positioned on the cruiser's bow, the formation utilised in the 26/27 April action. The ship in the foreground is again HMCS *Haida* (G 63). During the latter 'Tunnels' the Plymouth forces shifted from line ahead when they began their run along the enemy coast. (Department of National Defence)

this was his first 'Tunnel'. At 0140, when Force 28 was two-thirds of the way across the Channel, *Bellona*'s Type 272 detected a contact 'fine on the starboard bow, distance six miles.'[24] More contacts popped up, leading Norris to suspect they were E-boats, and he ordered evasive action to port. As the manoeuvre was carried out, 'fresh and separate contacts, close in, appeared about every 40 degrees during the turn.' About to order starshell, Norris decided 'the "enemy" plan of attack was "too perfect" and that we were in presence of some freak of nature, rather than E-boats.' At this point a 'more experienced' radar operator took over and reported that the 'freak of nature' was actually 'a low flying aircraft.' That mystery settled, Force 28 resumed its southern course.

At 0322, after they assumed their eastward sweep up the coast, it appeared they had found opposition torpedo boats when *Tartar*'s Type 276 picked up a surface contact bearing 141 degrees at eight miles. *Bellona* confirmed the contact, and at 0325 *Tartar* reported that her plot indicated its course was 180 degrees at nine knots. At 0327 *Bellona* illuminated the contact and Norris ordered the 'Tribals' to engage. Starshell 'disclosed what appeared to be an enemy Destroyer' and *Bellona* opened up with her main armament. *Tartar*, which also reported enemy ships, joined in. Unhappily, as the range dropped, a chagrined Norris reported: 'These "ships" quickly changed their form into small islets, lands, etc. Fire was ceased accordingly.'

The 'action' reveals several problems that could be encountered on 'Tunnels'. Regarding the first incident at 0140, Norris reflected that 'an experienced Radar operator would undoubtedly have been able to interpret these contacts as a low flying aircraft, but, unfortunately, the few more highly trained men cannot be on watch all the time.' One would have thought the most qualified rating would have been at the set – perhaps *Bellona* had not yet gone to action stations; nonetheless, radar was still in relative infancy, and false echoes and ground returns often impaired results. That, strong tidal currents and flawed navigation – Norris explained that an unusual southerly set meant their position was actually 'a good 3 miles to the southward of that estimated' – combined to cause the erroneous identification of the islets. As it was, Force 28 narrowly avoided disaster when the destroyers approached to within two miles of dangerous shoal waters before breaking off their high speed 'pursuit'. Accumulated experience would lessen such problems as sailors became more familiar with their equipment and the treacherous currents and rocky coastline of Brittany – more helpful would be the radio navigation aid QH-3, the naval version of Bomber Command's GEE, fitted later in the spring, which eliminated many of the navigational challenges associated with the Channel.

Kriegsmarine destroyers were not at sea that night, but were active when similar issues beset the next 'Tunnel'. On the night of 1/2 March, Force 28, comprising *Bellona* and five 'Tribals', was directed to sweep westward from Ile de Batz towards Ushant; intelligence revealed an opportunity – 'thought to be a good one'[25] – to meet enemy destroyers. The mission got off on the wrong foot when a radar breakdown forced *Bellona* to withdraw, and Cdr St John Tyrwhitt, D10 and CO of *Tartar*, only learned he was leading the mission as he took his 'Tribals' out of Plymouth. Well-seasoned, Tyrwhitt had commanded *Tartar* through hard fighting in the Mediterranean since taking command in June 1942; nonetheless, this was only his second 'Tunnel' and the first he led.

Events began to unfold as they approached the French coast. At 2343, when about 20 miles off Les Sept Isles, *Tartar*'s Type 276 detected a contact 10 miles ahead. A radar officer on the Plymouth staff who was on board *Tartar* – likely due to the issues of the previous mission – assured Tyrwhitt the contact comprised eight echoes. Tyrwhitt 'judged this to be a flotilla of large 'E' boats, there being no reason to suppose that the enemy had a similar number of large ships in the area.' Choosing to avoid such a clash, he altered to evade, and at 0130 Force 28 assumed a westward course down the coast. When that came up empty after half an hour, Force 28 turned back to the east. The next 45 minutes were fraught with confusion and doubt. Numerous small radar contacts popped up inshore; however, Tyrwhitt later related that none 'was large enough to suggest it being a ship.' At 0226 and 0227, *Haida*'s 'Headache' monitored enemy transmissions repeating call signs that had been used in the October 1943 actions and indicating torpedoes had been fired. That caused Tyrwhitt to alter course to seaward. When Force 28 turned to 075 degrees ten minutes later, *Haida* detected a radar contact three miles off the starboard bow, while *Tartar*'s 'Headache' overheard more torpedo firing orders. Force 28 again took avoiding action, but starshell fired southward illuminated only empty seas. Tyrwhitt had had enough, deciding 'By this time it was obvious that our presence was well known to the enemy and it was suspected that 'E' boats were attempting to attack.' 'Accordingly', he headed home. In fact, there were no E-boats lurking; rather, four torpedo boats under Kohlauf had been guided into contact by coastal radar and launched two unsuccessful torpedo attacks. Expressing relief that Kohlauf escaped undetected, *Kriegsmarine* analysts assessed, correctly, that the British failed to ascertain the composition of his force, thus missing an opportunity to exert their 'superior battle strength'.[26]

The 'Tunnel' caused consternation, and in his report to the Admiral Leatham judged the result 'disappointing'. He accepted Tyrwhitt's initial evasion of E-boats, though he considered evidence of their presence 'slender and unconvincing'. He was, however, critical of Tyrwhitt's subsequent decisions, when 'there were definite indications, by both radar and Headache, of the presence of an enemy force to the south and east of Force 28':

Senior Officer, 10th Destroyer Flotilla at the same time turned his force 90° to avoid torpedoes which he presumably thought must be coming from E-boats, but took no action to try to locate the enemy. This omission was unfortunate, as it seems certain that a starshell sweep to the south-east at 0227 would have revealed them. From the call signs used, it seems likely that these would have proved to be Elbings.

'Sufficient steps', Leatham emphasised, 'were not taken to verify the very uncertain information received in spite of the enemy's sting having apparently being drawn at 0226 or thereabouts.'[27] 'Disappointing', 'slender and unconvincing', 'took no action' and 'omission' are expressions no officer would want applied to a mission they had led.

In fact, ten days before Leatham submitted his analysis, Tyrwhitt had been replaced as D10 and appointed to the Admiralty. The reasons for this remain unclear, but it may well be that Leatham lost confidence in his leadership after the 1/2 March 'Tunnel' – Tyrwhitt may also been due for a rest; however, *Tartar* had just emerged from a three-month refit, which should have provided the required respite.[28] No matter the rationale, the flotilla grasped a clear message; be aggressive and think offensively. Tyrwhitt had twice turned away from suspected torpedo attacks without searching for the enemy. Although E-boats were elusive and dangerous, Leatham emphasised to his commanding officers that confronting them 'must not be allowed to colour their judgement to such an extent that action against the enemy is not taken.'

> Depending on the circumstances, which only the commanding officer can gauge, evasive action (e.g. to avoid a position of tactical disadvantage) may have to be taken, but it does not mean that the main objective is to be missed as a result of it. In fact to reach it, it may be necessary to fight a way through the E-boats, which are likely to be frightened of a Tribal bearing down on them with guns in action.

Risks had to be accepted, and in the aftermath of the command change, the flotilla became single-minded in its pursuit of Leatham's 'main objective'. *Haida*'s Cdr Harry DeWolf, the flotilla's No 2, later recalled 'my feeling at the time was that we had failed to go after some ships, some echoes, so from then on in the back of my mind was if we ever get an echo, we're going after it. So, if we found one, we did. That was just natural fear of criticism.'[29]

Pinnacle

Commander Basil Jones, RN, took over as D10 and CO of *Tartar* on 15 March 1944, and although he did not lead the flotilla in action until June, he had an impact on how they fought. A veteran of destroyer sweeps in the Mediterranean where he probably came to Leatham's attention, Jones had been slated to command an escort group when he was instead ordered to Plymouth. He had an opportunity to assess the challenges ahead when he sailed on three 'Tunnels' in the last week of March, two as Senior Officer. Jones thought the biggest question regarding sweeps down a 'well-known "tramline"' that typically resulted in a head-on encounter with the enemy turning away and launching torpedoes, was the formation to be used. 'Tunnel' forces had previously relied almost exclusively upon line-ahead while sweeping down the enemy coast, but Jones thought it 'vulnerable', probably because ships obstructed the radar and guns of those ahead. One cannot confirm who conceived the tactical solution but Jones, Captain Norris and Commander Morice, the Cdr (D) Plymouth, probably contributed; Cdr DeWolf recalled playing no part. Improved technology eased their task, and the cohesion derived through group training stoked confidence in different formations. To Jones, the fact that all ships had centimetric radar with revolving aerials, some with PPIs, lessened 'the risk of mutual interference' and made new formations 'feasible'. The one adopted for mixed forces saw a sub-division of destroyers positioned 1.5 miles, 40 degrees on each bow of the cruiser:

> The force would be manoeuvred into contact by the cruiser, who would then illuminate the enemy with her more powerful and long range starshells and release the destroyers to attack from either bow or quarter of the enemy. Continued illumination by the cruiser allowed all destroyers' guns to be used offensively, while the cruiser's own offensive and longer-range guns would tend to slow down, if not destroy, the enemy until the destroyers were in close action.
>
> All three parts of the formation had greater freedom of action in avoiding the enemy torpedo attack to be expected, and the radar equipment allowed continuous watch on all parts of the formation.[30]

To gain confidence with the tactics, the Plymouth strike force continued its high-speed manoeuvring and plotting serials whenever opportunity arose. Also, seven 'Tunnels' between mid-March and the final days of April, three including *Bellona* or *Black Prince*, added realism to the training.

These 'Tunnels' demonstrated how the operations had evolved. Force 26/28 was invariably supported by both coastal forces and patrol aircraft. Three or four MTBs were positioned beyond the extremities of the patrol route to report on opposition surface activity. Under strict instructions not to reveal their presence, to avoid friendly fire incidents, they were prohibited from launching attacks until the strike force cleared the area. Aerial reconnaissance took the form of a Coastal Command ASV radar-equipped patrol aircraft that swept the area prior to the main force's arrival – an activity that also provided cover for 'Ultra' intelligence. Additionally, the radar plot at Plymouth's combined headquarters kept Force 26/28 apprised of E-boat activity. One thing that had not much changed was that the sailors carrying out the 1944 'Tunnels' had the same mixed feelings as those who had experienced them the previous autumn.

Athabaskan's Leading Writer, Stuart Kettles, described the reaction to news that they were going on another sweep in late April:

> Yes, once again we were going to F.A.F.C. What under the sun does F.A.F.C. mean? Well, in sailors' language it wouldn't be polite to say but to give you a general idea, F.A.F.C. means "Foolin' around the French Coast", and that is what we had been doing for about six weeks.[31]

The fact they had yet to meet victory while engaged in such "foolin' around" was surely of little comfort.

The Final 'Tunnel'

Only one of the final seven 'Tunnels' run from mid-March resulted in an interception, and it only occurred because of a miscalculation by *Korvettenkapitän* Kohlauf. On the night of 26/27 April, Kohlauf was at sea with *T-29*, *T-27* and *T-24* supporting a convoy proceeding westward for Brest. There was no moon, and visibility on the starlit night was about two nautical miles. At 0106, when Force 26 – *Black Prince*, *Haida*, *Athabaskan*, *Ashanti* and *Huron* – was 18 miles off the coast, the cruiser's radar monitor indicated they were being held by enemy coastal radar. When informed of their presence, Kohlauf assumed they were heading to intercept the convoy to the west of his torpedo boats and held his westward course. In fact, Force 26 had not detected the convoy and soon turned on its prescribed west-east sweep – aware they had been detected by enemy radar, Captain Dennis Lees, Senior Officer of Force 26 in *Black Prince*, thought that 'there seemed to be little hope of an engagement'.[32] However, coastal radar failed to alert Kohlauf of Force 26's alteration to the east until 0201, at which point the two forces were on a collision course. At 0200 *Black Prince*'s Type 272 reported an echo ahead at 21,000 yards and, with the 'Tribals', tracked the contact as it steadily closed until, after nine minutes, it suddenly reversed course. Finally warned that Force 26 lay directly ahead, Kohlauf had wheeled around eastward to lead it away from the convoy. For this he was later criticised, the *4.Torpedoboot Flotille* war diary concluding that 'a headlong attack launched without changing the westward course might have brought a hard-fought-for success. In situations which appear to be hopeless the boldest decision leads in most cases to success.'[33] Perhaps, but a 'headlong attack' was contrary to *Kriegsmarine* doctrine, and Kohlauf probably did not consider his position 'hopeless'. As it was, his turn away sparked a chase up Channel with Force 26 nipping at his heels.

At 0220, with the range at 13,100 yards, Lees thought the enemy might succeed in pulling away even though Force 26 was surging ahead at 30 knots. He therefore signalled 'L.C.' to his destroyers:

> The particular significance of the signal 'L.C.' when used by Force 26 is that the destroyer sub-divisions proceed on the best course and speed to engage the enemy whilst the cruiser manoeuvres to achieve the best position for providing starshell illumination at the same time avoiding possible enemy torpedo zones and keeping clear of the melee.

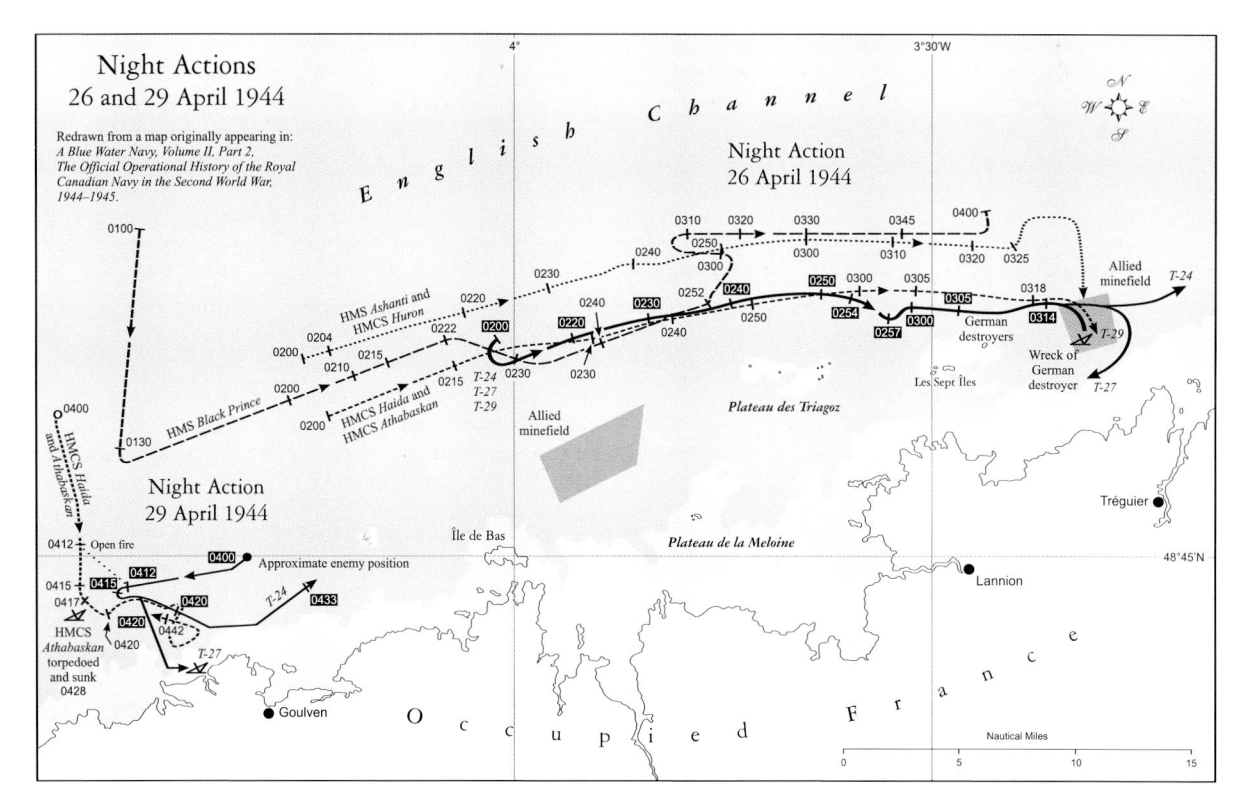

In perhaps her final pose, a battle-worn *Athabaskan* glides into Plymouth the morning after the action of 26/27 April, when she played a major role in sinking *T-29* and damaging *T-27*. Her loss two nights later with 128 of her crew, including CO Lt-Cdr John Stubbs, was the major warship loss suffered by Canada in the Second World War. (Authors Collection)

In a sombre exchange upon *Haida*'s return to Plymouth, Commander DeWolf and Admiral Leatham discuss the 28/29 April action that saw the loss of *Athabaskan* and the destruction of *T-27*. DeWolf, Canada's most respected sailor, had an exemplary career at sea and ashore and retired as Chief of the Naval Staff. (Author's collection)

Here, then, is a lesson from the *Charybdis* 'Tunnel': keep the cruiser clear of immediate danger, but continue to support the destroyers. And, unlike the destroyers accompanying *Charybdis*, the four 'Tribals' had worked together often, had matching capability and, almost instinctively, could implement tactics absorbed through hard training.

In contrast to the 3/4 October 1943 'Tunnel', when Kohlauf's torpedo boats outpaced the 'Hunts' and disabled *Ulster* and *Grenville*, this time the four 'Tribals' turned the tables. The bulk of the action fell to Force 26's 2nd sub-division of *Haida* and *Athabaskan* positioned on *Black Prince*'s starboard bow. After being turned loose, *Haida* opened fire at 0226 and the enemy responded two minutes later, but the forward gun armament of four radar-controlled 4.7s that was a feature of the 'Tribals' paid dividends. They scored hits on *T-27* within 10 minutes, causing her to break south towards the coast.[34] *Haida* tracked the move but DeWolf continued after the other two torpedo boats fleeing east. As *T-27* approached the loom of the coast she fired a spread of torpedoes as Force 26 swept by: all missed, but they caused *Black Prince* to turn away. For the rest of the night the cruiser shadowed the action from seaward, 'in case one of the enemy doubled back or one of our own destroyers required assistance.' *T-27* eventually escaped into Morlaix.

Finding the range on Kohlauf's *T-29*, at 0320 *Haida* and *Athabaskan* hit the torpedo boat, damaging the rudder, severing steam lines and cutting her speed. A hit on the bridge ten minutes later mortally wounded Kohlauf. By 0335, *T-29* lay stopped with flames engulfing her upperworks. In the meantime, *Ashanti* and *Huron* had pursued *T-24* to the east, scoring hits on her superstructure. Her speed was unaffected, and when she eventually escaped, the two 'Tribals' rejoined their companions. *Athabaskan*'s CO, Lt-Cdr John Stubbs, described the confusion typical of night action at sea:

> Although by this time burning fiercely, the Elbing maintained a constant fire of close-range weapons as we were circling her. *Huron* and *Ashanti* joined and there was a certain amount of dangerous cross fire although this was unavoidable. Fighting lights had to be switched on to avoid collision.[35]

T-29 was clearly doomed but, adhering to Leatham's call for aggressiveness, DeWolf stubbornly refused to withdraw until the enemy was sunk – incredibly, 15 torpedoes fired at close range at the drifting hulk all missed. Finally, the 0421 entry in *Haida*'s log book rejoiced 'ENEMY HAS SUNK!!'[36] The 'Tribals' turned for home, but *Ashanti* collided with *Huron* as they reformed; both made Plymouth but required docking. When all was done, Captain Lees observed: 'This minor engagement has done much to enhance the already high morale of the ships' companies engaged. They now realise how worthwhile are the days, weeks and months of dull solid, slogging working up practices.'[37]

The 25/26 April 'Tunnel' had a sequel. Two nights later, *Marine Gruppe West* attempted to slip *T-24* and *T-27* down the Channel to Brest for repairs. 'Ultra' intercepts warned of the move and super refraction caused by abnormal atmospheric conditions enabled Plymouth's coastal radar to track the torpedo boats as they steamed west. *Haida* and *Athabaskan*, in mid-Channel supporting

Operation 'Tunnel' 5/6 September 1943 – 26/27 April 1944

Date	Ships	Senior Officer	Contact
5/6 Sep 43	*Phoebe* (SO) *Grenville* *Ulster*	Capt CP Frend, RN	No
6/7 Sep 43	*Phoebe* (SO) *Grenville* *Ulster*	Capt CP Frend, RN	No
8/9 Sep 43	*Phoebe* (SO) *Grenville* *Ulster*	Capt CP Frend, RN	No
10/11 Sep 43	*Phoebe* (SO) *Grenville* *Ulster*	Capt CP Frend, RN	No
3/4 Oct 43	*Limbourne* (SO) *Tanatside* *Wensleydale* *Grenville* *Ulster*	Cdr CB Alers-Hankey, RN	Yes
14/15 Oct 43	*Limbourne* (SO) *Talybont* *Melbreak* *Tanatside* *Grenville* *Rocket*	Cdr CB Alers-Hankey, RN	No
15/16 Oct 43	*Tanatside* (SO) *Melbreak* *Grenville* *Rocket*	Lt-Cdr FD Brown, RN	No
16/17 Oct 43	*Tanatside* (SO) *Talybont* *Melbreak* *Grenville* *Rocket*	Lt-Cdr FD Brown, RN	No
17/18 Oct 43	*Tanatside* (SO) *Talybont* *Melbreak* *Wensleydale* *Grenville* *Rocket*	Lt-Cdr FD Brown, RN	No
22/23 Oct 43	*Charybdis* (SO) *Grenville* *Rocket* *Limbourne* *Talybont* *Wensleydale* *Stevenstone*	Capt GAW Voelcker, RN	Yes
19/20 Jan 44	*Tanatside* (SO) *Brissendon* *Talybont* *Haida* *Ashanti* *Athabaskan* *Iroquois*	Cdr B de St Croix, RN	No
1/2 Feb 44	*Haida* (SO) *Iroquois* *Ashanti*	Cdr HG DeWolf, RCN	No

a minelaying mission – a duty that would have gone to *Ashanti* and *Huron* – were vectored to intercept and surprised the torpedo boats off Ile Vierge. Pummelling from *Haida*'s accurate gunnery quickly forced *T-27* aground; however, a torpedo fired by *T-24* as she wheeled to escape struck *Athabaskan* astern, stopping the destroyer and igniting fires. Minutes later the 'Tribal' blew up in a massive magazine explosion. After chasing *T-24* eastward, *Haida* returned to rescue 44 sailors before daybreak forced her to head home. Others were rescued by the *Kriegsmarine*, but 128 perished. Perched on the Breton rocks, *T-27* was later finished off by torpedoes from MTBs.

Outcomes

The 25/26 April 'Tunnel' was the last of the series. Demands related to the imminent invasion of northwest Europe made the ships difficult to spare; plus, with the destruction of *T-29* and *T-27* and the damage to *T-24*, Plymouth had achieved the goal of whittling down enemy destroyer strength. In all, Plymouth had run 22 'Tunnels'. The enemy was encountered on four, with results that can be judged as mixed: the destruction of *T-29*, while *T-27* and *M-156* were dispatched as a result of damage sustained on 'Tunnels'. No German merchant ships were sunk. While *Marine Gruppe West* sometimes postponed sailings, at no point did they cancel any outright. Against that, Plymouth lost *Charybdis* and *Limbourne*, while *Athabaskan*'s destruction can be linked to a 'Tunnel'. In the cold calculus of war, Allied navies could make good such losses; the *Kriegsmarine* could not.

The odds were stacked against 'Tunnels'. *Marine Gruppe West* knew where the sweeps would occur and that they would come during darkness. Effective planning, forewarning from coastal radar, harbours in which to escape and the skill of Kohlauf's *4.Torpedoboot flotille* all thwarted interceptions. When they did occur, 'Tunnel' forces had to cope with the vagaries of high-speed, relatively close-quarters night action, under a forbidding, enemy-held coast. Plymouth Command and the Admiralty understood these challenges; nonetheless, the missions had to proceed. The Allies were pressing the offensive in advance of the invasion, and it was critical to improve their degree of sea control in the Channel. The risks associated with 'Tunnels' had to be accepted. If they were to wear down the opposition, particularly their destroyer strength – the acknowledged 'prize' – they had no choice but to engage the enemy on his own terms.

In early 1944, when the currents of the naval war allowed, the Admiralty reinforced Plymouth with the warships considered necessary for a permanent strike force. Just as important, they kept them there, and Admiral Leatham laid on the training and seasoning

Date	Ships	Senior Officer	Contact
4/5 Feb 44	*Tanatside* (SO) *Talybont* *Brissendon* *Wensleydale*	Cdr BJ de St Croix, RN	Yes
25/26 Feb 44	*Bellona* (SO) *Tartar* *Haida* *Huron* *Ashanti*	Capt CFW Norris, RN	No
1/2 Mar 44	*Tartar* (SO) *Haida* *Huron* *Ashanti* *Athabaskan*	Cdr StJ Tyrwhitt, RN	No
20/21 Mar 44	*Tartar* (SO) *Haida* *Ashanti* *Huron*	Cdr B Jones, RN	No
26/27 Mar 44	*Tartar* (SO) *Haida* *Ashanti*	Cdr B Jones, RN	No
28/29 Mar 44	*Bellona* (SO) *Tartar* *Haida* *Huron* *Ashanti* *Athabaskan*	Capt CFW Norris, RN	No

Date	Ships	Senior Officer	Contact
5/6 Apr 44	*Ashanti* (SO) *Huron*	Lt-Cdr JR Barnes, RN	No
6/7 Apr 44	*Ashanti* (SO) *Huron* *Athabaskan*	Lt-Cdr JR Barnes, RN	No
24/25 Apr 44	*Black Prince* (SO) *Haida* *Ashanti* *Athabaskan*	Capt DM Lees, RN	No
26/27 Apr 44	*Black Prince* (SO) *Haida* *Athabaskan* *Ashanti* *Huron*	Capt DM Lees, RN	Yes

Senior Officers

Capt CP Frend, RN (4)
Lt-Cdr FD Brown, RN (3)
Cdr CB Alers-Hankey, RN (2)
Cdr BJ de St Croix, RN (2)
Capt CFW Norris, RN (2)
Cdr Basil Jones, RN (2)
Lt-Cdr JR Barnes, RN (2)
Capt DM Lees, RN (2)
Capt GAW Voelker, RN (1)
Cdr HG DeWolf. RCN (1)
Cdr StJ Tyrwhitt, RN (1)

deemed necessary to enhance their performance. Where they fell down was in the provision of stable leadership. The 22 'Tunnels' were led by eleven different senior officers and, even after the formation of a permanent strike force, seven officers commanded the remaining twelve sweeps. Moreover, only six senior officers led consecutive missions. As results demonstrated, experience was critical to the outcome of 'Tunnels', yet no officer was given a real opportunity to learn the job. *Phoebe*'s Captain Charles Frend led the first four 'Tunnels', the most of any officer, but he, like *Limbourne*'s Cdr Byron Alers-Hankey and Lt-Cdr Frank Brown who led five in mid-October 1943, could never enter full stride. Perhaps the greatest blemish came with *Bellona* and *Black Prince*'s Captains Charles Norris and Dennis Lees, who played an instrumental role in training Force 26/28, but who each led only two 'Tunnels'. And, of course, there were the abrupt tests that confronted George Voelker and St John Tyrwhitt. War imposes exorbitant demands and long-range planning can be tricky; nonetheless, more should have been done to provide the 'Tunnels' with stable leadership.

The 'Tunnels' experience shows there is more to going on the offensive than flicking a switch. Even with the stability of a permanent strike force – buttressed by relentless training, accrued experience, and enhanced capability – offensive operations along an enemy coast

Participation of Individual Warships

Light Cruisers (4)

Phoebe	4
Bellona	2
Black Prince	2
Charybdis	1

Fleet Destroyers (9)

Ashanti	11
Grenville	10
Haida	9
Huron	7
Athabaskan	6
Ulster	5
Rocket	5
Tartar	5
Iroquois	2

'Hunt'-class Destroyers (7)

Talybont	7
Tanatside	7
Limbourne	3
Melbreak	4
Wensleydale	4
Brissenden	2
Stevenstone	1

are inherently difficult, particularly at night. The line between success and failure can be thin indeed, and fortune – always a factor at sea – can intercede on either side. Whether the 'Tunnels' can be judged an overall success is debatable, but achieving the desired level of sea control through maritime presence demanded the sweeps be undertaken and the risks ventured.

Acknowledgements:

The author wished to thank Trent Hone, Norm Jolin, Gil Lauzon and Vince O'Hara for commenting on drafts. Danny Covey, Jerry Proc and Tim Lewin contributed images. The maps, drawn by Chris Johnson, are courtesy of *Fighting at Sea: Naval Battles from the Ages of Sail and Steam* (Robin Brass Studio, 2008). Finally, Jim Walker shared information and sparked this study when he queried the circumstances surrounding the tragic loss of *Charybdis* on behalf of two elderly neighbours trying to comprehend the death of their brother, Able Seaman John F Jewes.

Sources:

This study is based upon records held at the UK National Archives (TNA), the US National Archives and Records Administration (NARA), Library and Archives Canada (LAC), and the Directorate of History and Heritage (DHH). Supporting monographs include:

O'Hara, Vincent, *The German Fleet at War, 1939–1945*, USNIP (Annapolis, 2004).
Smith, Peter, *Hold the Narrow Sea*, USNIP (Annapolis, 1984).
Whitby, Michael, 'Shoot; Shoot; Shoot: Destroyer Night Fighting and the Battle of Ile de Batz' in McLean, DM (ed), *Fighting at Sea: Naval Battles from the Ages of Sail and Steam*, Robin Brass Studio (Montréal, 2008).
Whitley, MJ, *German Destroyers of World War Two*, USNIP (Annapolis, 1999).
The section on HMS *Charybdis* at www.naval-history.net is also an invaluable source.

Endnotes:

[1] Director of Operations Division (Home), 4 July 1944, UKNA, ADM 199/532. For the Type 39 torpedo boats see Hervieux, 'The Elbing Class Torpedo Boats at War', *Warship* No 38, 95–102.
[2] After this 'Tunnel', strike forces were designated either Force 26 or 28.
[3] RN Fleet destroyers typically achieved speeds of at least 35 knots on trials, but operational conditions and impediments such as added topweight typically reduced speed. To reflect contemporary perceptions, this study relies upon the speeds mentioned in various reports and appreciations.
[4] Most destroyers with tripod foremasts had Type 271 search radar mounted in the former searchlight position amidships, causing a blind arc of some 30 degrees ahead. New, stronger lattice foremasts could support newer centimetric search aerials, enabling unobstructed all-round search.
[5] Donald, *Stand by for Action*, Seaforth Publishing (Barnsley, 2009).
[6] Kemp, *Friend or Foe: Friendly Fire at Sea, 1939–1945*, Leo Cooper (London, 1995), 23.
[7] Air Historical Branch, 'The RAF in Maritime War: Volume IV', DHH, 79/599.
[8] C-in-C Plymouth to Admiralty, 19 October 1943, Admiralty War Diary, NARA.
[9] Voelcker had a successful career in submarines, but *Charybdis* was his first wartime surface command.
[10] Cited in Smith, *Hold the Narrow Sea*, 190.
[11] Hill, *Destroyer Captain: Memoirs of the War at Sea, 1942–45*, William Kimber (London, 1975), 165.
[12] CO *Grenville*, 'Night Action with Enemy Destroyers', 22 October 1943, UKNA, ADM 199/1038.
[13] C-in-C Plymouth, 'Action with German Ships on the Night of October 22nd/23rd, 1943', 13 November 1943, UKNA, ADM 199/1038.
[14] Ibid.
[15] After being damaged in two major RAF raids directed against her at Cherbourg, *Münsterland* was finally sunk by the Dover batteries on 20 January 1944.
[16] C-in-C Plymouth, 'Action with German Ships on the Night of October 22nd/23rd, 1943'.
[17] DOD(H) minute, 26 January 1944, UKNA, ADM 199/1038.
[18] C-in-C Plymouth, 'Action with German Ships on the Night of October 22nd/23rd, 1943'; and Admiralty to C-in-C Plymouth, 10 February 1944, UKNA, ADM 199/1038.
[19] *Grenville* and *Rocket*'s sailors assumed the transfer was to prevent them from spreading news of the *Charybdis* tragedy; however, the transfer had been formalised in mid-October. Moreover, the four 'Hunts' remained at Plymouth.
[20] C-in-C Plymouth to Admiralty, 21 December 1943, Admiralty War Diary, NARA.
[21] Admiralty to C-in-C Home Fleet, 28 December 1943, Admiralty War Diary, NARA.
[22] A-scans displayed echoes as vertical spikes on a horizontal scale, whereas PPIs presented a bird's-eye view that enabled continuous monitoring of contacts.
[23] *Haida*, Report of Proceedings, 7 April 1944, DHH, 81/520 *Haida* 8000.
[24] CO *Bellona*, 27 February 1944, UKNA, ADM 199/532.
[25] DOD(H) minute, 4 July 1944, UKNA, ADM 199/532.
[26] CO *Tartar*, Report on Operation TUNNEL, 3 March 1944, UKNA, ADM 199/532; and *Kriegstagebuch (KTB) der Seekriegsleitung*, 2 March 1944, DHH.
[27] C-in-C Plymouth, '10th Destroyer Flotilla Report on TUNNEL Night 1st/2nd March', 25 March 1944, UKNA, ADM 199/532.
[28] Tyrwhitt did not suffer professionally, retiring as Second Sea Lord in the rank of Admiral.
[29] DeWolf to author, 5 September 1987.
[30] Basil Jones, *And So to Battle: A Sailor's Story* (Battle, 1975), 75.
[31] Kettles, 'A Wartime Log: A Personal Account of Life in HMCS *Athabaskan* and as a Prisoner of War', DHH, 74/458.
[32] CO *Black Prince*, Report on Action, 2 May 1944, DHH, 81/520 *Haida* 8000. Having previously commanded HMS *Calcutta* through tough fighting in the Norwegian, Dunkirk and Mediterranean campaigns, Lees was the most seasoned cruiser captain to work with the 'Tunnel' forces.
[33] *KTB, 4.Torpedoboot Flotille*, 'Evaluation of Actions Fought by 4th TB Flotilla, DHH, 81/520 *Athabaskan*.
[34] The map incorrectly shows *T-27*'s withdrawal southwards later than it actually took place.
[35] Lt-Cdr JS Stubbs, Report of Proceedings, DHH, 81/520 *Athabaskan*.
[36] *Haida*, Deck Log, 27 April 1944, author's possession.
[37] CO *Black Prince*, Report on Action, 2 May 1944.

THE IJN CARRIERS
SŌRYŪ AND *HIRYŪ*

Sōryū and her half-sister *Hiryū* were arguably the best of the medium fleet carriers completed between the wars, and served with distinction until they were sunk in the Battle of Midway. **Kathrin Milanovich** looks at the development of the design in the context of the London Treaty limitations.

After the failure of the Geneva Naval Conference in 1927, the Japanese Naval General Staff (NGS) set up a committee to study future requirements. More than 70 sessions took place between 24 December 1927 and 6 August 1928. It was determined that four aircraft carriers were required for national defence – not three as previously stipulated – and that these should be of the 'medium' type with a displacement of 13,500 tons standard and an air group of 36 aircraft (three twelve-plane squadrons of fighters, bombers, and reconnaissance aircraft), a powerful anti-aircraft armament, high speed (34 knots) and good endurance They would be supported by a number of 'auxiliary' carriers of 10,000 tons standard carrying 24 aircraft. The urgent need for new construction was emphasised, but with the new definition of the aircraft carrier in the London Arms Limitation Treaty of 1930,[1] the medium carrier provisionally included in the First Naval Armament Replenishment Programme of 1931 became a victim of the red pencil.[2] Due to the abandonment of this project and in preparation for the next replenishment programme, the IJN first turned to a hybrid type combining cruiser and carrier characteristics permitted by the London treaty.

The Design of a 'Flying Deck Cruiser'

The London Treaty not only placed a quantitative limitation on cruiser tonnage and the number of ships – for Japan a maximum of twelve category 'A' (8in/20cm) cruisers with a combined displacement of 108,400 long tons plus an allocation of 100,450 tons for category 'B' (6.1in/15.5cm) cruisers – but stipulated that 'not more than 25 per cent of the allowed total tonnage in the cruiser category may by fitted with a landing-on platform or deck for aircraft' (Article 16 para 5). Further to this, Article 3 para 2 stated: 'The fitting of a landing-on or flying-off platform or deck on a capital ship, cruiser or destroyer … shall not cause any vessel so fitted to be charged against or classified in the category of aircraft carriers.'

The fact that this detailed wording had been incorporated into the treaty at the behest of the US delegation meant that suggestions that the US Navy was considering the construction of a 'flying deck cruiser' (*kōkūjunyōkan*) could no longer be dismissed as rumour, but had to be considered a real danger. In order to oppose these ships a flying deck cruiser was designed by the NTD in 1932 as design number G6 (*kihon keikaku bangō*). However, the IJN constructors opted to prioritise 'carrier' over 'cruiser' characteristics.

The official trial displacement of the 240-metre design was to be 17,500 tonnes. Three superfiring 20.3cm twin turrets Model E (elevation angle 70°) were mounted on the forecastle deck. Immediately abaft the main guns was a 194.2-metre long flight deck which projected over the stern. A total of 70 fighters and torpedo-bombers (reserve planes included) were to be carried, and there were to be six 12.7cm/40 Type 89 twin HA gun mountings. The ship would have had a large island to starboard to house the command spaces, and the exhaust uptakes for the boilers would have emerged to starboard at about half the height of the hull and would have been inclined aft. Horsepower was to be 150,000shp (comparable to the latest cruisers of the *Mogami* class) for a maximum speed of 36 knots.

However, the US flying deck cruiser proposal remained unfunded, and its Japanese counterpart was abandoned before being inscribed into a building programme.[3] The design and construction principles currently in favour with the IJN suggest that this ship would have been top heavy with insufficient stability, and might also have suffered from a lack of structural strength.

The Second Naval Armament Replenishment Programme of 1934

Japanese foreign relations had deteriorated since the Manchurian Incident and grew steadily worse after leaving the League of Nations; Japan became increasingly isolated from the international community. The Japanese–American conflict over the Manchurian question was far from being resolved. The concentration of the US Fleet on the Pacific coast, the proposal of representative Carl Vinson from Georgia for the expansion of the US Navy up to the limits permitted by the arms limitation treaties (4 January 1932) and Roosevelt's Executive Order 6174 (for the construction of two carriers, nine cruisers, 20 destroyers and four submarines) were recognised in Japan as pointers towards an 'armed solution' to the Manchurian question requiring the urgent reinforcement of naval forces.

Prince Fushimi Hiroyasu, who had relieved Admiral Taniguchi Naomi as chief of the NGS on 2 February 1932, submitted a memorandum to Navy Minister Osami Mineo (NGS Secret Document No 154 dated 6 May 1933). Taking into account the international situation, and in addition to the expansion of the naval air force and the fleet during and after FY 1934, he proposed that Japan's numerical inferiority be compensated by an improvement in quality[4] in order to attain a level of national defence which permitted a solution to the crisis. Details were to be worked out by the Vice Chief of the NGS and the Vice Minister of the Navy.

However, irrespective of the general principle, 'Quantity before Quality' remained valid for the carrier because of its vulnerability to attack, and parity should be sought with the USN. This took into consideration not only the maintenance of the 'inferiority ratio' in the London Treaty *vis-à-vis* the USN and RN but also the results of a study according to which Japan was not able to compete with these two naval powers in naval construction.

The IJN currently had four carriers (*Hōshō, Akagi, Kaga, Ryūjō*), the US Navy three (*Saratoga, Lexington, Ranger*). The operational value of *Hōshō*, the prototype for this category, was limited, *Ryūjō* was a failure, and *Akagi* and *Kaga*, prior to their mid-1930s conversion, were essentially experimental. In contrast, the US Navy carriers were all modern, powerful ships. To make matters worse, Roosevelt's Executive Order 6174 permitted the construction of two more carriers (to be named *Yorktown* and *Enterprise*). The carrier which had been included in the First Naval Armament Replenishment Programme of 1931 had been stricken for financial reasons, but an additional factor was the change in the treaty provisions which meant that carriers of less than 10,000 tons were now counted in the total tonnage; moreover, the fighting power of such a carrier was low.[5] In view of the reinforcement of the US Navy and the increasingly tense relations with the USA, the IJN deemed it essential to maximise the total tonnage available and to build two medium-sized carriers. The 12,630 tons available could be increased to 20,100 tons by the decommissioning of *Hōshō* in accordance with Article VIII of the Washington Treaty, which stated that 'all aircraft carrier tonnage in existence or building on the 12th November, 1921, shall be considered experimental, and may be replaced, within the total tonnage limit prescribed in Article VII, without regard to its age.' In order to conform with treaty requirements the displacement of each vessel was therefore to be restricted to 10,050 tons.

On 14 June 1933 the Navy Minister received the outline of the new programme as NGS Secret Document No 199 dated 12 June. No fewer than 86 warships displacing 159,370 tons were to be built. For the two carriers the NGS required the following characteristics:

Displacement::	10,050 tons standard
Armament:	5 – 20cm LA guns on centreline
	20 – 12.7cm HA guns
	> 40 MG
Aircraft:	c100
Speed:	36 knots
Range:	10,000nm at 18 knots

Propulsion was to be by diesels if possible, and more than half the embarked aircraft were to be carried on the flight deck ready for take-off. The aviation facilities were be designed to speed take-off and landing operations; if necessary, the number of reserve planes was to be reduced in order to attain these goals.

Sōryū immediately after delivery at Kure on 29 December 1937. She was incorporated into the 2nd Carrier Division (*Kōkū Sentai*) and, although she is flying the flag of Rear Admiral Mitsunami Teizō, she remains incomplete. (IJN Department of Naval Aeronautics)

The Second Naval Armament Replenishment Programme, now reduced to 48 naval vessels of 137,380 tons, was submitted to the 65th Diet Session (28 December 1933–26 March 1934). It passed after much discussion and was promulgated on 20 March 1934 following the authorisation of the Emperor. Of the Yen 431,688,000 approved by the Diet for auxiliary shipbuilding expenses, ¥80,400,000 were for the carriers.

Revision of the Requirements

Design G6 had a displacement (standard) of *c*12,000 tons, whereas the proposed new design was to have a larger air group, more HA guns, a longer flight deck (and hangars) and only one fewer 20.3cm gun. Clearly the wishes of the NGS could not be fulfilled within the 10,500-tons tonnage restriction; in order to avoid the defects of *Ryūjō* a displacement of 22,000–25,000 tons would have been necessary.[6] The author is unable to provide a logical explanation for this, but Japanese historians have generally assumed that the requirements were established after consultation with the NTD (the usual procedure) and that the stated displacement was intended to conceal the true characteristics of the ship.[7] However, possibly due to resistance on the part of the NTD or perhaps because the impossible nature of the demands was acknowledged, new requirements were transmitted to the NTD early in 1933:

Displacement:	10,050 tons standard (peacetime)
	11,000 tons (wartime)
LA guns:	5 – 15.5cm (1 x III, 1 x II)
HA guns:	16 – 12.7cm (8 x II)
MG:	*c*40
Aircraft:	*c*70
Range	10,000nm at 18 knots

Preliminary design G8 was drawn up in the autumn of 1933, and became the initial design for *Sōryū*. It had some features of the contemporary British carrier *Ark Royal*, in particular a flight deck which projected beyond the stern, a large island bridge to starboard about one-third of the length from the bow, and a tiered arrangement of fire control directors and rangefinders, which were grouped centrally. One unusual feature was the tall vertical funnel abaft the island bridge.[8] The two turrets with the 15.5cm LA guns were to be mounted back-to-back on the forecastle (anchor deck) beneath the forward end of the flight deck,[9] and the 12.7cm HA guns – reduced to twelve – were to be located on sponsons in the after half of the hull with comparatively good arcs of fire.

As a carrier design G6 was markedly superior to G8, although the calibre, number and location of the main LA guns made for a much-reduced anti-surface capability. Moreover, whereas the armour protection for the G6 was comparable to a cruiser of the *Myōkō* class, in the case of G8 it was totally inadequate against an enemy cruiser. The protection of the ammunition magazines was to be sufficient to keep out 20cm (8in) projectiles, but other vital spaces had protection only against 12.7cm (5in) shell. The proposed mounting of LA guns is difficult to understand, as the IJN knew full well that no carrier was well-suited for surface engagement; on the other hand the aircraft it carried could cripple or sink any ship far outside the range of its guns. It was recognised that the carrier needed to be protected against submarine and aerial attacks by antisubmarine and anti-aircraft escorts, and the operation of carriers within a force composed of such ships had already been discussed in various publications.

It may be that the tacticians in the NGS were so impressed by the rumoured development of American flying deck cruisers that they felt a similar response was needed. If so, the error of the 20cm LA armament adopted for *Akagi* and *Kaga* under the Washington Treaty would have been repeated under the much tighter constraints of the London Treaty. The LA gun requirement automatically brought with it considerable costs in weight. Given Japan's quantitative inferiority in carrier tonnage and the recognition that numerical parity with the USN was required, it made sense to invest the tonnage available in an improved type of aircraft carrier rather than a generation of hybrids with limited proficiency in either the anti-surface or the carrier role.

As with other ships designed during this period, the excessive requirements of the NGS were reflected in the design defects which resulted; the preliminary design studies attempted to incorporate features which simply could not have been realised within the given tonnage.

At the end of 1933, even before the building program was authorised by the Diet, Kure NY received the (secret) instruction to prepare to build the first ship, and early in 1934 the manufacture of the frames, hull plating and armour plates was begun.

The *Tomozuru* Incident and its Impact

When the torpedo boat *Tomozuru* capsized on 12 March 1934 construction was temporarily halted. New stability calculations were performed on the basis of the report of the investigation committee; these proved that a viable aircraft carrier could not be built within the current tonnage limitation.[10] The NGS was compelled to accept that excessive requirements had disastrous effects upon ship design and construction techniques if naval architects responded to them, and a new carrier design had to be drawn up. It was decided to abandon the mounting of LA guns as inappropriate to this type of ship, and to build the revised *Sōryū* as a pure aircraft carrier.

The following parameters were now established:

Displacement:	10,050 tons
Aircraft:	18 (+6) fighters
	33 (+11) dive-bombers
	Total: 51 (+17)
Horsepower:	152,000shp (engines as *Kumano*)
Speed:	35 knots
Range:	7,800nm at 18 knots
Armament:	12 – 12.7cm HA guns (6 x II)

Protection for the magazines was to be provided against 20cm projectiles fired between ranges of 12,000m and 20,000m,[11] and for the machinery spaces and avgas tanks against 12.7cm destroyer shell.

The drawings for preliminary design G9 were completed at the end of October 1934. A comparison with the original G8 design shows considerable changes which reflect the lessons learned from the defective Japanese ship design and building processes of the late 1920s and early 1930s, an era during which technological progress was made without sufficient research and testing.

Table 1 provides a comparison between the characteristics of the G6, G8 and G9 designs. The tables reveal: a reduction in length, an increase in draft, a scaling down of the island bridge, a reduction in the lateral plan, a lowering of the centre of gravity (CG), an increase in the range of stability, a modified armament, and the adoption of curved funnels facing downward to starboard.

Kure NY began construction officially on 20 November 1934. When the later Vice Admiral (constructor corps) Niwata Shōzō was transferred to Kure as shipbuilding inspector in July of that year, the dock had just been readied for construction. Because there were no structural changes, most of the original frames and plating could be used. On 23 December 1935 the carrier, whose construction was treated as a military secret, was launched; she would be completed on 29 December 1937, one year later than scheduled.[12]

Following the initial failed conversion of *Akagi* and *Kaga* into aircraft carriers and the excessive topweight problems experienced with *Ryūjō*, Japanese carrier design reached its highest point thus far with the fifth ship *Sōryū*. However, this favourable result was attained only on the basis of experience with the earlier ships and after going through a design process which is hard to understand from a technical point of view. In her final guise *Sōryū* had the features typical of later designs, and became the first of a series of modern 'light fleet' carriers built for the IJN. However, the following deficiencies were acknowledged and accepted: insufficient protection, a comparatively small air group and limited range. These would continue to be features of the medium carrier types derived from *Sōryū*. Moreover, even this was achieved only by an increase in the standard displacement by almost 60 per cent.

Outline Technical Description

The hull form was adapted from the contemporary cruisers of the *Mikuma*/*Kumano* classes. In order to reduce weight and economise on shipyard labour, and also with a view to using the weight saved to improve fighting power within the required displacement, even structural members were of light construction, and electric welding was employed in place of riveting. Differentiation of frame spacing was used to increase longitudinal strength: frame separation in the bow and stern areas, where the greatest stresses were experienced, was 0.9m increasing to 1.2m amidships.

Following the 4th Fleet Incident of 1935 the structural

Table 1: Comparison of Designs G6, G8 and G9

	Flying Deck Cruiser G6	Carrier-Cruiser G8	Aircraft Carrier G9
Displacement (standard)	12,000 tons	10,050 tons	15,900 tons
Displacement (trial)	17,500 tonnes	18,000 tonnes	18,800 tonnes
Length pp	?	223.02m	206.52m
Length wl	240m	240m	222m
Beam	21.7m	23.4m	21.3m
Draught	6.3m	6.27m	7.62m
Complement	?	1,800 officers & men	1,100 officers & men
Propulsion machinery:			
Boilers	8 Kampon Ro Gō	8 Kampon Ro Gō	8 Kampon Ro Gō
Main engines	4-shaft Kampon geared turbines	4-shaft Kampon geared turbines	4-shaft Kampon geared turbines
Horsepower	150,000shp	150,000shp	152,000shp
Speed	36 knots	35.5 knots	34.5 knots
Range	10,000nm at 18 knots	10,000nm at 18 knots	7,680nm at 18 knots
Fuel, heavy oil	3,760 tonnes	4,000 tonnes	3,400 tonnes
Armament:			
LA guns	6 – 20cm/50 (3 x II)	5 – 15.5cm/60 Type 3 (1xIII/1xII)	none
HA guns	12 – 12.7cm/40 Type 89 (6 x II)	12 – 12.7cm/40 Type 89 (6 x II)	12 – 12.7cm/40 Type 89 (6 x II)
Light AA	–	28 – 25mm Type 96 (14 x II)	28 – 25mm Type 96 (14 x II)
Aircraft:			
Fighters	Total 70	24 Type 90	12 (+4) Type 96
Torpedo-bombers	(incl reserve)	48 Type 89	9 (+3) Type 97
Reconnaisance	–	–	9 (+0) Type 97
Dive-bombers	–	–	27 (+9) Type 96

Source: Shizuo Fukui, *Japanese Naval Vessels Illustrated, 1869–1945*, Vol 3: Aircraft Carriers, Seaplane Tenders & Torpedo Boat & Submarine Tenders, 69, 97, 331, 333.

Sōryū during trials off Tateyama on 22 January 1938. After being handed over to Yokosuka she ran trials on the measured mile between Iwafukuro and Ukishima. (IJN Department of Naval Aeronautics)

strength of all warships was reviewed. In the case of *Sōryū* the outer shell was reinforced, as was the double bottom; the upper hangar deck was significantly reinforced, with some strengthening of the lower hangar deck. Based on experience with *Ryūjō* it was also decided in the course of the enquiry to raise the freeboard at the bow and stern by one deck height to improve seaworthiness. These measures resulted in an increase in the trial displacement to 18,448 tonnes and in the draught to 7.47m.

The trend towards building out from the hull sides was reversed following the damage suffered by *Ryūjō* during the storm. The passageways were generally moved inside the hull. The fore part was reinforced by a grid structure and the after part by additional plating to provide resistance to green seas. The fitting of a gyro stabiliser was thought to be unnecessary because of the increase in displacement compared with *Hōshō* and *Ryūjō*. However, the bilge keels were given much greater depth (1.8 m) and length (one-third of the hull) compared to earlier ships.

The 216.9-metre flight deck was almost as long as the hull. Maximum flight deck width was 26m, reducing to 25m abeam the island and 16m at the forward end; the width of the projecting platforms at the stern was 38m. The round-down at the stern was retained because it had proven advantageous for landings. The flight deck incorporated a number of expansion joints to cope with hogging and sagging forces.

For the first time since *Hōshō* there was an island bridge. It was located on an extension of the flight deck to starboard about one-third of the ship's length from the bow. The island was very small and contained only the spaces and equipment necessary for navigation and the control of flight deck operations.

There were three aircraft lifts connecting the hangars and the flight deck. Dimensions were 11.5m x 16m (LxW) forward, 11.5m x 12m amidships and 11.8m x 10m aft. A windscreen was located forward of the first lift. The flight deck was crossed by nine arrester wires set at irregular intervals, the last of which was abaft the third aircraft lift; these were connected to Kure-Type Model 4 brake motors. Between the forward and centre aircraft lifts there were two fixed and one dismountable crash barriers. A second dismountable barrier was intended for installation forward of the windscreen near the fore end of the flight deck. It was planned to fit catapults but this was never realised.

Three retractable searchlights were fitted on the flight deck and a fourth was located on a pedestal at the forward end of the island; the searchlights were primarily to provide target illumination for the anti-aircraft guns at night, but could also be used to help returning aircraft locate their carrier. The hinged radio masts were mounted on sponsons on both sides of the flight deck; they were lowered during air operations.

The aircraft were stowed in one of two superimposed hangars. The upper hangar extended from the anchor deck to the boat deck aft; the lower hangar was shorter, the forward section being employed for officers' living quarters. The door in the lower hangar at the stern for the embarkation/disembarkation of aircraft was not fitted because it had proved to be the weak point for watertightness in the case of *Ryūjō*. Instead a watertight door two metres square was fitted in the upper hangar to allow for engine repairs and testing outside the confined space of the hangar. Aircraft were now embarked onto the flight deck when the ship was alongside or at anchor, and for this purpose a collapsible crane was fitted in a recess to starboard of the after aircraft lift.

The initial plans envisaged an air group comprising only fighters and dive-bombers. This was revised four

Table 2: *Sōryū/Suzuya* Comparison

	Sōryū	*Suzuya/Kumano*
Displacement (trial)	18,800 tonnes	13,440 tonnes
Horsepower	152,000shp	152,000shp
Speed	35 knots	35 knots

times before *Sōryū*'s completion, and in the final version torpedo-bombers were included. *Sōryū* was now to embark a total of 57 (+16 reserve) aircraft: 12 (+4) Type 96 fighters, 27 (+9) Type 96 dive-bombers, 9 (+3) Type 97 torpedo-bombers and 9 (+0) Type 97 reconnaissance planes.

Because it had been decided that the 'standard' carrier should have a speed of at least 34 knots, the machinery plant of the cruisers *Suzuya* and *Kumano* was adopted for both *Sōryū* and her half-sister *Hiryū*; the machinery differed only in the arrangement of the cruise turbines (see the description given for *Hiryū*). The maximum speed of *Sōryū* and the two cruisers was around 35 knots, as shown in Table 2.

In order to attain the same maximum speed as the cruisers with the same machinery but a much-increased displacement the body plan of the cruisers was adopted for the hull of *Sōryū* but waterline length increased from 198 metres to 222 metres – this increase in length was also important in order to maximise the length of the flight deck. V/L ratio was 1.30 for *Sōryū* compared to

Sōryū and *Ryūjō* in Tōkyō Bay around 1938. The port 12.7cm high-angle guns look almost white in this view. The aircraft in the foreground is a Type 95 'Dave' reconnaissance floatplane. (IJN Department of Naval Aeronautics)

1.38 in *Suzuya*, thereby reducing resistance through the water.

The arrangement of the machinery was a departure from previous practice. Each set of turbines and each of the boilers was located in its own space, making for eight boiler rooms and four engine rooms. This arrangement was to become standard for large IJN warships and was considered 'ideal'. The fuel tanks were outboard of the machinery spaces. The adjacent compartments fore and aft and parts of the double bottom were also used for the

Sōryū: Midship Section

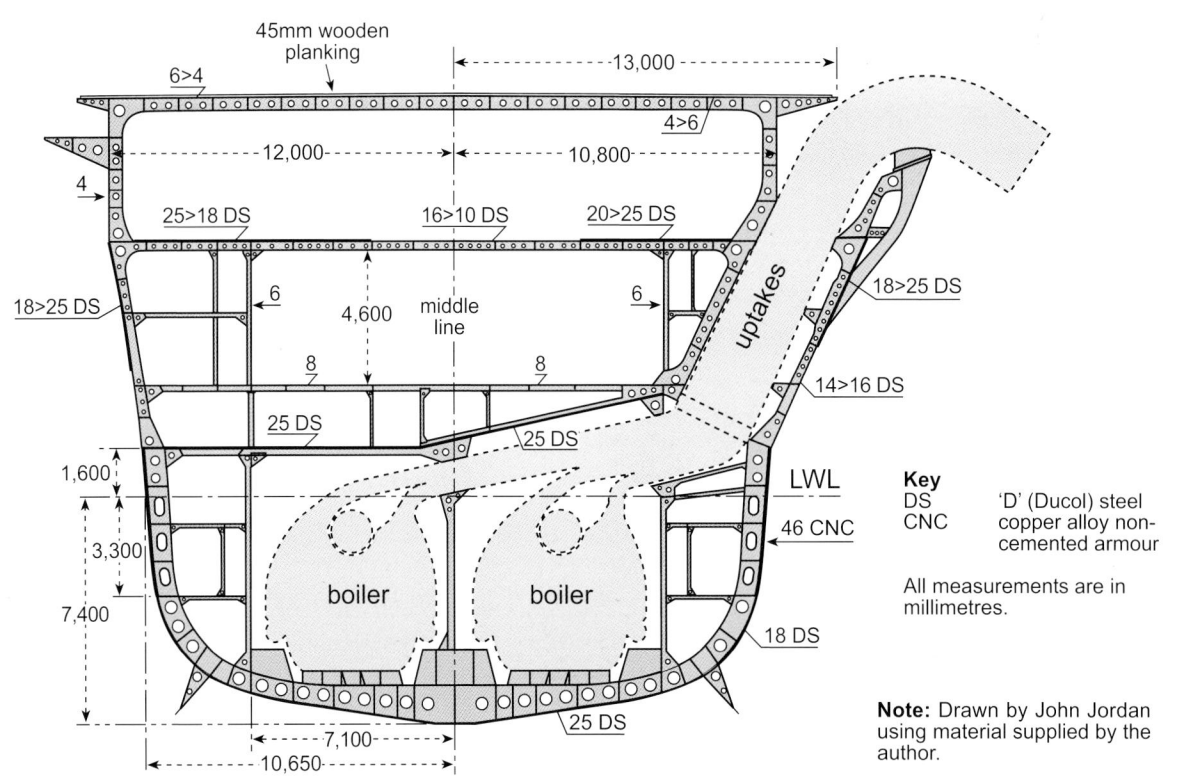

Key
DS 'D' (Ducol) steel
CNC copper alloy non-cemented armour

All measurements are in millimetres.

Note: Drawn by John Jordan using material supplied by the author.

stowage of 3,400 tonnes of heavy oil. Based on experience with *Ryūjō* the shape of the funnels, which emerged abaft the island, was revised: the uptakes were raised a little above the flight deck before being bent downwards towards the surface of the water and angled aft.

One problem with the ships built for the IJN during this period was the excessive angle of heel when the helm was put over. Although topweight was undoubtedly the main cause, it was thought that the type of the rudder exerted a certain influence. For this reason *Sōryū* was fitted with a pair of spade rudders inclined outwards; it was believed that this would reduce the angle of heel.

Six 12.7cm/40 Type 89 HA guns in twin mounts were located on sponsons on either side of the flight deck; the odd-numbered mounts were to starboard, the even numbers to port. Following experience with *Ryūjō* the flight deck was cut away around the positions of the mountings to minimise the projection of the sponsons and their supports from the outer shell. This also facilitated fire against targets approaching from the opposite side of the ship.

Three of the fourteen 25mm/60 Type 96 twin MG mounts were grouped together on a platform close to the bow below the flight deck to port. The remaining eleven mounts were on sponsons close to the HA guns: six to port, five to starboard. HA gun mounting No 5 and MG mounts Nos 1, 3 and 5 had completely enclosed shields as protection against the funnel gases. The fire of the HA guns was controlled by two Type 91 fire control systems (separate rangefinder and director); each of these systems controlled the three mounts on that side. There were Type 95 FC directors for each group of MGs.

Protection for the ship's vitals was relatively light, as specified by the NGS: 140 NVNC vertical armour with 40mm on the middle deck for the magazines, 35mm with a 25mm deck over the machinery spaces. The level of protection was acknowledged to be inadequate, but could not be revised because of the need to keep displacement within the prescribed limits.

Hiryū: An Improved G9

The provisional No 2 carrier in the Programme of 1934 (*Maru Ni Keikaku*) was to be laid down to the same design as *Sōryū* approximately one year after the name ship. However, the NGS required the design to be revised

View of *Sōryū*'s wake from the after end of the boat stowage deck, while making nearly 35 knots on speed trials in November 1937. Note the underside of the flight deck above, the heavy supporting brace between the main and flight decks, and the safety netting around the edges of the flight deck. (Naval History and Heritage Command, NHHC 73062)

WARSHIP 2022

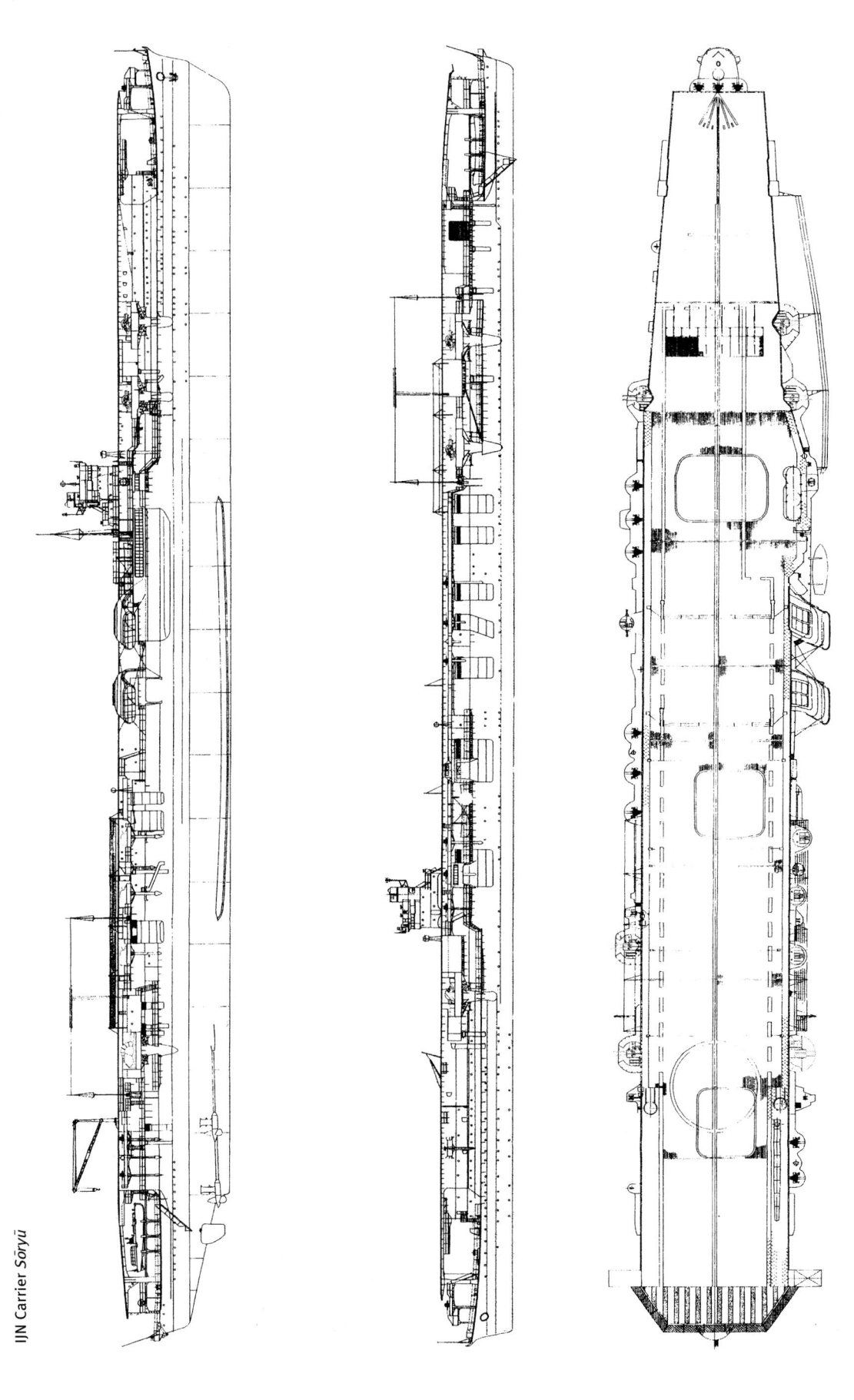

IJN Carrier *Sōryū*

(Drawn by Jürg Tischhauser)

THE IJN CARRIERS *SŌRYŪ* AND *HIRYŪ*

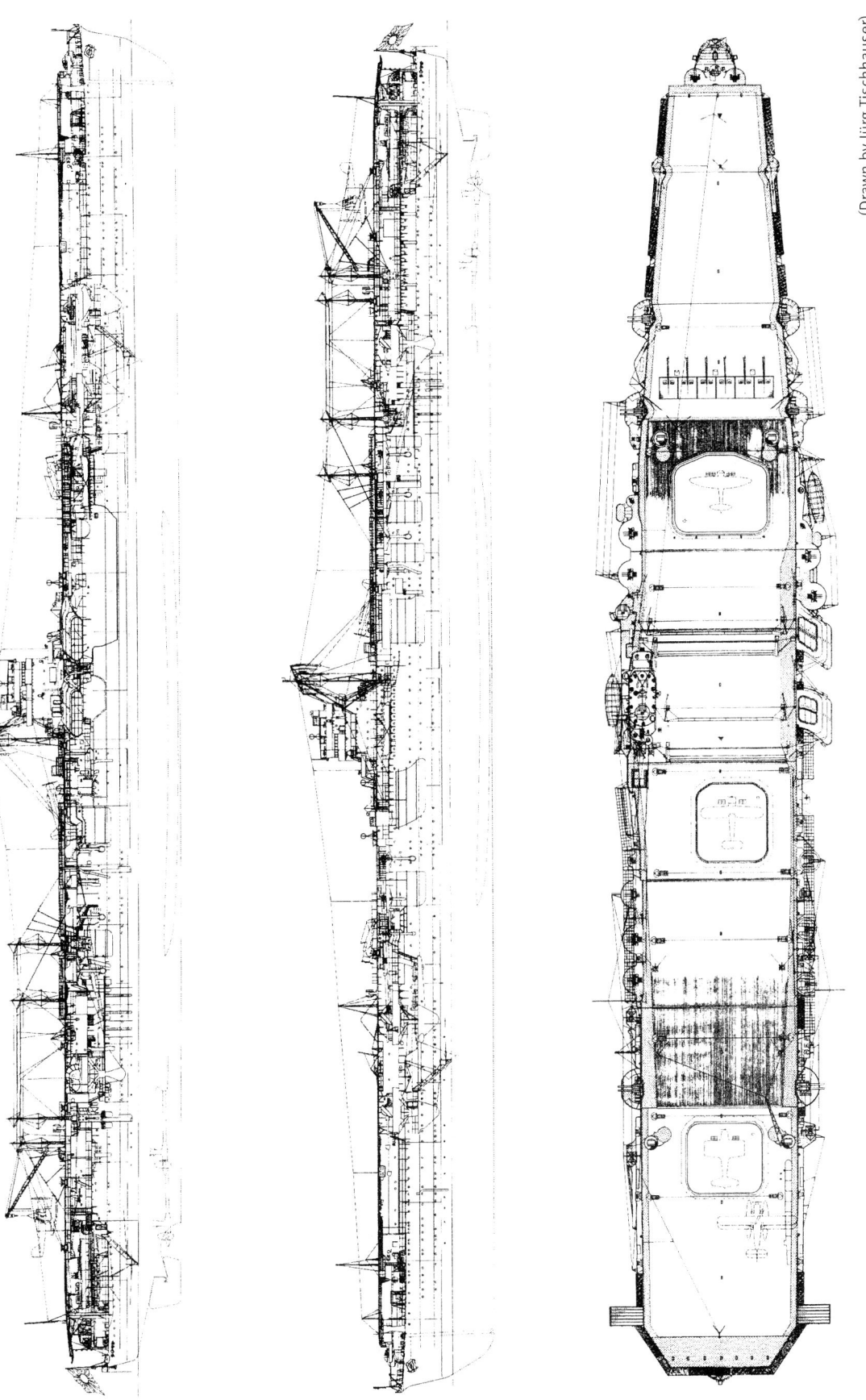

IJN Carrier *Hiryū*

(Drawn by Jürg Tischhauser)

to take into account the reconstruction of *Kaga* and the far-reaching modifications to *Sōryū* which had to be carried out after the *Tomozuru* and 4th Fleet Incidents. Also, there were inherent defects in the original design that had been recognised, foremost among which was the lack of protection which became the 'deadly weak point' of the medium-type carrier, but which had been accepted in order not to increase *Sōryū*'s displacement.

The revised design of *Hiryū* was given the number G10. The modifications can be divided roughly under two headings: improvements in the quality of the ship, and improvements in aviation facilities. Due to the requirement for new design studies and plans, the keel of *Hiryū* was laid at Yokosuka NY only on 8 July 1936. The hull was launched on 16 November 1937 and she was commissioned on 5 July 1939, 16 months behind schedule and more than two years after the in-service date envisaged when the Second Naval Armament Replenishment Programme was authorised.

The requirements relating to the design of the ship were:

– improvement in seaworthiness by raising the freeboard at the bow and stern
– improvement in manoeuvrability by changing the type of rudder
– reinforcement of the structural strength (rigidity) of the hull by reverting to riveting instead of electric welding for the longitudinal strength members and other key structures
– increase in the thickness of the shell plating and strength deck
– adoption of three different frame spacings depending on the distribution of the stress curve
– reinforcement of the magazine protection
– increase in endurance (not achieved).

The requirements relating to aircraft operations primarily comprised:

– increase in the width of the flight deck by one metre (27m *vice* 26m in *Sōryū*)
– change in the position of the island bridge (to port and at the mid-point).

Outline Technical Description

The requirement of the NATD for a 27-metre-wide flight deck meant an increase in topweight. Retaining the hull of *Sōryū* would raise the centre of gravity (CG), and the consequent reduction in the distance between the CG and the metacentric height (GM) would have compromised stability. The increase in weight would also result in a deeper draught, resulting in corresponding reductions in freeboard and speed. In order to avoid these adverse consequences certain dimensions of the hull had to be revised. Beam at the waterline had to be increased by 0.7m; the GM was now 1.81m compared to 1.52m for *Sōryū* and the stability range increased from 105.2 to 109.6 degrees. Draught increased to 7.84m. Overall depth was increased by 0.1m to 20.5m, but despite this there was a reduction in freeboard amidships by 0.31m to 12.57m at the official trial displacement. This in turn

Hiryū during the final stages of fitting out in the Yokosuka Navy Yard's Koumi Basin on 20 February 1939. The after aircraft lift is being lowered into place using a 200-tonne crane, and the funnel smoke tells us that the boilers have been flashed up. (IJN Department of Naval Aeronautics)

A starboard view of *Hiryū* running trials off Tateyama on 28 April 1939. Trials began soon after construction and the paintwork is stained. (IJN Department of Naval Aeronautics)

made it essential to raise the height of the bow, which was achieved by extending the upper hangar deck to the bow, where it became the anchor deck. Freeboard at the stern was likewise increased by 0.4m to improve sea-keeping.

The enquiry set up following the *Tomozuru* Incident had demonstrated clearly the need to improve longitudinal strength. The thickness of the outer shell plates of the double bottom was increased and the longitudinal strength members riveted, not welded as in the case of *Sōryū*. However, electric welding continued to be used for plating and hull members which did not contribute to hull strength in order to save weight. A further measure used to improve structural strength was the use of three different frame spacings. The 900mm spacing used in *Sōryū* for the bow and stern was reduced to 600mm, increasing to 900mm by adaptation to the stress curve; frame spacing amidships remained at 1.2m.

The shape of the bow had already been modified in *Ryūjō*, and both *Sōryū* and *Hiryū* had the 'S'-shaped bow that was a characteristic feature of Japanese warships of the period. The underside of the bow beneath the water was cut away to give a 30-degree angle, so that it approached that of the 'Maier bow', while the curved and projecting line above the waterline resembled a 'clipper' bow. This particular configuration was adopted because it was thought that Japanese warships would probably have to pass through a minefield laid using Type 1 mines in the path of the US main force during the decisive battle.[13]

The flight deck had the same length and a similar configuration to that of *Sōryū*. Influenced by the reconstruction of *Kaga*, the constructors of the NATD decided to increase the width of the deck by one metre, as the 25 metres of *Sōryū* abeam the island was considered insufficient. The supporting sponson for the island impacted on the form of the outer hull and also on the internal configuration of the upper hangar.

The first IJN carrier, *Hōshō*, had been completed with an island. It was removed shortly afterwards due to opposition from the pilots, and *Akagi*, *Kaga* and *Ryūjō* were completed with a flush deck. However, over the course of ten years not only had aircraft developed, but landing techniques and aviation facilities had improved and the area of the flight decks had increased. Moreover, the level of training of the air crews was incomparably higher. Island bridges had been widely adopted by foreign navies, and the advantages for conning the ship and control of flight operations were obvious and far outweighed the objections still expressed by some aviators. When *Kaga* was reconstructed it was decided to fit an island bridge to starboard at approximately one third of the ship's length.

Kaga was completed in June 1935 after a conversion lasting one year and trials conducted. One result was the requirement of the NATD for the bridge to be relocated amidships to port, and the NTD agreed with its conclusions. However, it was too late for a revision of the drawings for *Sōryū*, so the new location was first adopted for the conversion of *Akagi* and for *Hiryū*. Compared to *Sōryū*, *Hiryū*'s bridge was one deck higher to improve the view over the bow. While *Akagi* was being rebuilt objections to the new location of the island were lodged by the pilots, and the NATD gave way. However, the advanced building stage of *Hiryū* did not permit this revision, and the location of the island at the mid-point to port was retained for both *Akagi* and *Hiryū*, which were the only carriers in the world to have this arrangement. It was, however, decided that *Sōryū* would provide the template for future carriers.

The double hangar and aircraft lift arrangements were broadly as in *Sōryū*. Changes in the dimensions of the aircraft lifts reflected the latest developments in naval aviation: the forward lift was 13m x 16m (LxW), the centre lift 13m x 12m and the after lift 11.8m x 13m. The windscreen and the arrester wires were identical to those fitted in *Sōryū*, but there was an additional dismountable crash barrier. Other flight deck equipment, including the hinged W/T masts, was as in *Sōryū*, but there were four (*vice* 3) retractable searchlights.

The complement of aircraft was likewise identical to that of *Sōryū*, except that torpedo-bombers were included from the outset. The original plans prioritised the dive-bomber as a type, and in both ships the figure of 27 (+9) aircraft accounted for almost 50 per cent of the air group. There were two bomb hoists and one torpedo hoist to transport the air weapons from the magazines to the hangar decks. Stowage of Type 91 aerial torpedoes was sufficient for three sorties by the nine torpedo bombers.

Machinery

The eight main boilers were large Ro Gō Kampon *shiki* (NTD 'B' type) oil-burning small watertube boilers fitted with air preheaters and superheaters. They were the 'A' model with the superheater in the middle of each bank of tubes and the air preheater above. The boiler room was not pressurised, as double casings with air under pressure between were used. Steam pressure at the superheater outlet was 22kg/cm^2, steam temperature 300°C. One boiler produced 107 tonnes/hr of steam; this output was the highest of all the boiler types of the IJN.

The four main engines were Kampon type steam turbines. Each set comprised a high presssure (HP) turbine, an intermediate pressure (IP) turbine and a low pressure (LP) turbine driving through single reduction gearing. The astern turbines were enclosed within the LP turbine casings. Cruise turbines were installed in the forward engine rooms and could be coupled individually to the HP pinion by means of cruise reduction gearing and a jaw clutch. Steam from the cruise turbine exhausted into the HP turbine on the corresponding inboard shaft. There were no cruise turbines coupled to the inboard engines.

The main engines developed 153,000shp (152,000shp in *Sōryū*), 2/3,000shp more than those mounted in the battleships of the *Yamato* class. This output was surpassed only by the main engines mounted in the two carriers of the *Shōkaku* class and *Taihō*, which were rated at 160,000shp.

The main steam lines from the two forward boiler rooms (Nos 1 & 2) on each side of the ship were combined with those from boiler rooms Nos 3 & 4 respectively, and led aft on the outboard side to the forward set

Table 3: Trial Results of *Sōryū* and *Hiryū*

Condition	Displacement (tonnes)	Speed (knots)	SHP	RPM	Effective HP	Efficiency
Sōryū unofficial trial						
10/10.5 overload	18,603	35.286	160,325	339	–	–
10/10	18,644	34.989	152,098	333	–	–
8/10	18,691	33.646	122,361	318	–	–
6/10	18,733	32.162	93,147	288	–	–
Sōryū official trial						
10/10.5 overload[1]	18,621	35.217	160,326	339	–	–
10/10 full power	18,871	34.898	152,483	333	–	–
8/10	18,702	33.860	122,918	312	–	–
6/10	18,756	32.189	92,094	287	–	–
4/10	18,746	29.156	61,132	252	–	–
3/10	18,715	26.511	41,022	229	–	–
Standard[2]	19,039	18.006	15,533	182	–	–
10/10 final run	18,492	34.935	152,115	333	73,400	48.36%
Full astern	18,541	–	40,028	225	–	–
Hiryū official trial						
10/10 full power	20.346	34.28	152,733	326	–	–
Standard[3]	20,156	18.142	13,629	n/a	–	–
10/10 final run[4]	20,165	34.59	153,000	326	75,400	49.28%

Sources: Fukuda Keiji, *op cit*, 113–14, Shizuo Fukui, Vol 3.

Notes:
[1] In the overload operation the performance was increased by 5 per cent over the designed full power, hence 10.5/10 overload power.
[2] The standard speed of the major surface ships was increased initially from 14 knots to 16 knots, then to 18 knots. It was an operational value mainly determined by the requirement to outmanoeuvre enemy submarines; there was no direct relation to the so-called economical speed. The standard speed was also the basis for the endurance calculation.
[3] The most effective coefficient was attained with *Hiryū*'s standard speed: 7,200 effective HP and 53 percent.
[4] The speed difference at the 10/10 full power between the two ships was insignificant, despite the 1,600-tonne greater displacement of *Hiryū*, which had a superior propulsion coefficient.

Hiryū at Yokosuka on 5 July 1939. At 1300 Vice Admiral Araki Hikosuke, of the Yokosuka Navy Yard, handed the ship over to Captain Takenaka Ryūzō. Note the ceremonial Shinto trees and curtain forward of the bridge behind which the ceremony took place. (IJN Department of Naval Aeronautics)

of main turbines (wing shafts). The steam lines of the after boiler rooms (Nos 5 & 7 to starboard and Nos 6 & 8 to port) were led aft on the inboard side, being combined only after they had passed through the bulkhead of the after engine rooms. A cross-connecting line just aft of the boiler room bulkhead connected the main steam lines of all four shafts. The main steam lines were made as straight as possible with expansion joints between the fixed mountings and also between boiler stops and the point where the line joined the main steam line or one of the fixed mountings. A cross-connection to the auxiliary steam line was provided in each of the four engine rooms. The auxiliary steam line operated at the same temperature and pressure as the main. This permitted cross-connection of the two systems in each space so as to receive all steam for the auxiliaries fitted in the engine room from the main steam in each space if so desired.[14]

A cross-connection from port to starboard was provided only for the forward boilers. When operating under split plant condition the bulkhead valve was closed. Two mains ran aft on the centreline to the engine rooms with a connection to the auxiliary steam lines in each of the boiler rooms traversed. It was general practice not to use this line; instead, each boiler supplied its own steam, and the engine room received steam through the main/auxiliary cross-connection in each space. A peculiarity was the supply of steam to the bilge eductors from the local steam supply; in other ship types this was supplied by a line in one of the boiler or engine rooms.

There were four auxiliary exhaust systems. Two serviced the two forward boiler rooms on either side (Nos 1 & 3 and 2 & 4) and ran aft and outboard to each of the two forward engine rooms. The other two systems served the two after boiler rooms, one on each side, and ran inboard to the two after engine rooms. Between the

Table 4: General Characteristics

	Sōryū	*Hiryū*
Length wl	222.00m	222.93m
Length pp	206.52m	209.52m
Length oa	227.00m	227.00m
Beam wl	21.30m	22.32m
Depth (keel to flight deck)	20.50m	20.50m
Draught (mean)	7.40m	7.50m
Displacement, standard	18,000 tons	19,000 tons
Wetted surface	5,106.39m²	5,241.12m²
Cb (Block coefficient)	0.502	0.507
Cp (Prismatic coefficient)	0.590	0.592
Co (Amidships section coefficient)	0.851	0.856
Cw (Waterline coefficient)	0.728	0.720
Length/beam (L/B)	10.43	10.09
Draught/length (d/L)	0.334	0.338
Beam/draught (B/d)	2.878	2.933
Length/depth (L/D)	11.82	10.85
Depth/draft (D/d)	2.77	2.74

Sources: Fukuda Keiji, *Gunkan Kihon Keikaku Shiryō*, 8, *KZGG*, Vol 2, *op cit*, 240ff.

forward and after systems on each side there was a cross-connection in each of the forward engine rooms. There were also cross-connections between the systems in the two after engine rooms and the forward engine rooms. The bulkhead valve in the after engine rooms could be operated from either side of the bulkhead. There was also a connection between the lubrication oil pumps in the two forward engine rooms and those fitted in the after engine rooms. These arrangements ensured that as much of the machinery as possible could continue to operate in the event of action damage.

Fuel oil stowage was 3,750 tonnes – 350 tonnes in excess of that required for *Sōryū* to attain the required 8,000nm at 18 knots. A document relating to *Hiryū* states that endurance was 7,670nm; however, no reason is given for the 330nm reduction in range despite her carrying 350 tonnes more fuel.

The exhausts of the eight boilers were combined into two funnels as in *Sōryū* and similarly disposed. The downward-inclined funnel had been trialled in *Akagi* for the first time and modified on the basis of results with *Ryūjō*; it became standard for *Sōryū* and the later IJN carriers, with few exceptions.

Sōryū had two balanced rudders arranged side by side and inclined outwards 18.5 degrees to counteract the heel generated by centrifugal force when turning. During trials this rudder arrangement was judged successful. However, in the case of the destroyers *Ariake* and *Yugure*, which had the same rudder arrangement, it had been considered a failure, speed being reduced by 1.5 knots compared to their sisters.[15] It was only after the replacement of the two balanced rudders by one semi-balanced rudder that the desired result was obtained. Based on this experience the constructors assumed that the performance of the twin rudders in *Sōryū* could be further improved by adopting a suitable semi-balanced rudder. Following the required modifications to the stern *Hiryū* was therefore fitted with a single semi-balanced rudder on the centreline. However, contrary to expectation, this rudder had the effect of increasing the turning circle at small rudder angles. This example serves to illustrate the complexity of ship design, and demonstrates that results obtained with smaller hulls cannot always be successfully transferred to major vessels (and *vice versa*).

Armament

As in *Sōryū* twelve 12.7cm/40 Type 89 HA guns in six twin mounts were disposed on sponsons on either side of the ship. The difference was that, due to the revised location of the island there were two HA mountings on either side forward and one aft. The fire control system was the

Table 5: Weight Distribution (tonnes)

	Sōryū (trial)	%	*Hiryū* (trial?)	%	*Hiryū*[1]	*Hiryū*[2]
Hull	7,939	42.2	8,240	41.5	9.050	9.050
Armour	818	4.3	1,573	7.9	1.600	1.600
Protection	590	3.1	222	1.1	165	165
Fittings	1,527	8.1	1,565	7.9	1,350	1,350
Equipment, permanent	250	1.33	250	1.3	275	275
Equipment, consumable	469	2.73	513	2.6	–	–
Armament, guns	469.3	2.54	502	2.5	293.5	494.6
Armament, torpedo	136.4	0.68	128	0.6	90.7	100.5
Armament, electric	499.5	2.72	515	2.6	503.3[3]	503.3[3]
Armament, aviation	660.6	3.52	676	2.6	305.8	677.1
Armament, navigation[4]	–	–	–	–	13.8	13.8
Machinery	2,574.2	13.6	2,601	13.1	2,331	2,616.1
Fuel, heavy oil	2,270	12.1	2,500	12.6	0	3,750
Light oil (gasoline)	240	1.28	240	1.28	0	370[5]
Lubrication oil for engine	58[6]	0.31	58[6]	0.31	0	51
Lubrication oil for planes					0	36
Reserve feed water	100	0.53	100	0.5		
Ballast	201.8[7]	1.07	177[7]	0.89	0	0
Margin	–	–	–	–	93.2	93.2
Total	18,800	100	19,860	100	16,270	21,887

Sources: Fukuda Keiji, *op cit*, 42; position of centre of gravity data for *Sōryū* on page 67 largely agree.
Notes:
1   Official data book of 5 February 1939 light load condition.
2   Official data book of 5 February 1939 full load condition.
3   Includes 28.5 tons for W/T equipment.
4   Weight of nautical instruments included under 'Electric'.
5   Includes 10 tonnes for ship's boats.
6   Includes 24 tonnes for aircraft.
7   Must belong to another weight group depending on the official data.

new Type 94 that featured a combined rangefinder and director. One director was located on a pedestal forward of the bridge, the other on the open air defence control deck atop the island.

Close-range defence was reinforced by an additional three 25mm MG for a total of 31. There were five twin mounts (two at the bow and three to starboard aft of the funnels in enclosed shields) and seven triples (five to port, of which two were forward of the island and three aft, plus two to starboard forward of the funnels). *Hiryū* was the first IJN carrier to be fitted with triple mounts. The MGs were combined into five groupings of two/three mounts, each group being controlled by a Type 95 MG director. One director was on a platform at the bow; the remaining four were on sponsons at the four 'corners' of the ship. On an extension of the upper hangar deck near the bow was a 12.7cm loading training gun (*sōten enshū hō*) Type 3.

Protection

As in *Sōryū* most of the vulnerable parts of the ship (machinery spaces, avgas tanks, steering compartment) were protected against 12.7cm shell. *Sōryū*'s magazines had the thickest armour, designed to resist penetration by 20cm cruiser shell fired at ranges between 12,000m and 20,000m. Armour thicknesses were slightly increased in *Hiryū*; for detail see the midship section drawing on page 52. These measures not only improved protection but had the effect of lowering the centre of gravity, thereby improving stability. On the other hand the reinforcement of the armour protection was costly in terms of weight, and contributed to the 1,000-ton increase in displacement compared with *Sōryū*.

Boat Outfit

As in *Sōryū* the upper hangar deck was extended close to the stern to serve as an upper boat platform. There were eleven boats in total: three 12-metre motor boats, one 8-metre motor launch and one 6-metre service boat were stowed on the upper platform, two 12-metre motor launches and two 13-metre 'special transport' boats (landing boats) on the upper deck. An overhead boat crane running on transverse rails secured to the underside of the flight deck was employed for lowering and lifting the ship boats; the rails ran from one side of the flight deck to the other. Two 9-metre cutters were suspended from radial boat davits outside the hull; this type of davit had been adopted for the first time in *Sōryū*.

Endnotes:
[1] The Washington Treaty of 1922 had defined the aircraft carrier as having a displacement in excess of 10,000 tons; air-capable ships below this displacement were classified as 'auxiliary' vessels and their numbers were unlimited, leading the IJN to design *Ryūjō*. Under the London Treaty of 1930 *Ryūjō* had to be added to the total tonnage of the carriers, meaning that the IJN had now used up 68,370 tons (*Hōshō* 7,470 tons, *Akagi* and *Kaga* each 26.900 tons, *Ryūjō* 7,100 tons) of its 81,000-ton allocation. Only 12,630 tons remained.
[2] The size of air groups was also considerably reduced. In *Senshi sōsho* Vol 31, 401, it is stated: '… the proposed build-up of the carrier fleet … did not pass, and the reduction in the *Kōkūtai* (Air Groups) was greater than anticipated and presented the tacticians with serious problems …'.
[3] Editor's note: There were two fundamental 'political' problems with the G6 flying deck cruiser design: If the ship was classified as a 'cruiser', under the London Treaty she would have had to count against Japan's allocation for category 'A' cruisers, but this allocation would be fully taken up once the four cruisers of the *Takao* class were completed. Also, as a cruiser she would be permitted a maximum displacement of 10,000 tons standard. This suggests a 'paper' feasibility design.
[4] This had already been recommended in secret document No 215 of 1932. The principle of individual superiority resulted in the battleships *Yamato* class, the two carriers of the *Shōkaku* class and the destroyers and submarines of the Third Naval Replenishment Programme of 1937, and was continued with a second pair of 'super-battleships', the carrier *Taihō* and a second batch of destroyers and submarines in the Fourth Naval Completion Programme of 1939.
[5] When the treaty rules changed the NGS demanded a 50 per cent increase in *Ryūjō*'s air group, from 24 to 36 aircraft; this requirement was to bring about fatal consequences – see Hans Lengerer, 'The Light Carrier *Ryūjō*', *Warship* 2014, 129–45.
[6] When *Sōryū* was commissioned, displacement in the standard condition was 15,900 tons. Even though both the number of aircraft and the armament had been drastically reduced, and

Table 6: Stability

[Upper values = *Sōryū*; lower values = *Hiryū*]

	Trial	Full load	Light load	Suppl light load
Displacement	18,000t	20,285t	14,999t	15,789t
	20,250t	21,887t	16,670t	16,495t
Draught	7.52m	8.03m	6.52m	6.75m
	7.84m	8.21m	6.60m	6.90m
KG	8.35m	8.16m	9.41m	9.01m
	8.27m	8.08m	9.25m	9.17m
GM	1.63m	1.78m	0.69m	1.05m
	1.81m	2.01m	0.96m	1.03m
OG	0.73m	0.13m	2.89m	2.26m
	0.43m	-0.13m	2.65m	2.47m
Stability range	100°	110°	91.5°	98°
	109.6°	110.7°	96.3°	97.8°
GZ	1.96	2.01	1.14	1.49
	2.30	2.38	1.65	1.71
A/Am	1.77	1.60	2.28	2.16
	1.71	1.59	2.23	2.18
Ballast (water)	–	–	–	800t

Key:
KG   height of centre of gravity above base
GM   metacentric height
GZ   maximum righting lever
OG   distance of centre of gravity above base
A/Am ratio of lateral plan above/below waterline

Sources: Fukuda Keiji, *op cit*, 79; official data book *Hiryū* 5 February 1939.

A Type 97 'Kate' carrier attack aircraft takes off from *Hiryū* while en route to Midway 31 May–1 June 1942. The photo was probably taken from the oiler *Kyokutō Maru*. (IJN Department of Naval Aeronautics)

    the speed and range requirements lowered, 5,850 tons (roughly 57% of the originally planned 10,050 tons) were necessary to obtain an operable carrier. This is a clear indication of the over-ambition inherent in the original requirements for the G6 and G8 designs.

7 This argument does not really stand up to scrutiny. Anyone who knew anything about ship design would have recognised that no ship of this displacement could be built to fulfil these requirements.

8 No IJN carrier was ever built with this feature.

9 This suggests that the length of the flight deck in G6 (194m) was deemed inadequate. For the reconstruction of *Kaga* a full-length flight deck replaced the shorter tiered take-off decks of the original conversion.

10 Preliminary studies for a carrier mounting 15.5cm guns had demonstrated that a standard displacement of at least 14,000 tons was required.

11 The shorter range dictated vertical protection, the longer range horizontal protection (to resist penetration of the decks by plunging shell).

12 The delay was not uncommon with the ships of this programme. *Sōryū* should have been completed on 31 March 1937, but the capacity of the Japanese shipyards, which were currently occupied with stability improvement measures after the *Tomozuru* Incident followed by structural reinforcement after the 4th Fleet Incident, and also with the reconstruction of the older battleships and cruisers and of the carriers *Kaga* and *Akagi*, was insufficient to cope with the rapid expansion in their workload. Other examples of delayed construction were: carrier *Hiryū* 16 months, cruiser *Chikuma* 14 months, seaplane tender *Chitose* class 9 months, fleet tanker *Takasaki* 33 months.

13 During the battle of 10 August 1904 the Russian C-in-C reported that the crews of a Japanese destroyer division had laid mines in front of his force, and that the necessary changes of course had an unfavourable effect on his formation and resulted in a Japanese victory. In fact, although a Japanese destroyer division did indeed cross the path of the Russian force, the crews dropped bags of coal, not mines into the water. Akiyama Saneyuki (later Vice Admiral), who served in the Russo-Japanese War (RJW) on the flag staff of C-in-C Tōgō Heihachirō and whose tactical lessons were a major influence on the IJN, read this report and proposed the development of a mine that would be laid in the path of an enemy fleet. His proposal was accepted, and in October 1904 the first 'compound mine' was completed. It comprised four mines with cylindrical bodies, each of which was connected to a 100m cable. The electric fuze mechanism armed the mines 15 minutes after they entered the water and rendered them safe after 45 minutes. The deployment of this mine at the Battle of Tsushima failed mainly due to the bad weather. After the RJW, a special mine investigation committee carried out a series of tests and exercises in the course of which improvements were carried out in 1915/16. In the latter year the classification of mines was revised and the compound mine was classified as the Type 1 Mine. The pronounced cutaway of the bow of Japanese warships was intended to prevent a mine cable becoming entangled and the mines detonated by a friendly ship.

14 Normally only under battle conditions.

15 See Hans Lengerer, 'The Japanese Destroyers of the Hatsuharu Class' in *Warship 2007*, 99.

THE DEVELOPMENT OF THE SMALL CRUISER IN THE IMPERIAL GERMAN NAVY

PART III: THE GUNBOATS

Dirk Nottelmann continues and completes the story of the development of the German *Kleiner Kreuzer* ('small cruiser') begun in *Warship* 2020[1] by looking at the development of German gunboats which, in the absence of 'real' cruisers, often substituted for their larger counterparts, sharing identical tasks abroad and sometimes operating alongside them. Despite their important contribution their story has, as in the case of their counterparts in the other major navies, rarely been told.

When working through files and historical publications,[2] one particular question presents itself: What *was* a gunboat? There are at least as many answers as there are authors. Some answers are brief, some elaborate, but all seem to agree on a common baseline: gunboats were small vessels built either for one particular task (*eg* the Rendel 'flat-iron' gunboats) or for a variety of tasks.

This limited definition highlights the problem: such vessels elude any kind of unified, coherent description. Perhaps the best approach is to work from the other direction, and thus to regard a 'gunboat' simply as an armed vessel that cannot be placed in any other category.

A second question then arises: Who first employed the term? Going back to its roots, it is easy to find early examples of oared or sailing vessels regarded retrospectively as gunboats, ranging from the American War of Independence, through the invasion fleet of Napoleon I, to the vessels that formed the 'skerry fleet' of Sweden's famous naval architect Hendrik af Chapman.

However, it is perhaps surprising to find no official use of the term 'gunboat'. Instead we find, for example, the terms '*kanonslup*' or '*kanonjoll*' (af Chapman), which mixes the local designation of the boat or launch with the ability to carry one or more cannon. For the Royal Navy,

Adler, the last of the 'crossover' vessels between the gunboat, aviso and cruiser categories. (Author's collection)

Major and Preston, in their seminal work on the gunboat, note that '[t]he gunboat itself is harder to define, for the term has been loosely used for as long as guns have been mounted on ships'.[3] Probably the Russian Navy came closest to first using the term as an official classification – depending on the translation used. Thus at least one coin of 1812 of the type that traditionally accompanied the keel-laying of each Russian vessel has survived: its inscription reads '3-gun canon boat/launch' (*3 пўшечн Канонир Лодока*).[4] Only around the time of the Crimean War did the term 'gunboat' became more universally employed, with the construction of the new and specialised vessels represented by the British *Dapper* class.

The Beginning

Even before the universal introduction of the term, one of the recently-founded German navies had introduced it for its own system of classification. Indeed, surprising as it may sound, the history of the German navies since 1815 is largely that of what one might term a 'gunboat navy', even if its ambitious designation as a 'High Seas Fleet' at the pinnacle of the Imperial Navy's power might suggest otherwise.[5] Naturally, during these very humble beginnings in Prussia no one dared to dream of 'cruiser warfare'; the direct protection of the extensive and undefended shorelines of the Baltic had been the sole objective.

The broader naval protection of the loose-knit German Confederation that emerged from the Congress of Vienna of 1815 lay in the hands of Austria and, to a lesser extent, Denmark. In Prussia, the development of a local navy began with the acquisition of six Swedish '*kanonslups*' on 7 June 1815, in the wake of Sweden's handover of Pomerania and the island of Rügen to Prussia as part of the Vienna settlement. These boats were simply left behind, and Prussia benefited little from their acquisition, as they were worn out. Consequently, in March 1816, the construction of the first new building for Prussia, a 'covered gun-sloop', was begun: *Stralsund* was designated 'war-schooner' at her launch in 1817. Even more ambitious was a plan of 1818 to construct a first 'steam war vessel', modelled on the early Russian steam gunboats operating on the River Vistula. However, as with many subsequent plans, which included the first 'Fleet Foundation Plan' of 1820, nothing came of it in a relatively poor, army-focused and over-bureaucratic Prussia.

Except for acquiring a handful of sailing/oared 'gun boats', built and discarded individually in rapid succession, no real progress could be made out during the following twenty years. This situation changed for the better in June 1840 with the commissioning of the *Kanonenjolle Nr. 1*, which marked the true beginnings of the Prussian Navy, insofar as there would henceforth always be purpose-built war vessels included in the inventory lists of the Prussian state. Some historians[6] have thus seen this date as the true birth of the Prussian/German Navy, but the great leap forward came in the revolutionary year of 1848. Then, Denmark tried to incorporate the Duchy of Schleswig into its own territory, contrary to a century-old treaty which bound Schleswig to its neighbouring Duchy of Holstein – although both were under Danish administration. Consequently, both duchies declared their independence from Denmark and prepared for war, including the foundation of a navy of their own. They were supported by the first German National Assembly, gathered in Frankfurt to prepare the foundation of a constitutional Reich, which, faced with a blockade imposed on German ports and coastlines by the Danes, decided to build a truly national navy, the future *Bundesflotte* (Federal fleet).

Integrated into this fleet from the outset was the diminutive 'Hamburg flotilla', including the newly-completed oared gunboat *St Pauli*, together with some auxiliary vessels and, in February 1849, the navy of Schleswig-Holstein. This still-small navy won a certain renown for its innovations, not only by introducing the *Brandtaucher*, the first German submarine, but through the construction of the first European purpose-built 'screw gunboat', *Von der Tann*, publicly known at the time as 'the screw'. Although it may seem to be over-stating the case in view of the relative size of the respective navies, it may be argued that *Von der Tann* was an inspiration for the design of the British 'Crimean gunboats'.

The third navy to contribute to the *Bundesflotte* was that of Prussia, which placed an initial order for eighteen 'gun sloops' and two 'gun launches', but would never formally integrate these forces into it. Instead, Prussia's main contribution lay in the person of the commander-in-chief of the *Bundesflotte*, Prince Adalbert of Prussia.[7] In the eyes of the Frankfurt Assembly he had qualified himself for this position not only because of his prior experience in naval matters, but also because of a May 1848 memorandum regarding the foundation and future development of a German navy. Indeed, this memorandum would serve as the basis for the development of any future navy until its supersession by the plans developed by Alfred Tirpitz almost half a century later.

Prince Adalbert proposed the construction of a future fleet in three stages. The first would serve the operational demands of the current war, and would consist princi-

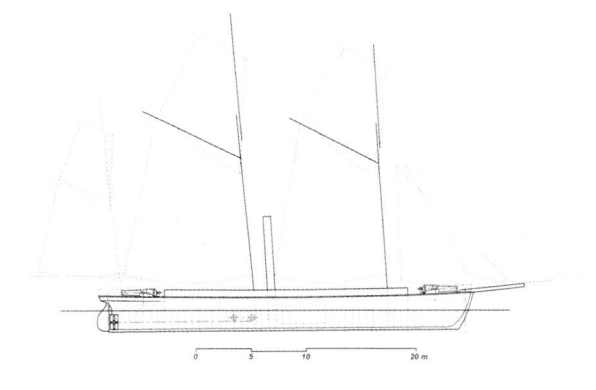

Von der Tann in her first, short-lived three-masted configuration. She would serve on in the Danish Navy until 1861. (Drawn by the author)

Table 1: Target Strengths

	1860	1865	1870	1875
Oared gunboats	40	40	–	–
Steam screw gunboats	20	35	39	52

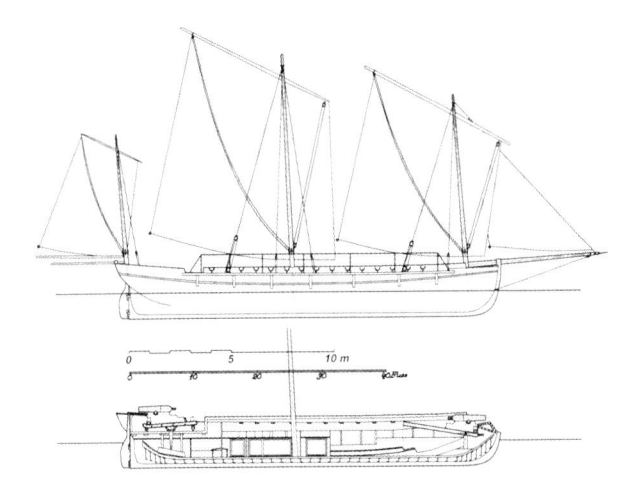

Prussian oared gunsloop No 6, which – unusually – was built of iron. Despite its being the most successful design of the type, the Navy did not continue down this constructional path, and subsequent vessels were built once again of wood. (Drawn by the author)

pally of coast defence vessels, gunboats, etc, with a few larger steamers to sortie locally against the blockading fleets. The second would add a considerable number of cruising vessels (frigates and corvettes) to the roster in order to project the power of the Reich overseas, aiming for comparability with at least the smaller European navies. Here, Prince Adalbert instituted a new term, which would have its impact far into the 20th century: the future naval forces had to be '*bündnisfähig*' – meaning of a size and relative power to be attractive to smaller powers as an ally. Only after fulfilling the objectives of the second stage would the third stage be implemented: the acquisition/construction of a number of ships-of-the-line, either to make Prussia/Germany more attractive as an ally, or to act as a respectable independent national force.

Raising Steam

The *Bundesflotte* was well on its way to meeting the requirements for the first stage when political events demanded its dissolution, leaving the Prussian Navy as the sole survivor and beneficiary.[8] However, its survival during the years 1852–53 was far from certain: influential circles spoke in favour of its dissolution, while others proposed a joint naval effort with the kingdom of Hanover to minimise costs. Eventually, two single events underpinned the decision to keep a navy alive: the acquisition of a stretch of wasteland on the banks of the Jade bay in July 1853 which, during the subsequent 16 years, would be transformed into the naval port of Wilhelmshaven; and, in April 1855, the signature of King Friedrich Wilhelm IV on a 'plan for the future extension of the Prussian Navy', commonly known as the 'Fleet Foundation Plan of 1855'.

As far as gunboats were concerned, this plan served initially to fix the existing numbers: 36 gun sloops and six gun launches, still under sail and oars. One year later, the Prussian Navy executed its first example of 'gunboat politics' abroad. Under the command of Prince Adalbert, the new paddle corvette *Danzig* was on a Mediterranean cruise when the Prince determined to undertake a punitive expedition against the Rif tribesmen of the Moroccan coast in retaliation for their raiding the becalmed Prussian sailing vessel *Flora* in 1852(!). This resulted in the 'Battle of Tres Forcas', where the Prussian landing party, under the personal leadership of Adalbert himself, was repulsed with casualties – the first dead of the Prussian Navy. Back home the Prince earned some harsh words from his cousin the King, despite the new navy having flown its flag on a distant sea and shown itself capable of power projection against a foreign shore.

At the end of the decade the political situation was becoming tense: to the north, fresh trouble with Denmark was looming; to the south, Prussia faced involvement in the conflicts accompanying the unification of Italy. Consequently, the thus-far reluctant Prussian Parliament urged the Admiralty to speed up naval construction and to present a costed 15-year plan to provide for the vessels needed. This plan included for the first time a request for steam gunboats and for the successive phasing-out of the older types. Target strengths are shown in Table 1.

Thus, in June 1859, Prince Regent Wilhelm[9] would publish an order for the construction of twenty steam-powered screw gunboats for a total cost of 1M Taler, taken – quite remarkably for Prussia – from the Army's mobilisation funds. However, the true background of these boats remains obscure, as was already the case 125 years ago. In the notes that accompany the plan they are simply referred to as 'steam gunboats following English patterns', but the actual situation was a little more complex. Even before the Prince Regent's order, chief naval architect Carl Elbertzhagen[10] had already been collecting information about steam gunboats built in Great Britain, France, Russia and Brazil, and distilling the

Table 2: Steam Gunboat Characteristics

	Jaeger class	*Camaeleon* class
Displacement:	237 tonnes	353 tonnes
Length (pp):	33.3m	38m
Beam:	6.5m	6.8m
Depth in hold:	2.5m	2.8m
Speed (designed):	9 knots	9–10 knots
Armament:	1 – 15cm Rk (24pdr)	1 – 15cm Rk (24pdr)
	2 – 12cm Rk (12pdr)	2 – 12cm Rk (12pdr)

most appropriate characteristics for Prussia's own requirements. Elbertzhagen would actually design two different types, both similar to British designs (see Table 2).

From the beginning it had been agreed to build these boats of wood in shipyards located along Prussia's Baltic coast, including nine yards that would produce an initial batch of fifteen boats of the smaller type. The four boats of the larger type – later augmented by another four – would have been built almost exclusively at the Royal Dockyard Danzig. The machinery would likewise be manufactured domestically, with tenders invited mainly from firms experienced in the construction of railway locomotives. These included Borsig of Berlin, Schichau of Elbing and what would become the later Oder-Werke and Vulcan concerns at Stettin; all would subsequently become important defence contractors. As the performance of the Vulcan engines constructed for the larger type proved disappointing during trials, the second batch of gunboats would receive engines imported from Penn & Son, Greenwich (UK). On 15 June 1861 the first Prussian steam gunboat *Jaeger* was commissioned for trials, followed a few weeks later by the first representative of the larger type, *Camaeleon*.

Early Conflicts

Still lacking appropriate cruising vessels, in 1863 the Prussian Navy despatched two of the larger gunboats, *Blitz* and *Basilisk*, to the Mediterranean and the Black Sea respectively, mainly to 'show the flag'. The following year saw the outbreak of the (second) Schleswig-Holstein war with Denmark – the so-called 'First Unification War'. The gunboats would have their share of action, their baptism of fire coming in the Baltic: under the leadership of Prince Adalbert, flying his flag in the aviso *Grille*, the 1st Gunboat Division (*Comet*, *Hay*, *Hyäne*, *Pfeil*, *Scorpion*, *Wespe*) had a brief encounter with the Danish ship-of-the-line *Skjold* and the steam frigate *Sjaelland* on 14 April 1864 east of Rügen. The outcome was inconclusive, but it was reported that the Danish assessment of the capability of the gunboats – which, they feared, might successfully avoid their relatively loose blockade, head north and bombard Copenhagen – resulted in changes in Danish tactics. This would influence the next encounter on 26 April between the Danish steam frigate *Tordenskjold* and the Prussian *Grille*, which had in company another four boats of the 1st Gunboat Division, led again by Prince Adalbert, off the north-western coast of Rügen. Again, this was a brief, inconclusive encounter, with the fast *Grille* conducting 'tip and run' attacks and the strongly armed gunboats in support.

In contrast, a major engagement – at least as far as this theatre of war was concerned – occurred in the waters off Heligoland on 9 May, when the Austrian Commodore Wilhelm Tegetthoff deployed his two frigates, *Schwarzenberg* and *Radetzky*, against two Danish frigates, *Niels Juel* and *Jylland*, and the corvette *Heimdal*. The Prussian paddle steamer *Preussischer Adler* and the gunboats *Blitz* and *Basilisk* (newly returned from the

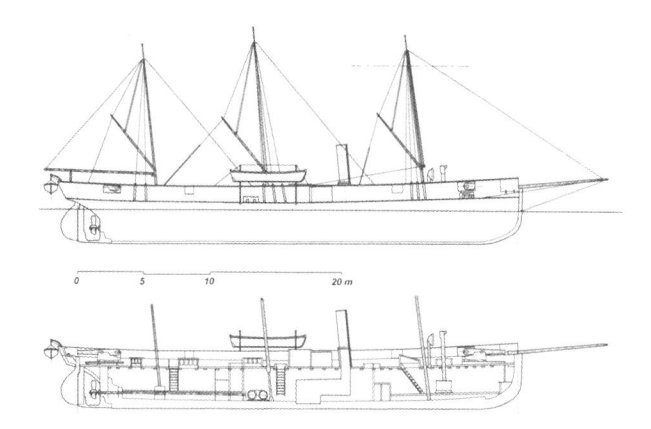

A Prussian second class gunboat of the *Jaeger* class. These vessels had relied mainly on steam instead of sails. Their low freeboard prompted their nickname 'sea-pigs'. (Drawn by the author)

Mediterranean) sailed in Tegetthoff's wake, one of the gunboats being credited with hitting *Niels Juel* with a shell forward at the waterline, causing considerable damage and thereby hindering her attempt to follow the retreating Austrians. The war proved to be the first and only highlight in the careers of these early gunboats; never again would a comparable number of them be in commission at the same time. However, it had shown the versatility of the type, which had been employed not only in purely coast defence actions, but had participated in a 'fleet' action and also in landing operations to regain some of the Danish-occupied islands in the North Sea. Anything appeared possible except, perhaps, true cruiser actions against vessels of commerce.

The gunboats' participation in the wars of 1866 and 1870–71 was sporadic. In 1866, just five of them remained in commission, largely due to a lack of engineering personnel. As the kingdom of Hanover, whose territory included a considerable part of the North Sea

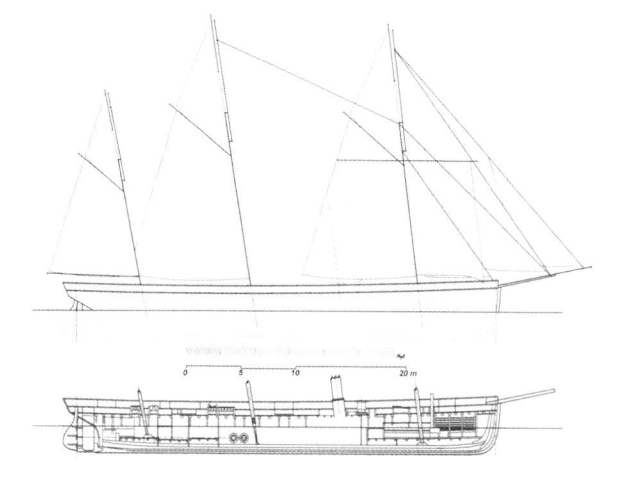

Prussian first class gunboat of the *Camaeleon* class; only slightly larger than their contemporaries, they were nevertheless sent abroad on several occasions, including *Meteor*'s deployment to the Caribbean. (Drawn by the author)

coasts and estuaries, supported the Austrian cause, they were employed in covering Prussian army operations against that state. The subsequent war of 1870–71 against France would be a mixed blessing for the gunboats. Given the French Navy's undeniable superiority in both seas off the German coastline, there were few opportunities for the gunboats to come into action. Six were employed as part of the Jade barrage off Wilhelmshaven, and three of the Baltic-based boats supported *Grille* during a brief sortie against parts of the French battle fleet northeast of Rügen. By far the most memorable naval event – for Germany – would actually occur far to the west: in the Caribbean.

The last of the gunboats, *Meteor*, had commissioned for the first time in September 1869, and had deployed across the Atlantic to monitor events during a revolution in Venezuela. On 28 July 1870, the commanding officer of the French 1,387-tonne aviso *Talisman* informed *Meteor*'s captain, Lieutenant Eduard Knorr, that he had heard rumours that their countries had already been at war for nine days. Knorr sought confirmation, which he eventually gained from American newspapers on 2 August. Henceforth, *Meteor* played hide-and-seek with stronger local French forces until the beginning of November, when Knorr was forced to coal at Havana, Cuba. Hard on his heels was the French aviso *Bouvet*,[11] slightly larger and better armed than *Meteor*. However, when *Bouvet* departed Havana on the 8 November, it was a matter of honour for Knorr to follow and to fight him as soon as the 24-hour neutrality rule expired.

On 9 November he left port to duel with the French aviso, which had waited for him outside neutral waters. The encounter was brief and intense, both vessels steering in attempts to ram one other. This resulted in a glancing blow in which *Meteor* came off worse, losing her main and mizzen masts, and coming to a temporary halt to avoid fouling her screw with the trailing rigging. However, *Bouvet* was unable to take advantage, as a 15cm shell from *Meteor* disabled her boilers and forced her to retreat under sail in the direction of Havana. Unfortunately, *Meteor* was still entangled in her rigging, and by the time the crew eventually managed to free her and bring her back to fighting trim, *Bouvet* had already reached Cuban (Spanish) waters. Under normal conditions this small encounter would rarely have been worth mentioning, but on the German side – given the lack of any other noteworthy naval events during the war – it was celebrated as a true victory, because *Bouvet* had seemingly fled the battleground. In a rare move, the gunboat was decorated with the Iron Cross,[12] and there were individual decorations for Knorr and his crew. The original *Meteor* would be discarded in 1877, but her name would become one of the 'traditional' names in the German Navy.

The First Gunboats for Overseas Service

Even before these events, in 1869, the question of replacements for the gunboats had to be addressed when *Crocodil* was found rotten beyond repair. For her replacement a completely new type of vessel would be introduced, which would become a classic illustration of the problematic nature of categorisation of ships in the German Navy for which it would become notorious until the Tirpitz era. The direct replacement of *Crocodil* would be *Albatross*, which was as a result designated as gunboat. However, the latter's sister *Nautilus*, under construction at the same time, was an addition to strength, and as such was classified as an 'aviso' (a classification ultimately merged into 'small cruiser'), eight of which had been included in the updated foundation plan of 1867. To make matters even more complicated, in 1872 *Nautilus* was re-designated a gunboat after several more of the older vessels had been condemned.

In any event, the projected mission of the pair was not far from that of true cruising vessels: they were officially designed for overseas service, especially as 'pirate hunters' in the South China Sea. Both vessels would spend much of their careers as station ships in South American waters and in the Caribbean, guarding German interests during the Carlist wars in Spain, as well as in Micronesia (*Albatross*), East Asia and East Africa (*Nautilus*). In 1884, both were officially re-classified as cruisers, but in 1888 they were taken out of frontline service to become survey ships. In this capacity *Albatross* served for a further ten years before being stricken in 1899; *Nautilus* had already been stricken in 1896. Both were subsequently sold, becoming coal barges. For

A well-known painting by Willy Stöwer, depicting the brief action between *Meteor* and *Bouvet*. Shown is the moment when the latter's boiler was disabled by the almost immobilised German gunboat. (Author's collection)

Table 3: The First Gunboats for Overseas Service

	Albatross class	*Otter*
Displacement:	713 tonnes	130 tonnes
Length (oa):	57m	31m
Beam:	8.3m	6.2m
Draught:	3.8m	1.6m
Speed (des):	10.5 knots	8.0 knots
Armament:	2 – 15cm Ringkanone (Rk)	1 – 12cm Rk
	2 – 12cm Rk	2 – 8cm Rk
	2 – 37mm Hotchkiss (1879)	

WARSHIP 2022

The newly-commissioned gunboat *Albatross* at the port of Danzig in October 1872. (Author's collection)

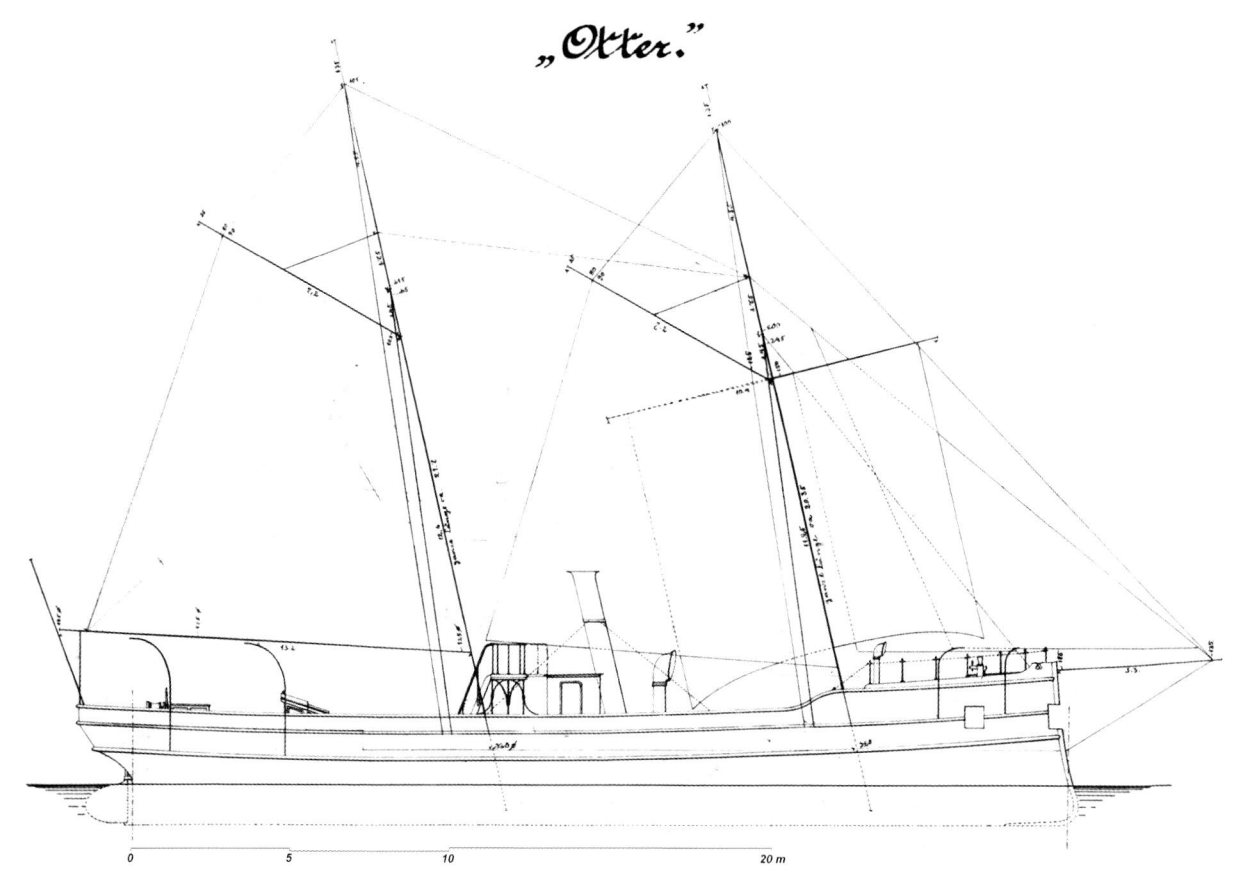

Sail plan of the diminutive *Otter*. It would indeed have been quite a task to sail this flat-bottomed vessel into Chinese waters. (Author's collection)

the main characteristics of these two vessels see Table 3.

After the unification of Germany in 1871 and the installation of *Generalmajor* Albrecht von Stosch as head of Admiralty and *de facto* Navy Minister in January 1872, the Navy once more required a plan upon which to build its future. Stosch took the still-valid plan of 1867 as a model to develop the so-called 'Memorandum regarding the development of the Imperial Navy and the resulting material and financial demands', publicly known as the 'Stosch Fleet Foundation Plan' of 1873. As regards gunboats, the plan demanded a strength of eighteen vessels, to 'represent and protect maritime trade on all seas.' [These words could, of course, been equally applied to cruisers!]

The first new addition to the gunboat force – although in practice a part-replacement for the old vessels that were being discarded during the first half of the 1870s – would, however, be an exception to the new type of gunboat established by the Stosch plan. Cautiously designated a 'pirate hunter' (for the China station), without any designation marking her as a replacement or as an addition to strength, the vessel had been designed speculatively by Schichau and offered to the Navy. The motivation for eventually buying her in 1877 remains unclear, because the vessel was so small that the Navy declined to send her round the north of Denmark to Wilhelmshaven, still less out to China; she thus proceeded from the Baltic to the North Sea via the Eider Canal (the precursor of the Kiel Canal). Consequently the boat, christened *Otter*, would never leave German waters, being attached instead to the artillery and torpedo school until her condemnation in 1907.

The Standard Gunboat for Overseas Service

Almost in parallel with the acquisition of *Otter*, the Navy had designed a new type of standard vessel as replacements for the condemned *Salamander*, *Scorpion* and *Tiger*. For the first time they were constructed of iron, which clearly would have disadvantages for station ships in tropical waters in terms of marine growth on the hull. Sheathing and coppering was not contemplated for fear of galvanic currents between sheathing and hull. The first two boats were built at the Imperial Dockyard Wilhelmshaven, to be christened *Wolf* and *Hyaene* during the first months of 1878, while the third boat was built at the Imperial Dockyard Danzig and named *Iltis* as early as September 1878, despite having been laid down in the subsequent fiscal year to her sisters. All three vessels would have long and successful careers: *Wolf* spent most of her career between 1878 and 1895 on the East Asiatic station, with a short trip home for overhaul, and a brief meeting with history in 1884 when she was present at the raising of the flag in the future German South West Africa. Her third and final stint abroad would be entirely in West African waters, between 1897 and 1905, mainly as a survey vessel; she was stricken from the list in 1906, but survived as a hulk until 1919.

Hyaene had a comparable career to *Wolf*, with a number of highlights. On her first voyage in 1882, she visited Easter island for a highly-regarded scientific survey of the ancient cultures of this remote island, remaining in Micronesia and Melanesia for some years. From 1886 she served on both East and West African

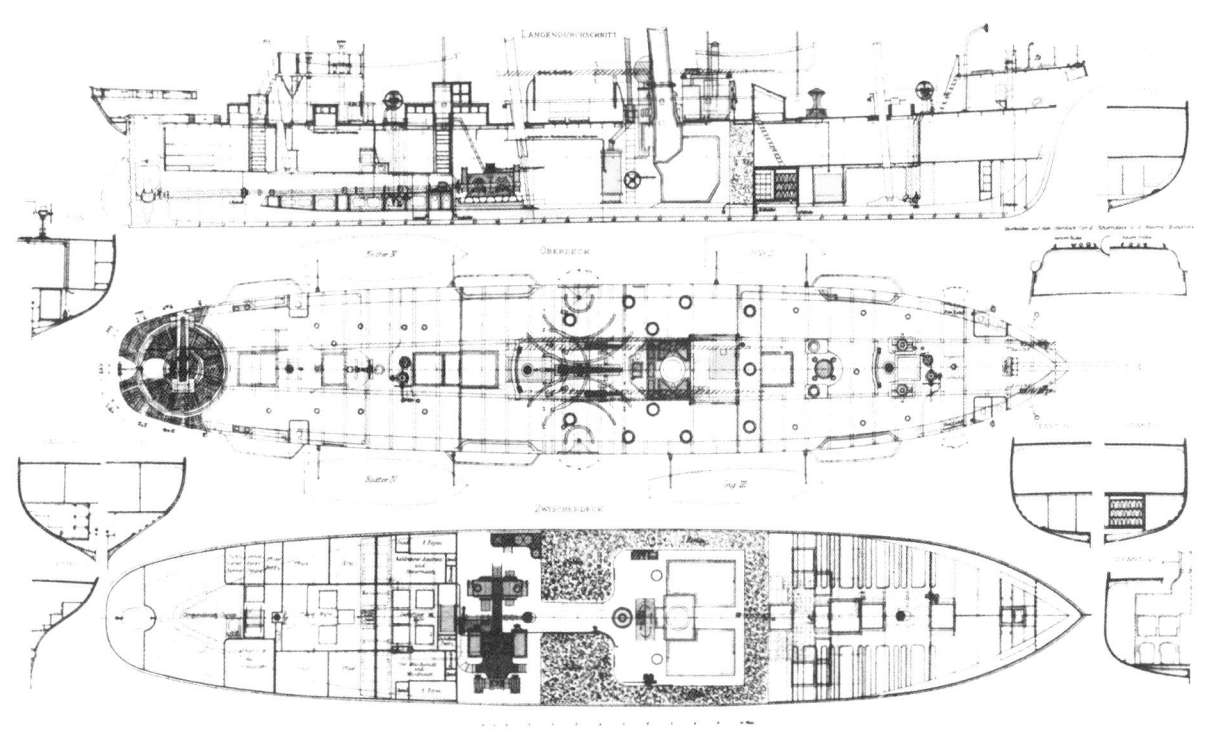

Sections and plan views of a gunboat of the *Wolf* class, taken from a set of similar drawings issued by the Imperial Admiralty at the end of the 1870s. (Author's collection)

stations, founding the town of Swakopmund in Namibia in 1892, and helping in 1893 to quell the so-called 'Dahomey uprising' in the Cameroons. Unlike her sisters, *Iltis* would spend her entire career between 1880 and 1896 in Chinese and western Pacific waters, with a short break for overhaul at home. She eventually won dubious fame by being washed ashore by a typhoon close to the northeastern promontory of the Shandong peninsula in July 1896. This event became famous in the German Navy because of the reported heroism of the doomed crew, who sang the *Flaggenlied* when facing death; this would become something of a second national anthem for the Navy. Of the crew of 82, 71 died.

Parallel to this event, *Hyaene* managed to squeeze between the millstones of domestic interests and naval politics. According to an 1896 memorandum concerning long-term planning for ship replacements (see below), the *Wolf* class would not be due to retire until 1908. Under normal circumstances, this would have been an indication of their soundness to survive for another twelve years. However, after the loss of *Iltis*, reports suddenly appeared that stated that *Hyaene* was being held together only by the rust of her plates and frames. The Navy thus succeeded in ensuring that when Parliamentary approval was being sought for a replacement for the lost *Iltis*, there was no objection when an 'Ersatz-*Hyaene*' was included in the same 1897/98 estimates, even though *Hyaene* was not due for replacement for a further decade. *Hyaene* was speedily re-designated as a survey vessel, to survive for another twenty years in the ranks of the Navy in spite of her allegedly dire state. Indeed, she not only outlived her replacement *Jaguar* (see below), but after her sale to a civilian owner in 1920 operated as a coaster under the name of *Seewolf* until her accidental destruction by fire in the port of Dieppe in 1924.

The Big Gunboats

The next class of gunboats is a perfect example of the reigning confusion regarding the proper designation for individual types of ships. The estimates of 1878/79 saw the addition of a further two vessels whose classification would be changed to that of 'aviso' during the design phase. This change was a consequence of their being given compound engines, constructed by their builder Schichau, then the leading shipyard in Germany for the development of such machinery. The installation of the more powerful and economical compound engines raised their designed speed from the customary 9 knots to 11 knots – sufficient in some minds to make it possible to employ them as reconnaissance vessels (avisos) for the battle fleet as well.

In order to retain this speed in service the hull was sheathed and metalled. This was a feature of ships for overseas service, and *Habicht* and *Moewe* were never employed as avisos, instead being sent abroad as soon as they had completed their trials, becoming 'gunboats' again in August 1881, albeit with the formal qualification 'of the *Albatross* class', implying a higher status than the run-of-the-mill smaller gunboats. However, the confusion regarding their correct designation lingered, as they officially became 'cruisers' from August 1884 to September 1893. After that date *Habicht* was again reclassified as a 'gunboat', while *Moewe* became officially a survey vessel.

Habicht was initially sent to the Australasian station in

Iltis (i), alongside at her customary winter berth at Tientsin (Tianjin), China. (Author's collection)

THE DEVELOPMENT OF THE SMALL CRUISER IN THE IMPERIAL GERMAN NAVY

One of the compound engines intended for either *Habicht* or *Moewe*. Schichau was clearly very proud of its achievement, as the image found its way into the lavishly-illustrated volume celebrating the centenary of the Schichau works in 1912. (Author's collection)

1880, returning in 1882. Deployed abroad again in 1885, she remained on the West African station during the subsequent years. From 1892 to 1896 she was back at Kiel for a thorough overhaul, only to be sent out to West Africa once more. Her final major employment would be during the Herero uprising in German South West Africa, beginning in 1904. In August 1905 she was ordered home for decommissioning and subsequent scrapping, which eventually took place in 1906. *Moewe* was likewise initially despatched to the Australasian station. Her second deployment, in 1884, took her first to the Gulf of Guinea, participating in the acquisition of the colonies of Togo and the Cameroons, and later to South West Africa.

A memorable occasion: on *Habicht*'s ultimate return to Kiel, Admiral Prince Heinrich could not resist to take her round the Kiel fjord under sail alone. In the distance, two representatives of 'modern' times: the battleship *Weissenburg* (left) and, at right, the coast-defence ship *Aegir*. (Author's collection)

Table 4: *Wolf* & *Habicht* classes

	Wolf class	*Habicht* class
Displacement:	490 tonnes	840 tonnes
Length (oa):	47.2m	59.2m
Beam:	7.7m	8.9m
Draught:	3.4m	3.5m
Speed (des):	9.4 knots	11.7 knots
Armament:	2 – 12.5cm Rk	(1 – 15cm Rk)
	2 – 8.7cm Rk	5(4) – 12.5cm Rk
	3 – 37mm Hotchkiss	5 – 37mm Hotchkiss

In 1889, she helped put down the so-called Arab uprising in German East Africa, but was called home at the end of the year. From 1890 to 1905 she would serve as a survey vessel, first in East Africa and later on the Pacific station. On 9 December 1905 *Moewe* was finally decommissioned, but remained for another five years as a hulk in the port of Tsingtau before been broken up there. A photograph taken during her demolition contradicts the widely-published claim that *Moewe* was sunk during the siege of Tsingtau in 1914.

Hay, *Adler* and *Eber*

The next 'gunboat' *Hay* barely merits mention because she was more of a sleight-of-hand than a proper addition to the ranks of the Navy. From the beginning, *Hay* had been planned as a tender to the gunnery training vessel *Mars*, since the existing tender, the former second class gunboat *Fuchs*, was worn out beyond repair. However, as there were no funds available for a new tender, the Navy opted to deceive Parliament by officially putting forward

Moewe hulked in Tsingtau dockyard. Ahead of her can be seen the gunboat *Tiger* and the torpedo boat *S 90*. (Author's collection)

a 'II class gunboat Ersatz-*Habicht*'[13] instead. *Hay* would remain in her role as tender from 1882 until her decommissioning in 1906; she was not, however, broken up until 1919, being employed in the interim as a target barge.

The subsequent *Adler* and *Eber* – which would ultimately share the same fate in March 1889 – were a different story. The first was ordered as a slightly-modified variation of the *Habicht* class in 1882 as the gunboat Ersatz-*Comet*, launched in 1883 as the 'aviso' *Adler* and commissioned for trials in 1884 as a 'cruiser'! Deployed for the first time in 1886, she spent her whole life on the Pacific station; the peak of an otherwise short and uneventful career was the raising of flags on two of the Solomon Islands in October 1886. On 13 March 1889 she was moored in the bay of Apia (Samoa) when a cyclone struck, throwing her onto a coral reef on the lee shore with a toll of 20 dead. The deteriorating wreck remained *in situ* far into the 20th century, but has now completely disappeared.

The estimates for 1885/86 saw, for the first time, the demand for a replacement of one of the gunboats of the second generation, Ersatz-*Albatross*. Since 1884 the colonial movement in Germany was in full swing and the Navy was being forced to meet the increasing consequent demands. One reaction was the introduction of purpose-built 'cruisers' for colonial service.[14] In order to provide a clear distinction between the 'gunboat' and the forthcoming 'cruisers', the size of the former would henceforth be considerably reduced, albeit with the profile adjusted to provide for a degree of uniformity across the German 'colonial' fleet. This meant, for example, the introduction

Table 5: *Adler & Eber*

	Adler	Eber
Displacement:	880 tonnes	582 tonnes
Length (oa):	61.8m	51.7m
Beam:	8.8m	8m
Draught:	4m	3.8m
Speed (designed):	11.3 knots	9–10 knots
Armament:	5 – 12.5cm Rk	3 – 10.5cm Rk
	5 – 37mm Hotchkiss	4 – 37mm Hotchkiss

One of only two known images of *Eber* (i), taken in home waters. (Author's collection)

The remains of *Adler* on the reef in Apia harbour in March 1889. A number of images of the wreck have survived, taken as she deteriorated over the coming decades. (Author's collection)

of a ram bow – which would be completely unnecessary against anything other than a Chinese junk or an Arabian dhow. Christened *Eber* at her launch at the Imperial Dockyard Kiel in February 1887, she sailed in November of that year for the Pacific to relieve the vessel she had been 'legally' built to replace, *Albatross*. During her remaining life of eleven months, she had no opportunity to show her merits before being struck by the same cyclone as *Adler* at Apia. She suffered even more severely, being driven onto another reef and breaking up: 73 officers and crew died, only five being saved.

The Colonial Gunboat

After the completion of *Eber*, there would be a hiatus of ten years during which no gunboat was laid down. This can be traced back, *inter alia*, to the 1884 reclassifications, when the larger gunboats had become 'cruisers'. The previous year had seen the replacement of Stosch by *Generalleutnant* Leo von Caprivi, and the new administration had reluctantly been forced into making provision for colonial 'cruisers' by the dawn of Germany's colonial era. This strongly contradicted Caprivi's main objective: the consolidation of the existing navy by concentrating on the protection of the German coastline.

For parliamentary purposes, the term 'cruiser' (without further qualification – contrasting with the terms 'cruiser-frigate' and 'cruiser-corvette' adopted in 1884 for ocean-going vessels) was applied broadly to vessels intended for overseas service in preference to the vague term 'gunboat'. However, Caprivi was very cautious initially, putting forward two cruisers 'A' and 'B' (later *Schwalbe* and *Sperber*)[15] that were only slightly larger than the *Habicht* class. This would be a problem insofar as another requirement – originating directly with the service – was for larger crews for station vessels to allow for shore parties to undertake local policing tasks in the colonies without degrading the ships' capabilities at sea. However, this proved to be impossible in vessels hardly larger than gunboats, and in 1893 the Kaiser had ordered yet another re-designation of the existing vessels which resulted in greater clarity. From now on, a vessel below a design displacement of 1,000 tonnes would be called a 'gunboat'. Only five vessels remained to be affected by the change: *Habicht*, *Wolf*, *Hyaene* and *Iltis*, plus the station ship at Istanbul, the old paddle aviso *Loreley*.

In 1892, a schedule for the development of the Navy through to 1900 was proposed by the *Reichs-Marine-Amt* (RMA – German Navy Office, the responsible authority for fleet planning). This stipulated the replacement of *Habicht*, *Wolf*, *Hyaene* and *Iltis* by additional cruisers of the *Bussard* class, six of which had already been ordered as follow-ons from the two units of the *Schwalbe* class. As was quite common during this period, the proposal disappeared without trace. Next, it was the turn of the *Oberkommando* (High Command) to present its own proposal for the next six years, with the goal of having nine operational gunboats at the end of the period. This would require a further five vessels to be built – one of the five boats above, not specifically identified at that time, was already considered due for replacement. The Kaiser took this as the cue to request a new plan from the RMA, to be presented in May 1896. Regarding the gunboats, it was anticipated that the *Wolf* class would remain in service until 1908.

All these plans came to naught with the wrecking of *Iltis* off Shandong. An immediate replacement would be inserted into the estimates for 1897/98, as was the first instalment for building an *Ersatz-Hyaene*. It was these very estimates that saw the final downfall of Admiral Friedrich Hollmann as head of the RMA and the subsequent appointment of Tirpitz to that post; despite the many cuts made, the gunboats survived and were approved in April 1897.

It is unclear whether chief naval constructor Alfred Dietrich had prepared rough sketches for new gunboat designs before the loss of *Iltis* in July 1896, but this is suggested by the short time-lag between this event and the December 1896 date on the earliest surviving drawings for the layout of the machinery. What is known, however, is that the basic requirements for any new design included a considerable increase in the power of the main engine(s) to permit the easy ascent of the River Yangtze, in the context of Germany's growing mercantile interests in that region. The profile of the new gunboat was to bear a strong resemblance to contemporary larger vessels, including the main armament, which reflected that fitted in recent avisos. Four of the modern 8.8cm/30 C/90 quick-loading guns were fitted in pairs on the forecastle and poop, supported by six 37mm machine cannon. Drawings were signed off on 3 June 1897, and on 1 October the contract for Ersatz-*Iltis*, subsequently *Iltis* (ii), and Ersatz-*Hyaene*, subsequently named *Jaguar*, was signed with Schichau, Danzig. Composite construction was employed, with a strong steel frame supplemented by additional steel strength-pieces, clad in wooden planking with anti-fouling metallic sheathing for the submerged part. There was no protection except for the conning-tower, which was made from 8mm special steel to protect against bullets or splinters.

Regarding the tasks for which these vessels were being built, it is fortunate that the fighting instructions for their employment has survived in part – quite a rarity among the surviving documents of the Imperial German Navy. From these, it becomes quite clear that the days of the dual-role gunboat/cruiser were over, following the evolution of the design of the small cruiser of the *Gazelle* class. The fighting instructions clearly stated (*inter alia*): 'SMS *Iltis* has been built as a gunboat for political service abroad. The vessel has been designed on the premise that it will not be a true fighting unit, even though it may be employed against soft naval targets, as well as similar targets on shore. This is why its offensive weapons are on the light side, omitting a ram or torpedoes and carrying only small-calibre weapons. … It is impossible to use the vessel as a ram.[16] Ramming should only be used as a final measure of desperation against unprotected targets…'[17] Overall, it was understood that, at outbreak of hostilities, gunboats would be decommissioned to serve as 'depots' to arm and man pre-designated merchantmen as auxiliary cruisers. This measure was confirmed in the spring of 1914, when the East Asiatic Squadron commander, Vice Admiral Maximilian Graf von Spee, on the verge of departing for an extended cruise in the Pacific, ordered that on mobilisation his four gunboats (*Iltis*, *Jaguar*, *Tiger* and *Luchs*) should be decommissioned immediately.

Altogether, three pairs of these very successful vessels would be built, each pair slightly different from the others. Consequently, we will look on each pair separately, only summarising their main data in common form at the end.

Ersatz-*Iltis* was laid down on 27 November 1897, her sister Ersatz-*Hyaene* on 3 January 1898. Despite the fact that only very few replacement vessels ever received the name of their predecessor, Ersatz-*Iltis* would unsurprisingly be one of the exceptions at her launch on 4 August 1898; the second ship received a newly-introduced name, *Jaguar*, on her launch on 19 November. *Iltis* (ii) ran her trials from December 1897 to February 1898, departing on 6 February for the China Station. She would remain there until her scuttling in September 1914, cruising the station zone from southern China to the shores of Sakhalin in the far north – and, naturally, the Yangtze basin. Its river port of Hankow would serve as a long-

The German station ship *Loreley* (i) at Istanbul-Tarabya. (Heidrich collection)

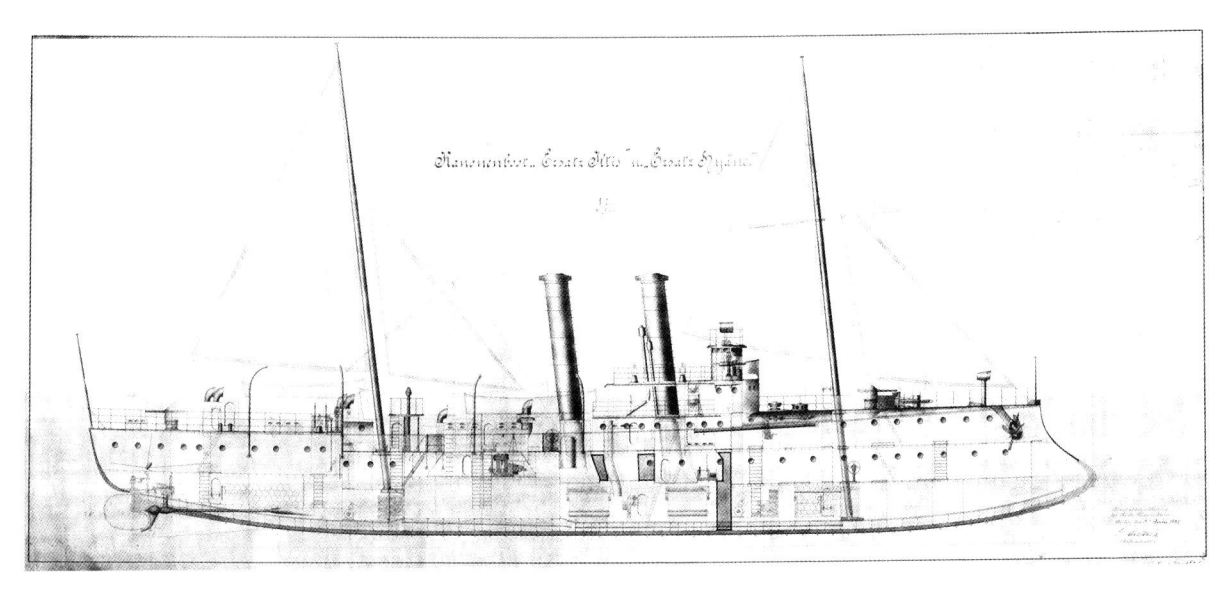

One of the preliminary sketches for the *Iltis* (ii) class that accompanied the specifications sent to the shipyards to elicit tenders; it was signed by Chief Naval Architect Alfred Dietrich on 3 June 1897. (Author's collection)

term base for *Iltis* and some of her half-sisters. The climax of events for *Iltis* would be her participation at the storming of the Taku fortresses on 17 June 1900 by a multi-national force in the wake of the Boxer uprising. The riverine supporting force comprised the gunboats (in order of their position in the Peiho river): *Algerine* (British), *Iltis* (German), *Lion* (French), *Bobr* (Russian); *Korietz* (Russian) and *Gilyak* (Russian). According to interviews held afterwards, the Chinese defenders felt *Iltis* to be their most dangerous adversary on the basis of her 'modern' profile: apart from *Gilyak*, she was the only vessel not fully rigged, and was therefore the target for many of the guns. By the end of the day, *Iltis* had received a dozen direct hits and numerous near-misses that caused splinter damage. Her casualties included Cdr Wilhelm Lans (severely wounded) and Lt Hans Hellmann (the first German naval officer to die in combat) among a total of seven dead and eleven wounded. The other vessels had received fewer hits but some had suffered more casualties (*Korietz* had eight dead and 45 wounded and *Gilyak*, which suffered two major hits, seven dead and 21 wounded).[18]

At the beginning of the First World War, *Iltis* lay in Tsingtau dockyard undergoing a major refit that included opening her up for the removal of her boilers. Her main guns and elements of her crew would be used to help equip the auxiliary cruiser *Prinz Eitel Friedrich*,[19] which would have a seven-month career as a commerce raider, sinking eleven ships before being interned in the USA in March 1915. The hull of *Iltis* was secured to that of her half-sister *Luchs*, and they were scuttled as a pair on 28/29 September south of Yu-nui-san lighthouse, together with former cruiser (now gunboat) *Cormoran*.

Jaguar would follow her sister to the Far East in June 1899, fulfilling much the same role. The highlight of her career was raising the German flag on the former Spanish islands of the Carolines, the Marianas and Yap in 1899.

Iltis in Taku roads for the burial of her dead. Seen from this distance, her 'wounds' are barely visible. (Author's collection)

She also made two long deployments to the western Pacific that ended in Samoa. For these voyages her auxiliary rig was enhanced, including rigging a bowsprit to reduce coal consumption under sail. Having completed a refit in June 1914, she was the 'fittest' of the boats recalled to Tsingtau as war-clouds gathered. This may have been the reason why *Jaguar* was selected to form part of the fortress's defences after having spent a short time out of commission at the beginning of August. She thus received back two of her 8.8cm guns, the other pair having been used to arm *Prinz Eitel Friedrich* (see above). To protect her screws from becoming entangled in fallen rigging under fire, her mainmast was removed and the W/T aerials led instead to the stern, while her foremast received a spotting top. She frequently duelled with Japanese field guns from the Tsang-kou deep in Kiautschou Bay, firing some 2,200 rounds from her two guns and being hit in turn by one 12cm/15cm projectile which caused only insignificant damage under the forecastle. However, on 7 November she was scuttled near her sisters as the German forces surrendered.

The subsequent pair of vessels was built to a slightly

The only known image depicting *Jaguar*'s conversion in August 1914. Here she is shown engaging Japanese batteries from the Tsang-kou deep inside Kiautschou Bay. (Author's collection)

modified design, most changes being external: the most obvious, as noted above, was the replacement of the ram bow with a straight stem, with the stern changed to a counter form. Despite this, the waterline length of both sub-classes remained identical and the underwater forms remained largely unchanged. The second significant variation was the change of main armament, from four 8.8cm QL to two 10.5cm QL guns. It is not wholly clear why this alteration was made, and some contemporary journals continued to record a four-8.8cm battery even after the ships had been laid down. Undoubtedly, the mounting used for the 8.8cm/30-gun, designed in 1889, was approaching obsolescence and still worked on the upwards-sliding Vavasseur principle for checking the recoil. When the second batch of the new gunboats was designed, the new 10.5cm 'cruiser gun' had been developed which, for the first time, operated on axial recoil. The temptation must therefore have been great to use this more modern gun, even at the cost of halving the battery and risking the loss of half the main armament should one of the two guns fail. Tactically, there was little difference between the two guns, as both calibres lacked armour-piercing ammunition; the limitations set out in *Iltis*'s fighting instructions thus remained valid for the larger gun. As noted below, the change had consequences. A third variation was that both main engines were mounted vertically abreast in a single engine room instead of horizontally in line. The engines were manufactured by Schichau, building on their experience in mounting high-speed low-profile engines in torpedo boats.

The first of this new pair, designated Ersatz-*Wolf*, was laid down in 1898 on the slipways of the Imperial Dockyard Danzig, and was named *Tiger* at her launch on 15 August 1899. *Tiger* commissioned on 3 April 1900 and left Kiel on 16 June for China. For the next fourteen years no special events were recorded, except that the standard area of operation was extended to the south to embrace Indonesia, Siam and the Philippines. Like her sisters and half-sisters, *Tiger* participated in the Chinese civil war(s) until, at the beginning of July 1914, she was once more ready to depart Tsingtau to ascend the Yangtze. Held back due to the events in Europe, *Tiger* decommissioned on 1 August to support the fitting out of *Prinz Eitel Friedrich*, surrendering her two main guns, some of her machine cannon and part of her crew.

The second ship of the second batch, Ersatz-*Habicht*, followed her sister by three months in nearly every respect. Christened *Luchs* at her launch on 18 October 1900, she likewise deployed to Chinese waters, despite

The brand-new gunboat *Tiger* in 1900, still in home waters. The revised profiles adopted for her bow and stern are evident, as is the largely-unchanged main body of the hull. (Heidrich collection)

Dockyard model of *Panther*, showing the extended deadwood aft, which precluded the installation of a balanced rudder. (Author's collection)

originally having been intended for the East American station – events in China were of higher priority and would remain so until the end. In August 1914, her crew would make up the bulk of *Prinz Eitel Friedrich*'s complement, including the commanding officer, Commander Thierichens. *Luchs* shared the fate of her sister *Iltis* (see above).

Although not covered by the Fleet Laws, since all the old gunboats now had formal replacements in hand, subsequent vessels of the new series were classed as additions to strength, beginning with gunboat 'A'. The new sub-class was slightly larger than the preceding vessels, with a displacement about 100 tonnes greater. Besides this, the only significant variation was an extension of the deadwood aft to improve course keeping. 'A' was laid down in July 1900 and was christened *Panther* on 1 April 1901. She would become, perhaps, the best-known German gunboat (see below), although her career did not begin well. Sent to the Caribbean in July 1902, she was ordered to neutralise the rebel Haitian gunboat *Crête à Pierrot* in the port of Gonaïves. The latter was better armed than *Panther*, but this potential advantage was

Eber (ii) in 1911, in Agadir roads, as relief for *Panther*. (Heidrich collection)

Table 6: The Colonial Gunboats

	Iltis class	*Tiger* class	*Panther* class
Displacement:	894 tonnes	894 tonnes	977 tonnes
Length (pp):	62m	62m	62m
Beam:	9.1m	9.1m	9.7m
Draught (mean):	3.3m	3.3m	3.1m
Speed (designed):	13.5 knots	13.5 knots	13.5 knots
Armament:	4 – 8.8cm QL	2 – 10.5cm QL	2 – 10.5cm QL
	8 – 37mm MK	8 – 37mm MK	8 – 37mm MK

offset by her rebel status; her crew was poorly trained and morale was low.

On *Panther*'s arrival, the Haitian ship was abandoned, except for the Haitian 'admiral' Hammerton Killick and four others, who ignited the after magazine to destroy themselves and their ship. An hour later, *Panther* opened fire to sink the burning hulk, only to find that after five and 24 rounds respectively both of her 10.5cm guns had become unserviceable. The forward gun encountered a mechanical failure, while the recoil forces of the after gun proved too strong for the scantlings of the deck, and the planking began to come apart. This problem would reappear a few months later during a bombardment of the Venezuelan fortress of San Carlos, resulting in *Panther* having to be docked at Newport News for permanent repairs. As a result of *Panther*'s experience, the Admiral Staff had become sufficiently alarmed to have work undertaken to modify the mounting, resulting in the 10.5cm/40 C/04.

From 1907 *Panther* was based on the West African station. While on the way home for a refit in June 1911, she was diverted to Agadir in Morocco against the background of French intervention following a rebellion, arriving on 1 July. This move was the (in)famous '*Panthersprung nach Agadir*' (*Panther*'s leap to Agadir) that brought the world a little closer to war. The resulting crisis was ended by a Franco-German agreement in November, although *Panther* left for Germany on 25 July, being relieved on the spot by her sister *Eber*. Following her refit at Danzig, she returned to Africa in January 1912, spending two more years on station and returning for a further refit in March 1914. After the declaration of war on Russia, *Panther* joined the Baltic Coast Defence Division, tasked primarily with patrolling and defending the barrages of the outer Kiel bight. In August 1917 she became the flagship of the Ærøsund flotilla, and was decommissioned following the Armistice, on 18 December 1918. She returned to service as a survey vessel in July 1921, finally paying off on 15 December 1926. She would remain in reserve until stricken on 31 March 1931; she was sold for scrap on 10 November and broken up at Wilhelmshaven.

Her sister 'B' would be the first and only gunboat to be built at AG Vulcan's Stettin yard; launched on 6 June 1903 as *Eber*, she commissioned on 15 September of that year. After two months of trials she was laid up, as there was no requirement for her service until 1 April 1910. It was only then that she was sent out to West Africa, to support *Panther* on that station. On 3 August 1914 she sailed into the Atlantic from German South West Africa to find a suitable vessel she could equip as an auxiliary cruiser. One was found on 28 August at the island of Trinidada, near the coast of Brazil, when the brand-new liner *Cap Trafalgar* appeared from Montevideo. After fitting out the liner with her guns, ammunition and part of her crew, *Eber* was decommissioned on 31 August and sailed as a merchantman to the port of Bahia. When Brazil joined the war, the former gunboat was scuttled at her berth there on 26 October 1917.

The Last of the Line

One last vessel needs to be mentioned, even if she was

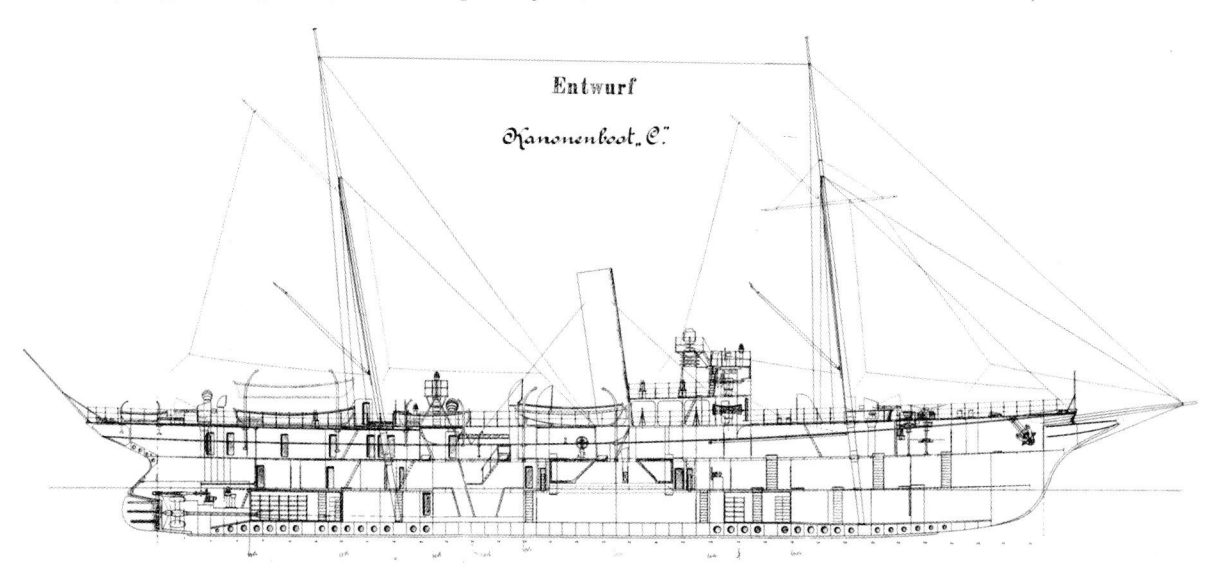

Published here for the first time: the initial project for 'Gunboat C' from 20 July 1904. In contrast to the final project, the four 10.5cm guns of this particulary handsome vessel were distributed in the same way as in the *Iltis* class. (Author's collection)

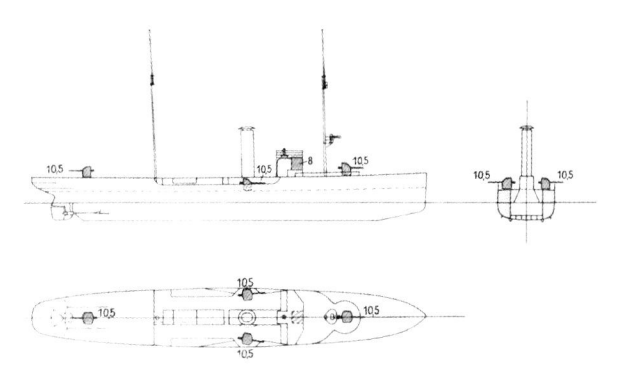

The final, very different project for 'Gunboat C', which was launched as *Meteor* on 18 January 1915. (Author's collection)

never commissioned as a gunboat. A new proposed gunboat, 'C', had featured in the estimates for 1904, 1905 and 1906, but had been rejected each time, not least because of the example of *Eber* lying idle. It would re-appear in 1912, as the eventual replacement of *Iltis* appeared on the horizon. Consequently, the estimates for 1913 contained a first instalment for the vessel, which was finally approved. In August 1913 the order went again to Imperial Dockyard Danzig, which laid down the keel on 26 February 1914. The ship was to be another enlargement of the basic design initiated by *Iltis* (ii), mainly to allow the installation of two more 10.5cm guns in the waist, a long-time requirement following the experiences of *Panther* in the Caribbean. Due to the outbreak of war, work was assigned a low priority and the ship was launched on 18 January 1915 to clear the slipway, being named *Meteor*. She would ultimately be completed as a survey vessel in 1924 – one of the most renowned ever – but that is another story. Principal characteristics were:

Meteor (unfinished)
Displacement: 1,150 tonnes
Length (oa): 71.1m
Beam: 10.2m
Draught: 3.8m
Speed (des): 12 knots
Armament: 4 – 10.5cm QL

Endnotes:

[1] Some readers may have anticipated that this part of the series would cover the *Brummer* and *Pillau* classes, which because they were outside the development mainstream were omitted from Parts I and II. These ships are covered in detail in A Dodson and D Nottelmann, *The Kaiser's Cruisers 1871–1918*, Seaforth Publishing (Barnsley, 2021).

[2] This article relies heavily on D Nottelmann and L Wischmeyer, *Das Kanonenboot ILTIS (II) – seine Vorgänger und Nachfolger. Ein technikgeschichtliche Dokumentation*, Arbeitskreis historischer Schiffbau, *das logbuch* – Sonderheft Nr.19, (Düsseldorf, 2018). This is the only book written on German gunboats in any language and contains full technical details of the vessels covered in the present article, which focuses on operational histories and, in particular, on the interaction between the gunboats and 'true' cruisers. Sources consulted include: E Gröner, *Die deutschen Kriegsschiffe 1815–1936*, JF Lehmanns Verlag (München 1937); *Deutsche Kriegsflotte – Bd V: Kanonenboote*, Reichs-Marine-Amt (Berlin, 1907–1915); *Übersicht über den Fortgang der Neubauten S.M. Schiffe – Abgeschlossen Ende 1917*, Reichs-Marine-Amt (Berlin, 1918).

[3] J Major and A Preston, *Send a Gunboat: The Victorian Navy and Supremacy at Sea, 1854–1904*, Conway Maritime Press (London, 2007), 14.

[4] VL Aleksandrov (ed), Lenko Izdat, *Admiralte?skie verfi: korabli i gody: 1704–1925*, Gangut (St. Petersburg, 1994), 67.

[5] Almost every navy had a strong coast defence aspect and/or colonial ambitions, requiring 'gunboats' to be employed either offensively or defensively.

[6] For example, Paul Koch (1872–1931), one of the most renowned historians of the early German navies.

[7] For Adalbert and his career, see J Duppler, *Prinz Adalbert von Preussen*, Mittler Verlag (Herford, 1986).

[8] Most importantly, Prussia would receive the paddle corvette *Barbarossa* and the former Danish sailing corvette *Gefion* (ex-*Eckernförde*, ex-*Gefion*).

[9] The younger brother of King Friedrich Wilhelm IV; from 1861, King Wilhelm I of Prussia, and later Kaiser Wilhelm I of Germany.

[10] Carl Alexander Elbertzhagen (1814–1881), master shipwright since 1839. He had served since 1855 as director of the Royal Dockyard Danzig, and since 1856 as head of department of naval construction at the Admiralty; he retired in 1872.

[11] *Bouvet* (1865), 748 tonnes, 10 knots, two 16cm, four 11.4cm guns.

[12] There would be only three similar instances later: the submarine *U 9* and the cruiser *Emden* would receive the Iron Cross for their exploits during the First World War, while the gunboat *Iltis* (ii) received the even higher decoration *Pour le Mérite* for its service during the taking of the Taku forts in 1900 (see below).

[13] This designation was even more confusing as this was old *Habicht* (i) of 1860, condemned back in 1877.

[14] See Part I of this series, *Warship 2020*, 103.

[15] Ibid.

[16] The adoption of the ram may have been Dietrich's personal preference, as his successor immediately discarded it for the succeeding small vessels.

[17] Nottelmann and Wischmeyer, *ILTIS*, 51ff.

[18] The true level of damage incurred on the bombarding fleet was – consciously – minimised for a long time by certain naval historians, witness the following account (with the source deliberately omitted here): 'A force of cruisers and gunboats shelled the Taku forts and easily silenced their obsolete guns, although the Russian gunboat *Gilyak* and the British sloop *Algerine* sustained some slight damage'.

[19] The armament of *Prinz Eitel Friedrich* has been subject to a wide range of speculation, including one particular account posted on the internet (http://www.dekalbhistory.org/documents/DHCNEWSLETTER_summer-2009.pdf) stating that a 10.5cm gun displayed in front of the courthouse in DeKalb, GA, USA, came from the ship. This is clearly an error, as the gun is a 10.5cm/35 C89 on an MPL/91 mounting, a type only surviving in Tsingtau on *Cormoran*. The gun would be mounted in the auxiliary cruiser *Cormoran* (ii, ex-*Ryazan*), which would subsequently be scuttled on 7 April 1917 after internment at Guam. In fact *Prinz Eitel Friedrich* mounted four 10.5cm from *Luchs* and *Tiger*, four 8.8cm from *Iltis*, and two 8.8 cm from *Jaguar*. There is a slight possibility, of course, that one of the *Cormoran* guns was salvaged and later installed on USS *DeKalb*, ex-*Prinz Eitel Friedrich*.

THE BATTLESHIP
JAURÉGUIBERRY

Generally regarded as the most successful of the battleships of the 'Fleet of Samples' (*Flotte d'échantillons*), *Jauréguiberry* went on to give distinguished service in the eastern Mediterranean during the First World War. In the latest of his series of articles, **Philippe Caresse** looks at the origins and service career of this unusual vessel.

Jauréguiberry was the first French battleship to be ordered from the private shipbuilding industry. She was the third ship listed under the 1890 budget. Together with her later half-sister *Bouvet*, she would be considered one of the better designs of the *Flotte d'échantillons*.

The plans were drawn up by the naval architect Amable Lagane, Technical Director at the Forges & Chantiers de la Méditerranée shipyard. On 8 April 1891, Minister of Marine Edouard Barbey signed the contract for the construction of a battleship to be named *Jauréguiberry* at their shipyard at La Seyne, opposite Toulon naval dockyard.

One of the important innovations made by Lagane was the adoption of electricity to power the training of the main guns – previously the French Navy had used hydraulics. Another was the introduction of the twin turret for the 138.6mm QF guns – the other four ships of the series all had their 138.6mm guns in single turrets. At that time, the *Conseil des Travaux* had strong reservations about the twin turret, which could be disabled by a single hit, meaning that a quarter of the QF armament would be out of action. Lagane was also forced to defend his choice of electricity to power the main gun turrets; the *Conseil Supérieur* gave its approval only after heated discussion.

As for the award of the ship to private industry, the latter had an execrable reputation in the eyes of French Corps of Constructors. Despite their opposition, they would be compelled to admit that *Jauréguiberry* turned out to be a far better ship that her half-sister *Carnot*, built on the opporite side of Toulon roads on the Mourillon slipways in the heart of the naval dockyard. As it was, *Jauréguiberry* was one of only two ships of the series to participate in the Great War, and the only one to survive it intact.

General Characteristics

The order for the battleship *Jauréguiberry* was placed on 8 April 1891. Construction was supervised by naval engineer Opine.

The hull was of steel throughout, and had a total weight of 3,688.94 tonnes. Bilge keels were fitted in May/June 1905.

General Characteristics

Length oa	111.05m
Length wl	109.70m
Length pp	108.60m
Beam (wl)	22.15m
Depth of keel	7.94m
Draught	7.75m fwd, 8.45m aft
Freeboard forward	8.08m
Normal displacement	11,888 tonnes
Full load displacement	12,081 tonnes
GM	1.15m
Complement	607 officers & men

Jauréguiberry: Midship Half-Section

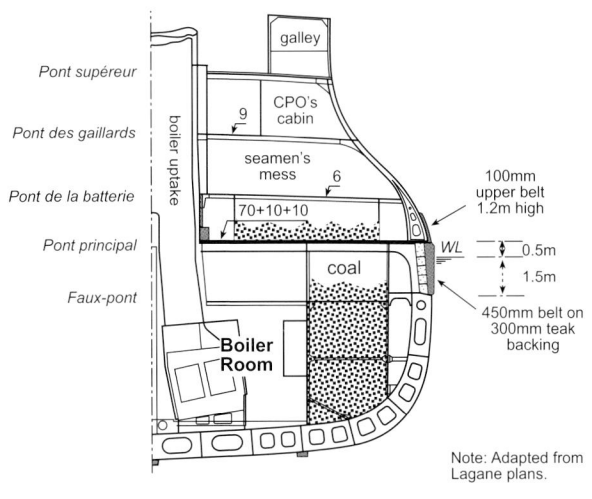

Note: Adapted from Lagane plans.

© John Jordan 2018

The upper part of the main armoured belt (*cuirasse épaisse*), which was of nickel steel, had a maximum thickness of 450mm amidships, reducing to 300mm at the bow and the stern. The lower edge was graduated from 250mm to 150mm. Height was 2 metres amidships, of which 1.5m was below the waterline, reducing to 1.81 metres at the stern. An upper belt (*cuirasse mince*) of 'special' steel ran the full length of the ship over the cellular layer; it was 1.2 metres high amidships, rising

THE BATTLESHIP *JAURÉGUIBERRY*

Jauréguiberry
Profile & Plan

GA Plans

Note: Adapted from Lagane plans.

© John Jordan 2018

305mm turret
wardroom
POs' mess
steering compᵐᵗ
305 shell room
aft perpendicular
305 magazine
47/37 hoist
138 mag
ENGINE ROOM
47/37 magazine
65/45 magazine
AFTER BOILER ROOM
CENTRE BOILER ROOM
FORWARD BOILER ROOM
305 magazine
138 mag
47/37 mag
47/37 hoist
TT
sick bay
general store
casualty station
wine hold
handᵍ room
305mm turret
47 hoist
navigation bridge
CT
47/37 hoist
Admiral's bridge
37 platform
47 platform
galleys
bakery
snr offrs' wardᵐ
staff office
Admiral's quarters
37 platform
47 platform
47/37 hoist
fore perpendicular

Cl (*Cloison*) watertight bulkhead
WTC watertight compartmt

Plate-forme de cale

WTC
shell rooms
blk pwdr
305 magazine
138 magazine
138 handing rm
port engine room
std engine room
blk pwdr
274 shell rm
submerged torpedo rms
274 shell rm
274 magazine
274 magazine
aft port boiler rm
aft std boiler rm
ctr port boiler rm
ctr std boiler rm
fwd port boiler rm
fwd std boiler rm
138 magazine
305 magazine
305 shell room p&s
wine hold
cable locker p&s
138 handing room p&s

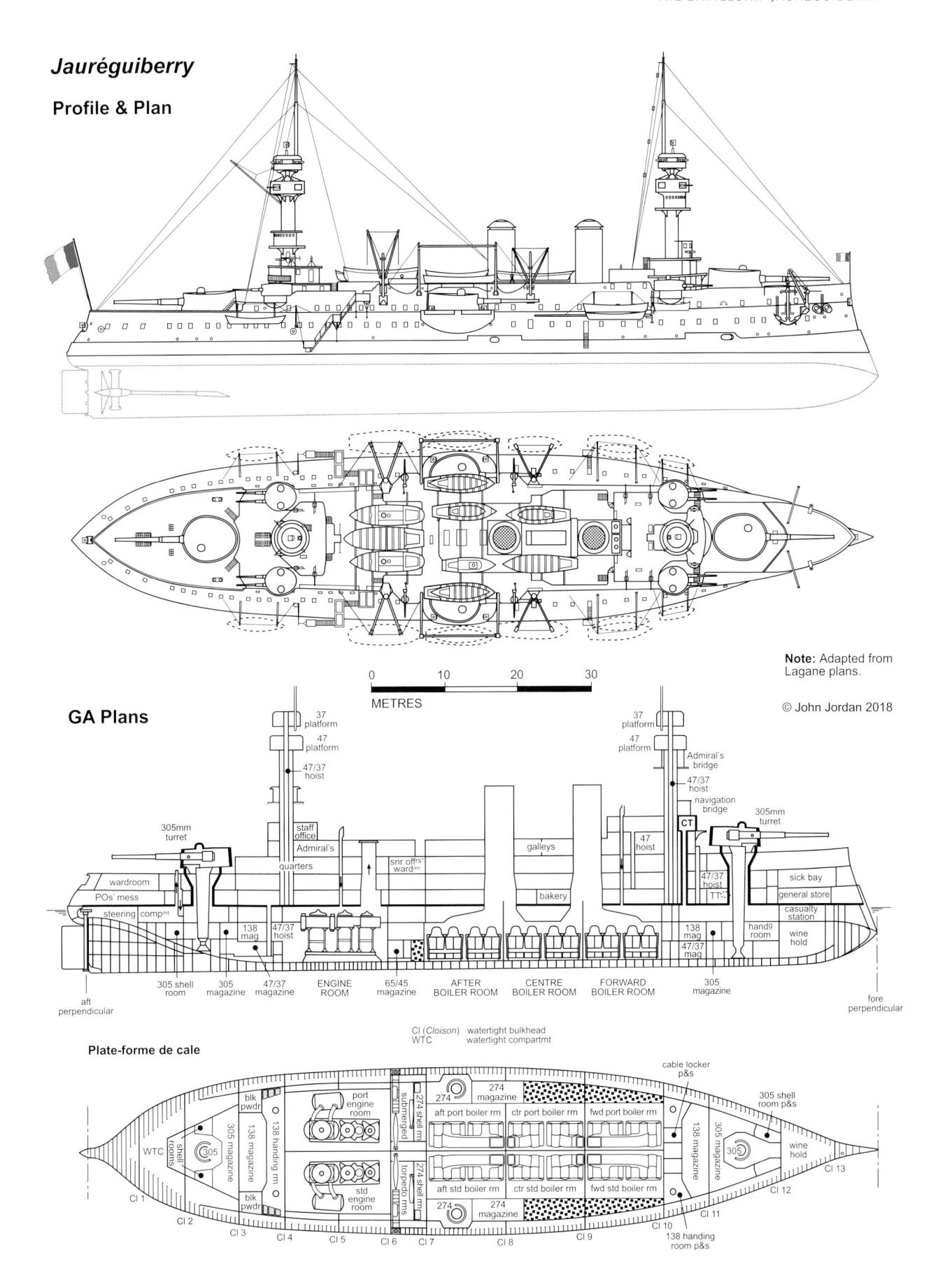

Cl 1 Cl 2 Cl 3 Cl 4 Cl 5 Cl 6 Cl 7 Cl 8 Cl 9 Cl 10 Cl 11 Cl 12 Cl 13

sharply at the bow to a height of 1.67 metres, and had a maximum thickness of 100mm amidships.

The main armoured deck comprised iron plating 70mm thick on a double layer of 10mm steel. The armoured coamings around the penetrations for the boiler uptakes and ladderways were of 350mm steel.

The conning tower, which weighed 28.64 tonnes, had 250mm walls, and the communications tube was formed with hoops of 220mm steel. A relatively small navigation bridge was located atop the conning tower.

The total weight of protection for the hull was 2,589.83 tonnes.

Armament

Jauréguiberry was the first battleship built for the French Navy to have her turrets powered by electricity. Two Sautter & Harlé dynamos provided the powet to train the turrets; they were rated at 800–1,200A and supplied a 160-Volt circuit. The electrical installations were designed by engineers Lagabbe and Savatier.

The battleship was armed with two 305mm 45-calibre guns Mle 1887 and two 274.4mm 45-calibre guns Mle 1887 in enclosed turrets in the traditional lozenge arrangement favoured by the French Navy. The 305mm turrets were mounted fore and aft, the 274.4mm turrets on the beam. The 305mm Mle 1887 fired a cast iron shell of 292kg, its 274.4mm counterpart a cast iron shell of 216kg. These would later be complemented by steel AP and SAP shells weighing 340kg and 255kg respectively. The guns were on trunnion-type mountings of identical design, with recoil cylinders and recuperators, and were capable of 10 degrees elevation. They would be fitted with a hydraulic breech flushing system in April 1901 and new optical sights in April 1907.

Both types of turret were built by the same contractor, and protection was identical: 370mm for the turret walls, 70mm for the roof, up to 200mm for the armoured hood, which was of cast steel, and between 370mm and 280mm on the barbette. The weight of the 305mm turret was 93.70 tonnes, that of the 274.4mm turret 79.10 tonnes.

Unlike the other four battleships of the *Flotte d'échantillons*, *Jauréguiberry* had her 138.6mm QF guns in twin turrets, mounted at the four corners of the ship. The turrets were armoured with 100mm plating on the sides, 30mm on the roof, and between 100mm and 80mm on the barbettes. Fouvasseur and Groult night sights were fitted in April 1912. The 138.6mm guns were supplemented by four 65mm QF guns in open mountings on either side of the boat deck amidships.

The total weight of protection for the turrets was 1,372.26 tonnes.

The number, calibre and disposition of the anti-torpedo guns for French battleships of the period varied considerably, and the complement of ATB guns was often revised

Armament

Two 305mm 45-cal Mle 1887 BL guns in two single turrets
Two 274.4mm 45-cal Mle 1887 BL guns in two single turrets
Eight 138.6mm 45-cal Mle 1891 QF guns in four twin turrets
Four 65mm 50-cal Mle 1891 QF guns in open single mountings
Twelve 47mm 40-cal Mle 1885 QF guns in open single mountings
Twelve 37mm 20-cal Mle 1885 revolver cannon
Six 450mm torpedo tubes

Calibre	Shell weight	Muzzle velocity	Firing cycle
305mm	292kg CI	815m/s	0.6rpm
	340kg APC	780m/s	0.6rpm
274.4mm	216kg CI	815m/s	1rpm
	255kg APC	780m/s	1rpm
138.6mm	35kg	730m/s	4rpm
65mm	4kg	715m/s	8rpm
47mm	1.5kg	650m/s	9–15rpm
37mm	0.5kg	435m/s	20–25rpm

* Cast Iron (CI) shell

Calibre	Angle of elevation	Range
305mm	+10°/-5°	12,500m
274.4mm	+10°/-5°	11,800m
138.6mm	+15°/-7°	9,400m
65mm	+20°/-15°	5,400m
47mm	+20°/-20°	4,000m
37mm	+20°/-20°	2,000m

A fine close-up of one of the twin 138.6mm turrets. (Author's collection)

THE BATTLESHIP *JAURÉGUIBERRY*

Plan of the 274.4mm single turret and its ammunition supply arrangements.
(Forges et Chantiers de la Méditerranée, courtesy of Philippe Caresse)

during trials and at subsequent refits. The original Lagane plans of *Jauréguiberry* show eight/twelve Hotchkiss 47mm guns Mle 1885 and twelve 37mm revolver cannon. The 47mm guns were mounted as follows: two/four in each of the military masts (middle platform), two on either side of the lower bridge deck, and two on either side of the after superstructure. Eight of the 37mm revolver cannon were mounted on the upper platforms of the military mast, two on the upper bridge deck, and two on the after superstructure. The ship also carried two 65mm Mle 1881 field guns for the landing party.

Jauréguiberry was fitted with no fewer than six torpedo tubes: four trainable above-water and two submerged. Two of the trainable tubes were mounted on pivots on the main deck abaft the forward 305mm turret; they had a command of 1.90m, and training arcs were between 16 degrees and 90 degrees from the ship's axis. The second pair was located amidships, just forward of the 274.4mm wing turrets; they had a command of 1.65m, and training arcs were 37 degrees either side of the beam. The four above-water tubes, which fired through a raised section of the upper belt, would be disembarked in 1906.

The submerged tubes were on the lower platform deck amidships, abaft the wing turrets; they were 4 metres below the waterline and were at a fixed angle of 90 degrees from the ship's axis. The two Thirion compressed-air pumps were rated at 500 litres per hour and 750l/h respectively.

The torpedo was the Mle 1892, a Whitehead torpedo built under licence at Toulon. It had a length of 5.05m; weight was 530kg including a 75kg explosive charge, and it had a range of 800m at 27.5 knots. Thirteen torpedoes were carried.

Fire Control

Six large-model fire control tables were provided for the torpedo tubes: two on the deck above the forward above-water tubes, two above the midships a/w tubes, and two in the conning tower. Communication with the torpedo compartments was by voice-pipe, with a bell for firing orders; the torpedo firing mechanisms were mechanical.

Two of the eight 60cm Mangin searchlight projectors were mounted on fixed pedestals on the upper platforms of the two military masts (*ligne haute*). The lower projectors (*ligne basse*) were on mobile carriages which ran on rails: one in the bow, one in the stern, and a pair fore and aft of the 274.4mm wing mountings. All the projectors were controlled remotely via electric cables, and all but the pair abaft the 274.4mm turrets could be retracted within the hull for protection from the elements.

Between January and February 1908 a single 2-metre Barr & Stroud pedestal rangefinder was mounted atop the navigation bridge. In December 1912 a second Barr & Stroud rangefinder was embarked and the rangefinders were linked to the conning tower by a Germain hydraulic transmission system. In May 1913 two R/F reception positions were fitted on the after bridge and linked to the rangefinders by a Germain transmission system.

In March 1917 a fire observation position was installed on the upper platform of the military foremast.

Machinery

Steam for the main engines was supplied by 24 boilers of the Lagrafel & d'Allest type built by Forges & Chantiers de la Méditerranée, mounted back-to-back on either side of the centre-line and distributed between six boiler rooms. Each of the boilers had a single rectangular furnace, and maximum pressure was 15kg^2; steam was first raised on 24–25 October 1895. There were twelve vertical ventilators manufactured by FC Méditerranée for the boiler rooms, each with a capacity of 30,000m^3 per hour. The two identical funnels housing the uptakes for the boiler exhausts each had a cross-section of 16.18 m^2 and a height of 23 metres above the gratings.

The boilers supplied steam for two vertical triple-expansion (VTE) engines, likewise built by Forges & Chantiers de la Méditerranée. Their installation was supervised by the shipbuilder's own engineer Moritz. The engines were installed side by side in two independent engine rooms divided by a centre-line bulkhead. There were four Elwell steam-powered ventilators with horizontal Compound motors for the engine rooms, each with a capacity of 30,000m^3 per hour.

The two outward-turning four-bladed propellers were of manganese bronze. At 11 knots the turning radius of *Jauréguiberry* was approximately 520 metres.

Machinery

Boilers	24 Lagrafel & d'Allest
Engines	two VTE
Propellers	two 4-bladed 5.70m
Rudder	non-balanced, 20.00m^2
Designed power	14,440CV
Maximum speed	17.67 knots
Coal	893 tonnes
Endurance	4,752nm at 12.15kts
	2,688nm at 15.88kts
	1,486nm at 17.67kts

Equipment

Searchlights	eight 60cm Mangin 45A projectors
Boats	two 10-metre steam pinnaces
	one 7.6-metre White steam launch
	one 11-metre pulling pinnace
	four 10-metre pulling cutters
	two 9-metre pulling cutters
	four 8.5-metre whalers
	one 7-metre whaler
	two 5-metre dinghies
	two 3.5-metre punts
	two 5.6-metre Berthon canvas boats
Anchors	two Marrel 13.1-tonne bower anchors
	one Marrel 6.13-tonne sheet anchor

At 16 knots the angle of heel was 5.30 degrees.

There were four main Thirion bilge pumps each rated at 600 tonnes per hour, two Thirion pumps rated at 45 tonnes per hour, and two Thirion pumps with a capacity of 30 tonnes per hour, together with six pumps in the boiler rooms each rated at 20t/h.

Electrical power for the ship was supplied by four dynamos designed and manufactured by Sautter & Harlé; they were rated at 400–500A and served an 83-Volt circuit.

The messdecks were heated by 99 steam-powered stoves supplied by Grouvelle & Arquembourg. There were also coal-fired stoves in the CO's appartments and the officers' wardroom.

Equipment

The two 10-metre steam pinnaces, the 11-metre pulling pinnace and two of the 10-metre pulling cutters could be fitted with a 37mm QF gun.

The larger boats were handled by four pairs of 'Y'-frame type cranes. The first pair was mounted between the funnels, the second between the after funnel and the mainmast. There were two pairs of davits above the 274.4mm wing turrets, and further pairs of light davits for cutters and whalers abeam the bridge and abreast the after 138.6mm turrets. Power for handling the boats was provided by two 2-cylinder steam winches supplied by Stapfer de Duclos; the power of these is not given in the *devis de l'armement*.

The anchor chains were 60mm in diameter; each had a length of twenty-four 30-metre shackles. The main capstan was a 2-cylinder steam model supplied by Caillard Bros, and was rated at 33,000kg. The auxiliary capstan was likewise a 2-cylinder model, but the power is not specified in the *devis de l'armement*. Two jet anchors each of 3.40 tonnes were also embarked; the chains had a diameter of 30mm and a length of four shackles

There was sufficient stowage for 90 days of provisions for a crew of 650 men; a complete ration was calculated at 1.65kg. A total of 52 tonnes of drinking water was embarked in iron tanks, sufficient for 20 days at sea. There was also stowage for 34,500 litres of wine and 2,500 litres of spirits. The galleys were fired by wood; 29 cubic metres were stowed in the wine hold forward.

Launch and Entry into Service

On 27 October 1893, during the Franco-Russian festivities, President Carnot, Minister of Marine Admiral Henri Rieunier and their entourage made a visit to the Russian squadron of Rear Admiral Avellan at Toulon. They were received on board the battleship *Nikolai I* and the armoured cruiser *Admiral Nakhimov*, then made their way to La Seyne to attend a ceremony which had long been planned. The correspondant of the journal *le Yacht* reported on the event:

> In the afternoon *Jauréguiberry* was launched. The gates opened at 1145 and crowds of spectators flocked in. The

The launch of *Jauréguiberry* on 27 October 1893 at La Seyne. (Author's collection)

Jauréguiberry fitting out. (Author's collection)

decoration of the battleship, the shipyard installations and the viewing stands were magnificent. At the stern to port, close to the sea, was the presidential platform; it was extended towards the bow to accommodate the most important guests. There was a second platform to starboard, and beneath were pontoons loaned by the Navy which accommodated the massed spectators. The launch operation went off smoothly and was a complete success. The signal to begin the launch was given by the widow of the late Admiral Jauréguiberry, the sponsor of the ship. The presence of so many high-ranking naval officers and civilians, together with that of the Russian sailors, significantly enhanced the impact of the occasion.

Following her successful launch, the hull of *Jauréguiberry* was towed to the fitting-out quay, which ran parallel to the slipway. Work on the ship made steady progress, and the installation of the propulsion machinery, built by Forges & Chantiers de la Méditerranée at their Marseille works, was supervised by the company's chief engineer Moritz. Engineers Lagabbe et Savatier supervised the installation of the electrics which they had conceived and designed.

The first trials took place on 24–25 October 1895. The ship's CO, *Capitaine de vaisseau* (CV) Foret, took up his post on 23 January 1896 in order to supervise the final stages of fitting-out. On 12 March the ship sailed for her first sea trials. Unfortunately, while manoeuvring, one of *Jauréguiberry*'s propellers became entangled around a buoy, and the tug *Samson* had to send a diver to free the shaft. He reported that one the propeller blades had been

Jauréguiberry steaming at more than 17 knots during her power trials. (Author's collection)

THE BATTLESHIP JAURÉGUIBERRY

Jauréguiberry in Villefranche. roads. (Author's collection)

slightly bent, but this was not serious enough to prevent the battleship's sortie.

On 9 April *Jauréguiberry* ran her 24-hour trials, achieving speeds in excess of 17 knots. The following day, at about 1725, one of the boiler tubes split while the furnace door was open, and a mix of steam and flame shot out into the stokehold. The ship's captain quickly took the necessary measures to control the fire: the boiler was isolated and shut down. Nine stokers were seriously injured, three of whom died shortly after the incident with three more hospitalised at Saint-Mandrier. A subsequent enquiry decreed the replacement of all the welded tubes in ships equipped with the Lagrafel & d'Allest boiler.

Jauréguiberry was docked in May and ran her first speed trial with natural draught on the 20th of the same month off Toulon. Seven days later she embarked on her fuel consumption trials.

On 30 March 1897 there was a serious incident when the compressed air vessel atop one of the torpedo tubes burst as the tube was being put in place. The ship underwent an emergency docking to seal the door of the affected torpedo tube.

Jauréguiberry was commissioned on 15 May 1897 and was assigned to the 1st Battle Division of the Mediterranean Squadron two days later. During her work-up it was noted that in a seven-metre swell, with a head wind, she had a maximum heel of between 11 and

Casualty evacuation exercise on board *Jauréguiberry*. (Author's collection)

87

Seen here at sea, *Jauréguiberry* was regarded as one the more successful ships of the *Flotte d'Echantillons*. Note the relatively low superstructures compared to her half-sisters. (Author's collection)

Jauréguiberry at her moorings at La Pallice, on the west coast of France. Behind her, on the left, can be seen the battleship *Masséna* and, on the right, the coast defence battleship *Bouvines*. (Author's collection)

Jauréguiberry and the old battleship *Formidable* in the port of Brest. The upperworks of *Jauréguiberry* are concealed beneath a large canvas awning rigged to facilitate maintenance work. (Author's collection)

17 degrees; she also took on a 1/2-degree list when the boats were not hoisted simultaneously on both sides of the ship.

In addition to the customary sorties to the anchorage at Les Salins d'Hyères, *Jauréguiberry* was docked in Missiessy No 1 dock 2–16 April, then remained moored in Toulon roads from 17 April to 2 May. On the 14th, she left for Golfe-Juan, Bastia, Santa Manza, Ajaccio and Endoume, entering the National Basin at Marseille on 3 June. She was at La Ciotat on the 9th and moored at buoy No 17 at Toulon the following day. After visits to Bizerte, Ajaccio, La Badine and Golfe-Juan, she would be at sea on 24 August for a general inspection of her gunnery prowess. Between 20 September and 24 November, the date on which she was docked in Missiessy No 2, she carried out gunnery trials and torpedo launches. *Jauréguiberry* left the dock on 7 December to continue her work-up. She was at Villefranche on 1 February 1899 and at Golfe-Juan one month later. She was readied for combat during the Fashoda crisis, but remained at her moorings at Toulon apart from a brief sortie to Les Salins. Peace having broken out between Britain and France, the Mediterranean Squadron visited Villefranche and Corsica; *Jauréguiberry* was then docked in Missiessy No 1 from 21 April to 23 May. During July, the squadron moored off Barcelona and Fort-Mahon. Towards the end of the year, the battleship embarked on a cruise to the eastern Mediterranean, visiting Piraeus, Jounieh, Jaffa, Beirut, Tripoli, Lattakia, Alexandretta, Messina, Rhodes, Smyrna and Salonika.

Jauréguiberry was again in the anchorage at Les Salins from 20 February 1900, followed by the customary visits to the coasts of Provence.

The 21 June saw the start of the annual grand manoeuvres: the battleship was at Mers el-Kebir 26–27 June, then sailed for the Atlantic for joint manœuvres with the Northern Squadron. These were followed by port visits to Douarnenez, Brest, Cherbourg, culminating in a naval review in the presence of President Emile Loubet. After returning to Brest, the fleet reentered Toulon via Royan on 14 August.

The Prewar Years

When *Jauréguiberry* entered service, her qualities were quickly appreciated by her officers and men of all ranks. The following remarks appear in the *devis de campagne* of CV David dated 1900:

> *Jauréguiberry* is a vessel with remarkable qualities. She manoeuvres admirably and is a good sea-boat, riding the swell rather than plunging into it like most of our battleships. She has a moderate roll, at least in the

Mediterranean. However, what is most appreciated in the design of the ship is that in addition to her graceful lines, movement around the ship is a simple matter due to the symmetry of the internal layout and also because of the spaciousness of the compartments, which means that the personnel always has adequate room to work.

The ship would be even better if she had a medium battery of 16cm guns rather than 14cm. Despite this, the height of command of the guns and the ease with which they are trained and operated, both electrically and manually, mean that that she could hold her own against any battleship afloat. She is as well protected as she could be. The launch of torpedoes from the submerged tubes is safe and precise, so she can rely on her torpedoes if an enemy vessel approaches too closely. The speed attained over two hours in trials, with only 90 per cent of her designed horsepower, was 17.25 knots.

Until 18 May 1901, *Jauréguiberry* would be stationed on the coast of the Var, around Toulon, as well as Golfe-Juan and Villefranche. She then sailed for Ajaccio and Porto-Vecchio (Corsica).

The 21 June saw the start of the annual grand manoeuvres: the Northern Squadron joined the Mediterranean Squadron for visits to Algiers and Ajaccio. The fleet would be at Toulon on 11 July, and *Jauréguiberry* visited Aigues-Mortes on the 18th, Le Lavandou on the 22nd, Saint-Tropez on the 24th, la Ciotat on the 26th and Ajaccio again on the 28th, returning to Toulon on the 30th.

On 22 August, *Jauréguiberry*, *Bouvet* and *Charles Martel* (2nd Battle Division) sailed for the Atlantic. They headed for Dunkirk, where they greeted the arrival of Tsar Nicholas II of Russia and his entourage. A severe storm when crossing the Bay of Biscay forced the ships to take shelter in Basque Roads; they arrived at Cherbourg on the 31st, entering Dunkirk on 15 September. The festivities were crowned with success, and the naval force sailed for the Mediterranean on the 24th, encountering heavy seas when passing through the Gibraltar Strait; it arrived back in Toulon roads on 1 October. The end of the year passed off without incident.

On 20 January 1902 there was a serious accident when a compressed air cylinder on one of the torpedo tubes burst, killing one man and injuring three. Other than that *Jauréguiberry* pursued the customary activities undertaken by a fleet unit until March 1904, the only notable event being the visit of President Émile Loubet to North Africa; the fleet then called in at Cartagena and the Balearics during the return voyage.

On 24 February 1904, part of *Jauréguiberry*'s crew was transferred to the battleship *Suffren*; the ship herself was to join the Northern Squadron with a reduced crew. She left for Brest on 25 March, arriving on 1 April. Incorporated into the 1st Division alongside her half-sisters *Masséna* and *Carnot*, she took part in exercises with the fleet in July, and suffered moderate flooding in her starboard bow compartments after striking a rock in the approaches to Brest. While attempting to avoid the rock *Jauréguiberry* was almost rammed by *Carnot*. Repairs were made during a docking at Pontaniou in December.

In May 1905 *Jauréguiberry* was struck to starboard by a 381mm exercise torpedo launched by the torpedo boat *Sagaie*, which resulted in some flooding in the steering compartment. The ship had to be docked at Le Salou and remained immobilised for three weeks.

Jauréguiberry at Portsmouth in August 1905. (Author's collection)

Commanding Officers

CV Foret	23 Jan 1896 – 30 Jan 1898
CV Daniel	30 Jan 1898 – 30 Jan 1900
CV Berryer	30 Jan 1900 – 30 Jan 1902
CV Aubert	30 Jan 1902 – 5 Nov 1902
CV Campion	5 Nov 1902 – 23 Oct 1903
CV Marius	23 Oct 1903 – 15 Nov 1903
CV Rabouin	15 Nov 1903 – 31 Jan 1905
CV Bomfay	31 Jan 1905 – 11 Apr 1907
CV Nicol	11 Apr 1907 – 12 Jan 1909
CV Le Cannellier	12 Jan 1909 – 20 Jul 1910
CV Gervais	20 Jul 1910 – 20 Jan 1912
CV Paillet	20 Jan 1912 – 20 Jan 1914
CV Tirard	20 Jan 1914 – 1 Apr 1914
CV Beaussant	1 Apr 1914 – 28 Nov 1915
CF Benoist d'Azy	28 Nov 1915 – 16 Jun 1917
CC Vincent	16 Jun 1917 – 30 Mar 1919

Flag Officers

CA Adam	25 Feb – 3 Mar 1904
CA Ramey de Sugny	1 Mar – 18 Mar 1913
CA Darrieus	1 Apr 1914 – 24 Mar 1915
CA Guépratte	1 Apr – 20 May 1915
CA Darrieus	19 Aug – 15 Nov 1915
CA de Spitz	6 Oct – 24 Nov 1916

Notes:

CA	*Contre-Amiral*	Rear Admiral
CV	*Capitaine de Vaisseau*	Captain
CF	*Capitaine de Frégate*	Commander
CC	*Capitaine de Corvette*	Lieutenant-Commander

On 11 July an important reception was accorded to the Royal Navy,[1] principally on board the battleships *Jauréguiberry* and *Formidable*. In return, the French Northern Squadron was welcomed in Portsmouth on 7 August and *Jauréguiberry* was moored alongside at South Railway Jetty. At 1030 on the 9th King Edward VII reviewed the French fleet. On 14 August, when leaving Portsmouth at the end of the festivities, the battleship collided wih the British steamer *Arana*; she was slightly holed but not seriously. Damage control teams were sent on board from the torpedo boats *Forbin* and *Bélier*.

In January 1906 stability trials were successfully carried out. On 8 March, a crew member of *Jauréguiberry* received minor injuries when a hawser parted while the ship was trying to avoid a collision with the armoured cruiser *Léon Gambetta*. On 3 July the annual grand manœuvres took place and the ship returned for a time to the Mediterranean, undertaking visits to the ports of North Africa.

A detailed bow view of *Jauréguiberry* taken before 1908. Note how close the forward 305mm turret was to the bow. (Author's collection)

Jauréguiberry at her moorings. She has her original livery of black hull and buff upperworks. (Author's collection)

Jauréguiberry at anchor in her post-1908 overall grey-blue paint scheme. (Author's collection)

On 15 January 1907 there was a major reorganisation of the battleships in favour of the Mediterranean, and *Jauréguiberry* was assigned to the 2nd Division of the Reserve Squadron. Visits were made to the ports of the Rhône estuary and the *départements* of Var and Alpes Maritimes over the next two years.

From 15 December 1909 *Jauréguiberry* was particularly active, visiting Algiers on the 17th, Mers el-Kebir on the 22nd, Tangier on 5 January 1910, Gibraltar on the 8th, Cadiz on the 13th, Lisbon on the 19th, Vigo on the 31st, La Palice on 11 February, the Isle of Aix on the 15th and Quiberon on the 23rd. She was at Cherbourg from 13 March to 9 May, then returned to Toulon to participate in the summer manœuvres. On 9 August *Jauréguiberry* was declared the best gunnery ship of the squadron during firing exercises against the decommissioned coast defence ship *Fulminant*. On the 14th the ship was at Brest, and on 29 September began a major refit at Cherbourg which focused on the retubing of her boilers; on 16 October *Charles Martel* replaced her in the 2nd Division.

Jauréguiberry recommissioned on 22 February 1911. On 1 August the 2nd Squadron (Brest) was rechristened the 3rd Squadron. On the 12th *Jauréguiberry* departed

Jauréguiberry prior to casting off; the photo was taken before 1908. (Author's collection)

Jauréguiberry in refit at Cherbourg in 1911. (Author's collection)

The secondary 138.6mm guns are trained, the boats deployed, and there is intensive activity on board *Jauréguiberry* during a visit to an anchorage in France. (Author's collection)

Jauréguiberry at Cherbourg in 1912. (Author's collection)

for Toulon to take part in the great naval review of 4 September in the presence of President Armand Fallières. After the catastrophic loss of the battleship *Liberté* to a magazine explosion, followed by a public funeral for the victims which took place on 4 October, the battleship sailed for Algiers, then Mers el-Kebir, before returning to Brest on the 24th. *Jauréguiberry* was again docked at Cherbourg for refit from 10 November. Following trials in the Iroise Sea, the battleship visited Le Havre and Dunkirk 11–15 June 1912. On 16 October the 3rd Squadron left for Toulon via Algiers, with a view to concentrating the French battle fleet in the Mediterranean. Visits were made to Golfe-Juan, Sidi-Abdallah (Bizerte) and Villefranche, followed by a naval review 8–9 June 1913 in the presence of President Poincaré.

Four days later, when casting off from her moorings, one of the battleship's propellers became entangled in a chain and the shaft support was damaged. *Jauréguiberry* had to be docked for repairs. On 3 July, during gunnery practice in the approaches to Bonifacio (Corsica), *Suffren* and *Gaulois* hammered water tanks cemented in place to the west of Vacca; the fire of *Saint Louis* was impressive, but it was *Jauréguiberry* that was singled out for the effetiveness of her gunnery.

The 3rd Squadron was dissolved from 11 November, and *Jauréguiberry* joined the Training Division (*Division d'instruction*), serving as host ship for trainee electricians. On 2 December she participated, along with the armoured cruiser *Pothuau*, in the destruction of the old battleship *Hoche*, which had been modified as a target and was towed by the armoured cruiser *Jules Michelet*. On this occasion the two ships trialled the new graphical plot and fire control computer devised by Captain Le Prieur. *Jauréguiberry* was at Les Salins at the end of the year, and again in July 1914.

The Great War

Faced with an increasingly tense international situation, on 1 August 1914 the C-in-C of the *Armée Navale*, Vice Admiral Boué de Lapeyrère, created the Special Squadron charged with protecting the troop transports due to make the passage from Algeria. The squadron comprised *Jauréguiberry*, flying the flag of Rear Admiral Darrieus, *Charlemagne*, the armoured cruiser *Pothuau* and the protected cruiser *D'Entrecasteaux*. On 3 August, these ships participated in the sortie of the fleet towards Algiers. The presence of the German battlecruiser *Goeben* in the western Mediterranean having been confirmed, Darrieus was ordered to head for Oran to provide cover for a convoy of seven troopships. After a safe passage to Sète, *Jauréguiberry* and *Charlemagne* were briefly at Toulon on the 8th. The following day Darrieus' force was reinforced by *Bouvet*, and conducted similar escort missions 10–13 between Ajaccio and Marseille. From the 21st, *Jauréguiberry* and *Bouvet* were off Barcelona to monitor mercantile traffic. On 21 September, Admiral Darrieus was instructed to head for Malta to ensure the safety of two convoys of 36 ships, transporting 25,000 men from Port Said to Marseille. From 17 October *Jauréguiberry* and *Bouvet* patrolled the waters off Genoa and intercepted 267 steamers, boarding 57 and seizing six. On 30 November *Jauréguiberry* was docked at Sidi-Abdallah, Bizerte, and from 18 December the ship was left with several light vessels to ensure the monitoring of traffic.

On 1 March 1915 the 3rd Squadron was reconstituted with the battleship *Saint Louis*, then the cruiser *D'Entrecasteaux* (1st Division) together with *Jauréguiberry* and the coast defence battleship *Henri IV* (2nd Division).

Faced with the threat of a Turkish assault on the Suez Canal, the British and French governments decided to base a naval force on Port Said with a view to blockading the coasts of Turkey and Syria. *Jauréguiberry* arrived at Port Said on 20 February. Three surveillance sorties were made on 24 and 26 February and 7 March. The battleship would then be at Alexandretta from 9 to 20 March.

Meanwhile, on 10 August 1914 *Goeben* and *Breslau* had passed though the Dardanelles, and the Ottoman Empire entered the war on the side of Germany. The Royal Navy deployed three battle divisions and a cruiser division to the eastern Mediterranean. Admiral Boué de Lapeyrère duly despatched Rear Admiral Guépratte at the head of five battleships.[2]

There were several abortive attacks on the Ottoman forts at the mouth of the Dardanelles and, in the attempt to force the straits on 18 March 1915, *Bouvet* was sunk by a mine and *Suffren* suffered serious damage. As for *Gaulois*, only a deliberate grounding on a sand bank on a neighbouring island prevented her from capsizing. In the course of the same day, the British lost the battleships *Ocean* and *Irresistible*. After this drama, with *Suffren* and *Gaulois* requiring extensive and urgent repairs, Guépratte sent a telegram to the Minister of Marine and the C-in-C *Armée Navale* which read:

> ... During the absence of *Suffren* I intend to hoist my flag in *Charlemagne*, which is undamaged. In order to sustain the honour of the flag I require the assignment of other battleships to my command, including *Saint-Louis* and *Jauréguiberry*, with instant effect.

Henri IV and *Jauréguiberry* duly arrived at Mudros on 1 April, and Guépratte hoisted his flag in the latter ship that same day, releasing *Charlemagne*, which headed for a maintenance period at Sidi-Abdallah. Despite the arrival of the new ships, the admiral was reluctant to expose *Jauréguiberry*, which was conceptually similar to *Bouvet*, whose end had been both tragic and alarmingly rapid.

The squadron remained inactive for some time due to the inclement weather prevailing in the Aegean. On 24 April a Franco-British expeditionary corps was landed at Gaba Tepe and Morto Bay. The big ships were to give fire support to the troops on shore. For this operation, General Albert d'Amade had embarked on the flagship to monitor the development of the military situation alongside Admiral Guépratte. *Jauréguiberry*, *Henri IV*, the

Admiral Guépratte and General d'Amade on the bridge of *Jauréguiberry* during the landings of 24 April 1915. (Author's collection)

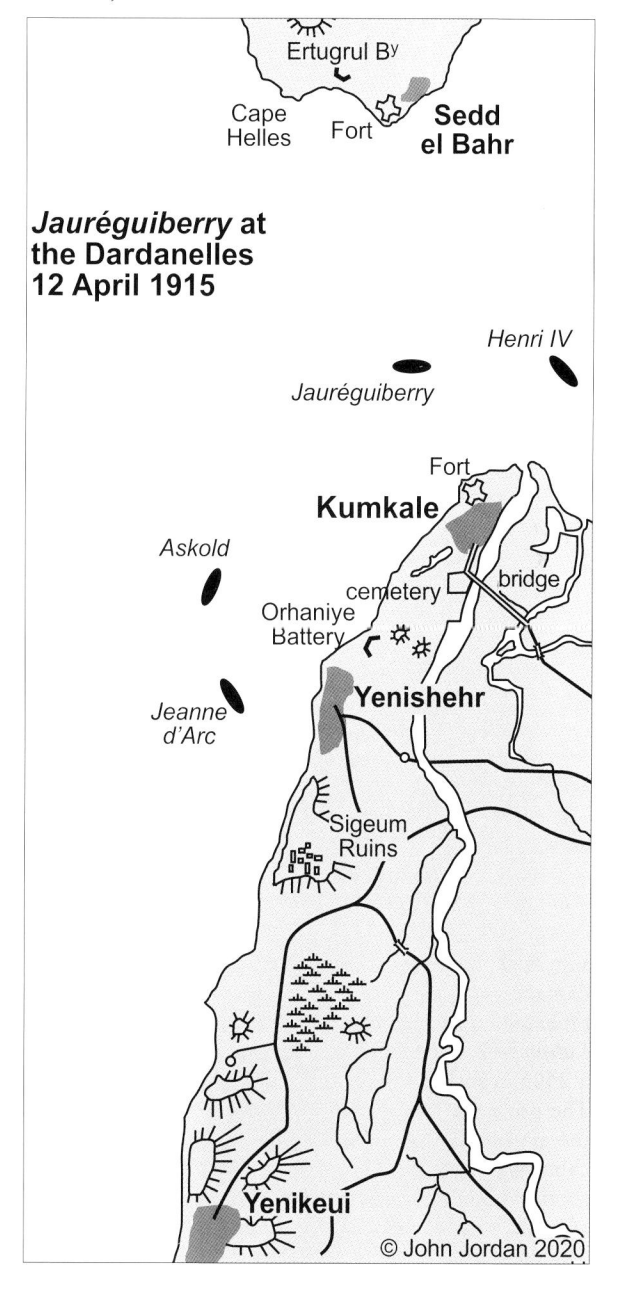

Jauréguiberry in 1915. Note the Barr & Stroud 2-metre rangefinders atop the bridge. (Author's collection)

armoured cruisers *Jeanne d'Arc* and *Latouche-Tréville* and six torpedo boats opened a sustained fire that same day against the Turkish fortifications to facilitate progress by the 6th Regiment of Colonial Infantry.

Despite the occupation of Kum Kaleh and against the advice of Guépratte and Admiral de Robeck, the British General Hamilton ordered the French troops to be reembarked on the night of 26–27 April. During this action, *Jauréguiberry* put up a luminous barrage south of the French lines, and fired on the enemy trenches and a battery of enemy howitzers which threatened the beach. The operation ended at 0500 on the 27th, by which time 500 Ottoman prisoners had been taken.

A new raid was carried out on 30 April, and the battleship was struck by a shell which caused only slight damage and no casualties. The bombardments were repeated on 5 May when *Jauréguiberry* relieved *Latouche-Tréville* off the Keraves defiles. At about 0600 a 240mm shell struck the battleship on her quarterdeck. The projectile entered via a louvred hatch and burst on the starboard side of the 2nd deck. The admiral's day cabin was devastated, and splinters shattered the bathroom bulkhead; two officers' cabins were destroyed together with a fresh water tank, and the armoured deck was depressed by 8mm.

During the second fortnight in May, the Dardanelles Squadron was formed with the 3rd Division comprising the fleet battleships *Patrie*, *Suffren*, *Saint-Louis*, *Charlemagne* and *Jauréguiberry*, and the coast defence battleship *Henri IV*.

On 20 May Guépratte lowered his flag in *Jauréguiberry*. Seven days later the battleship, on the orders of Admiral de Robeck, was again off the entrance to the Dardanelles, zigzagging at speed, and firing on the batteries on the Asia Minor side of the strait. However, after the torpedoing of the battleship HMS *Majestic* that same morning by *U-21*, these bombardment operations were suspended.

On 15 June *Jauréguiberry* was replaced by *Gaulois*, which had recently undergone repairs at Toulon naval dockyard. The following day, she headed for Port Said, arriving on the 19th. The Suez Canal Company immediately put its personnel and its facilities at the disposition of the ship to carry out repairs. At the same time, *Jauréguiberry* disembarked two 65mm and two 47mm guns intended for the patrol boat *Indien*.

On 13 August *Jauréguiberry* conducted two test firings of her forward 305mm gun against the railway station at Caiffa. On 1 September the landing parties of the battleship and the armoured cruiser *Jeanne d'Arc* occupied the

Jauréguiberry as a hulk for the School of Mechanicians and Stokers. (Author's collection)

Island of Fuad; there was no opposition. On 30 October the battleship was assigned to the defence of the Suez Canal; this mission involved a minimum of movement, the boiler tubing being on its last legs.

On 19 January 1916, faced with an attack on the Canal, *Jauréguiberry* was moored at the 42-kilometre mark then, on the 25th, off Ismailia. However, the assault by Turkish forces failed to materialise, and she returned to Port Said at 1600 on the 27th.

At 0600 on 25 November *Jauréguiberry* sailed for Malta; she was due to be refitted at Toulon. On the 26th a periscope was sighted, but no torpedo tracks. Due to the threat of submarine attack, *Jauréguiberry* remained at Malta for 23 days, then returned to Port Said on 26 December. She was moored at 'Post 3 Red' and served as headquarters ship for the *Division de Syrie*.

On 2 January 1917, an order was received for the decommissioning of *Jauréguiberry*; only two 65mm and two 47mm were to be manned. A long period of inactivity awaited the ship, but guns of various calibres would later be utilised for the defence of the Suez Canal.

Jauréguiberry returned to Toulon on 6 March 1919 to be deactivated. From the 30th, she would become a hulk housing the School of Mechanicians and Stokers. On 20 juin 1920, she was stricken from the lists, but continued to perform her support role for the mechanicians. On 23 February 1934 the hull of *Jauréguiberry* was put up for sale and was purchased by the Matériels Navals du

Photo of a trainee mechanician on board the old battleship *Saint-Louis*. Behind him can be seen the superstructures of *Jauréguiberry*. (Author's collection)

Midi company for 1,147,000 francs. She was broken up for scrap metal during the months that followed.

Editor's Note:
This article was translated from the French by the Editor. The drawings are based on the official plans of *Jauréguiberry* held at the Centre d'Archives de l'Armement.

Endnotes:
[1] The battleships *King Edward VII* (Vice Admiral May), *Victorious* (Rear Admiral Bridgeman), *Majestic*, *Magnificent*, *Illustrious*, *Prince George*, *Mars* and *Commonwealth* were present, together with the cruisers *Doris* and *Amethyst*.
[2] *Suffren* (flagship), *Bouvet*, *Gaulois* and *Charlemagne*; *Vérité* was initially assigned to the squadron, but departed for Toulon, via Malta, on 18 December 1914.

POSTWAR RADAR DEVELOPMENT IN THE ROYAL NAVY

In this latest in a series of articles on technical developments in the Royal Navy during the postwar era, **Peter Marland** describes the evolution of radar.

Previous articles published in *Warship* have chronicled the development of fire control, weapons, command & control and sonar in the postwar Royal Navy. The present article describes the development of the radar sensors used by RN surface ships across the post-war period. It complements the earlier articles and builds on the excellent work of Kingsley,[1] who documented the wartime effort on radar, but brings this forward to the turn of the 21st century. Despite the UK managing without having to import technology from other countries, it also charts something of a roller-coaster ride, with outstanding innovation interspersed by periods of stagnation or procurement disorder.

During this period, the RN shared knowledge with the other Allied 'five-eyes' community, and also had bilateral relationships with countries such as the Netherlands. For the bulk of the period development was driven by the Government's own research laboratory (ASWE). Research & Development projects led to progressive development of X, Y and Z models that transitioned over to industry for volume production, or to the somewhat less-demanding Cardinal Point Specification (CPS) regime of the mid- to late-1980s.

The wartime naval radar work was led by the Admiralty Signals & Radio Establishment (ASRE) at Haslemere, but this migrated to the new Portsdown site that opened in autumn 1948, which was later renamed the Admiralty Surface Weapons Establishment (ASWE).

Nomenclature

Individual RN equipments are described by a range of titles or different 'Nomenclature' schemes. Type numbers cover radio, radar and sonar transmitters, while three-letter Outfits refer to aerials, receivers and all else. Annex A gives more detail; there are a few gaps where numbers were issued for planning purposes but the project was subsequently cancelled.

It should be noted that there has been a change in the radar band descriptions, from the immediate postwar system to the current designations used by the ITU and NATO (see Fig 1). This also indicates the encroachment of modern civil communications equipment into formerly military allocations of the spectrum.

Radar Parameters

The principal radar parameters set by the transmitter are frequency, peak pulse power, pulse length, and pulse repetition frequency. The aerial shape also defines the vertical coverage diagram (VCD); for some examples, see Figure 2.

The range performance is crucially dependent on the target's radar signature or 'echoing area', usually expressed as metres squared (m²) or, for large targets, as decibel milliwatts (dBm). The simplified figures in Table 1 mask wide variations due to target aspect, from corner reflector 'highlights' or glint, to 'fading' during signature nulls.

The majority of maritime radars use 'pulse' waveforms,

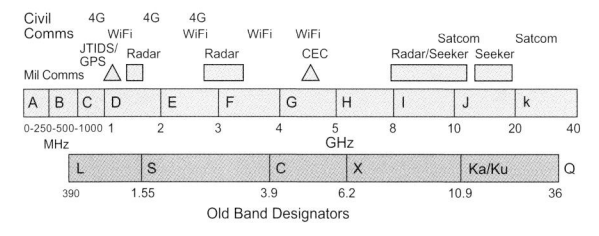

Fig 1: Electro-Magnetic Spectrum Definition
(Graphic by John Jordan using material supplied by the author)

Table 1: Typical Radar Cross Sectional (RCS) Areas (m²)

Major vessel	500
Jumbo jet	100
Nimrod	40
Fast Patrol Boat	30
Large fighter	10
Medium fighter	4
Small fighter/Helicopter	2
Bicycle	2
Man	1
Air flight missile	1–0.1
Bird	0.01

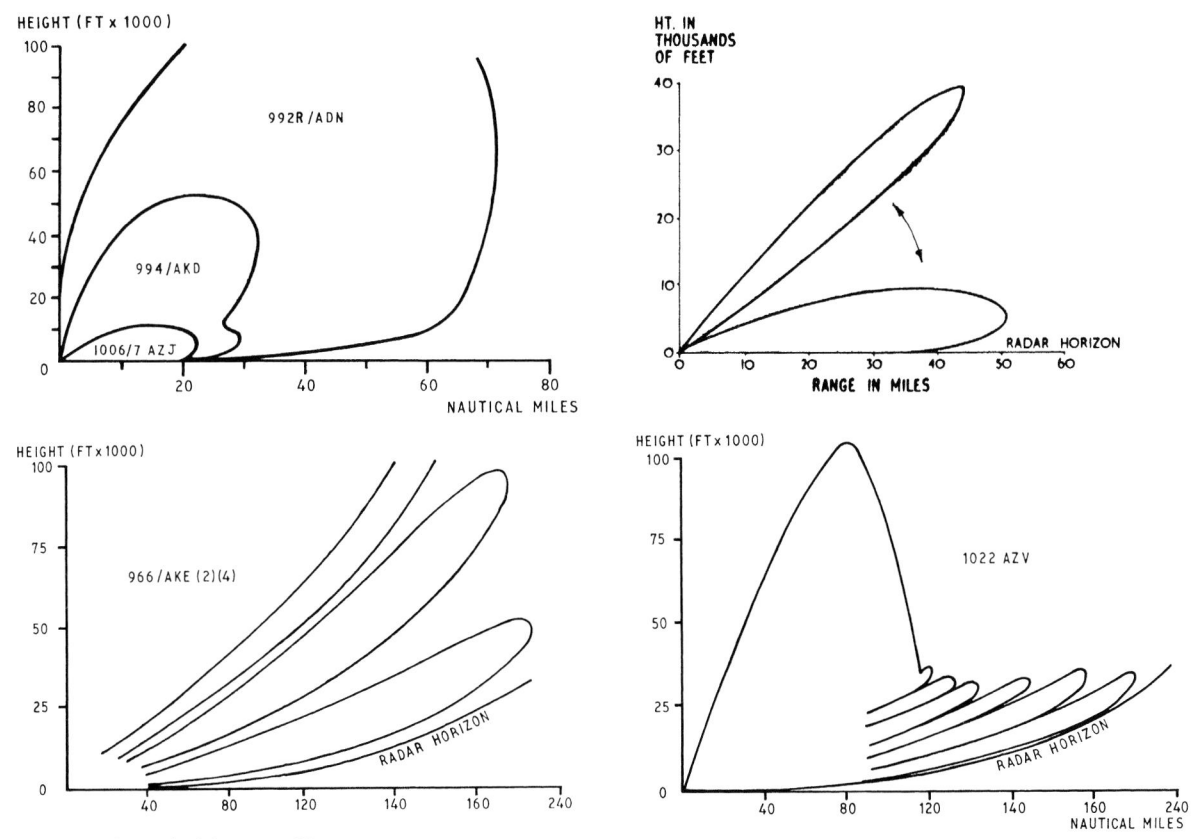

Fig 2: Sample Vertical Coverage Diagrams
[A] Surface Search/Navigation (Type 1006), and 2D air search and TI to higher elevations (Types 992/994).
[B] Pencil Beam Heightfinder (Type 278): scans in elevation or 'nods'.
[C] Metric Air Search with lobes (Type 965/966).
[D] D band Air Search: solid cover, with some surface reinforcement (Type 1022). (Author's collection)

though Continuous Wave (CW) is sometimes found in Low Probability of Intercept (LPI) work or for missile guidance. Basic fixed frequency waveforms are generated by magnetrons, but Travelling Wave Tubes (TWT) can generate more sophisticated waveforms such as swept or frequency-modulated 'chirp' pulses, compressed in the receiver to give better range definition. Other techniques include side lobe suppression, jamming assessment of background levels, Moving Target Indication (MTI), and the use of pulse repetition frequency (prf) stagger to avoid 'second time around' echoes.

Postwar Position

The immediate post-war period saw the RN consolidate around the following radars (for consistency, current NATO band names are used):

– Type 974 (I band, high definition surface warning and navigation)
– Type 293 (E/F band, 2D surface search and target indication)
– Type 277 (E/F band, heightfinder) and Types 980/982/983 (E/F band, fighter control radars)
– Type 960 (metric band, long-range air warning)
– Types 274, 275 and 262 (E/F and I band, fire control)
– Wartime IFF (Identification Friend or Foe) Mk II, III and IV, prior to arrival of IFF Mk 10.

Target Indication and Fighter Direction

One postwar area in which the RN led the US Navy was in Target Indication (TI) radars, able to point weapon systems at incoming targets, as the Gun Direction Officer (Blind) prioritised targets. He worked in conjunction with the Fighter Direction Officer, who was handling the outer air battle with a Combat Air Patrol (CAP) of friendly fighters. Table 2 shows the progressive development of TI radars.

One reason for the change in performance across the Type 293 versions (apart from receiver sensitivity and feeder loss) was the vertical beamwidth. Early Types 276 and 293X/M used aerial outfit AUJ/AUR with a particularly wide vertical beamwidth (65–70°); later sets 293P and Q with outfits AQR and ANS were optimised for range, with a smaller vertical beamwidth (35–40°) but a wider antenna, to focus more energy on low-angle detection at range, thus sacrificing high-angle cover. The RN never satisfactorily resolved volumetric cover (needed for fighter direction) apart from the large carrier-fitted

Table 2: Gun Direction Systems

Radar Type	276, 293X/M/P	293Q	992	993	992Q
In service	1943/44	1950	1959	1963	1966
Peak Power	400kW	500kW	2MW	600kW	1.75MW
Ae Rot Rate	7.5, 10, 15rpm	5, 10, 15rpm	0–30rpm/90rpm	24rpm	15/30rpm
Range vs 4m² Hunter	12–19kyds	25kyds	30kyds	30kyds	60kyds
GDS	GDS2	GDS2*	GDS3	GDS5	WDS6›
Channels	3 or 5	3	6	3	Parent AIO

Type 984, and failed with a series of smaller multi-purpose sets, eventually using separate azicator and nodding heightfinding radars to achieve the same goal.

UK TI arrangements did not include Threat Evaluation and Weapons Assignment (TEWA) in terms of prioritising air targets, although the later systems such as GDS3 did include a degree of hard-wired preferences for allocating directors in emergency 'alarm' conditions. The UK did introduce TEWA algorithms in Type 984 and CDS for allocating fighters to air targets, and by October 1950 the UK was monitoring early US work by Bell Telephone Laboratories (BTL) on the Mk 65 GDS using a digital computer. There was a notional UK GDS6 concept for an advanced (=elaborate) system; but later UK WDS (WDS6 onwards) were fitted to ADAWS and CAAIS ships as an adjunct to the parent command system.

The sequence of new radars in the E/F band to generate volumetric cover is detailed in Table 3.

Fighter Direction (FD) Radar[2]

The RN had observed and used US fighter direction sets such as SP that had dish aerials and helical scan patterns giving volumetric cover at the end of the Second World War. The first attempt at a UK FD set was Type 295 intended for a 2MW magnetron, with Type 294 as a fallback with the existing 500kW transmitter and a simplified (PPI beam only) scheme. Type 295, like 293, had a horizontal cheese antenna for the plan display function; it was about 12ft wide and had a 17in separation between the plates (293P had 7.5in separation), giving a high beam inclined upwards at 5 degrees. This was supported by an additional vertical aerial array with multiple horn feeds for heightfinding on receive (this may have contributed to the postwar Type 984 heightfinding scheme and displays). The heavy antenna (7 tons) used a complex pedestal that was tri-axially stabilised in roll and cross-roll to 6 minutes accuracy from its own dedicated metadyne set and Stabiliser M Mk 1.

This was followed by Type 990, which also would have required the 2MW magnetron, using an 8ft stabilised parabolic dish antenna for a long-range horizon scan – similar to the fixed low-cover warning beam eventually implemented in Type 984 (see below). The necessary 2MW magnetron did not become available until much later, and Type 295 was replaced by a combination of Types 980 and 981:

Type 980 was based on the Type 295 PPI beam, but stacked three such antennae together, one above the other, onto the 295 pedestal to overcome the lack of power (only 500kW, not 2MW). The innovation was the use of a high-speed (1,200rpm) phase changer to make the overall beam scan in elevation, and may have led to the scan mechanisation associated with Type 984. Type 980 became 982 but underperformed due to glint and fading. (The underpinning theory of probabilistic signal distribution in response to a target echoing area with nulls and in the presence of specular noise was not fully understood at the time.)

Type 982 then morphed into 982M, which introduced a slotted waveguide feed as part of the AKR trough aerial, and subsequently became 986 when the office equipment (transmitter and receiver) were replaced by common range equipment from the new 993. Type 982M/986 had a basic pedestal with a support mast (high to look over other radars in light carriers, lower in AD frigates with stability concerns).

Type 981 was the companion heightfinder with a nodding AQT aerial, a 14ft 6in diameter paraboloid cut to 5ft wide. Type 981 became 983, and then 987 with the same 993-derived office equipment as 986.

These radars acted together, with the azicator finding

Table 3: Generic Families E/F Band Radar + Aerial Outfits

Function	1944: 400 kW transmitter	1962: 600 kW transmitter
Combined	293 with ANS	993 with AKD, then high powered 992Q with ADN
Warning/TI Heightfinder	277 with AUK, then ANU dish	278 with same ANU (= AZR in ADA DLG)
Fighter Direction	982 with AQS stacked 3 reflector, vertical scan	982 failed; only specialised FD radar was 984
Azicator	982M with AKR azicator trough	986, same AKR aerial*
Heightfinder	983 with AQT nodding heightfinder	987, same AQT aerial*
Fire Control	275 with AUS	successors were I band radars

*based on 993 office equipment

POSTWAR RADAR DEVELOPMENT IN THE ROYAL NAVY

Fig 3: E/F Band Fighter Direction Radars

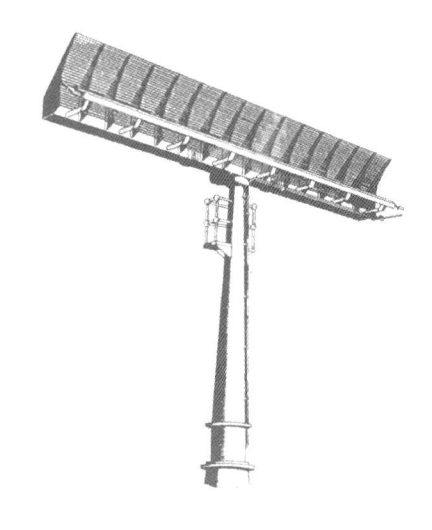

[A] Type 982 AQS reflector: 12ft wide, 5ft high; 19° vertical coverage; weight 7 tons. Same tri-axially stabilised pedestal as Type 983 AQT.
(HMS *Collingwood* CHC website, courtesy of Clive Kidd)

[B] Type 983 AQT reflector: 14ft 6in high, 5ft wide. Three motion stabilisation: azimuth, elevation and cross level/rotation about the sight line.

[C] Type 982M AKR reflector: 26ft 6in wide; maximum weight 3.5 tons. Slotted waveguide feed gave vertical cover up to 22° and more efficient/better distribution of RF power in fan-shaped Cosec2 beam. Not stabilised.

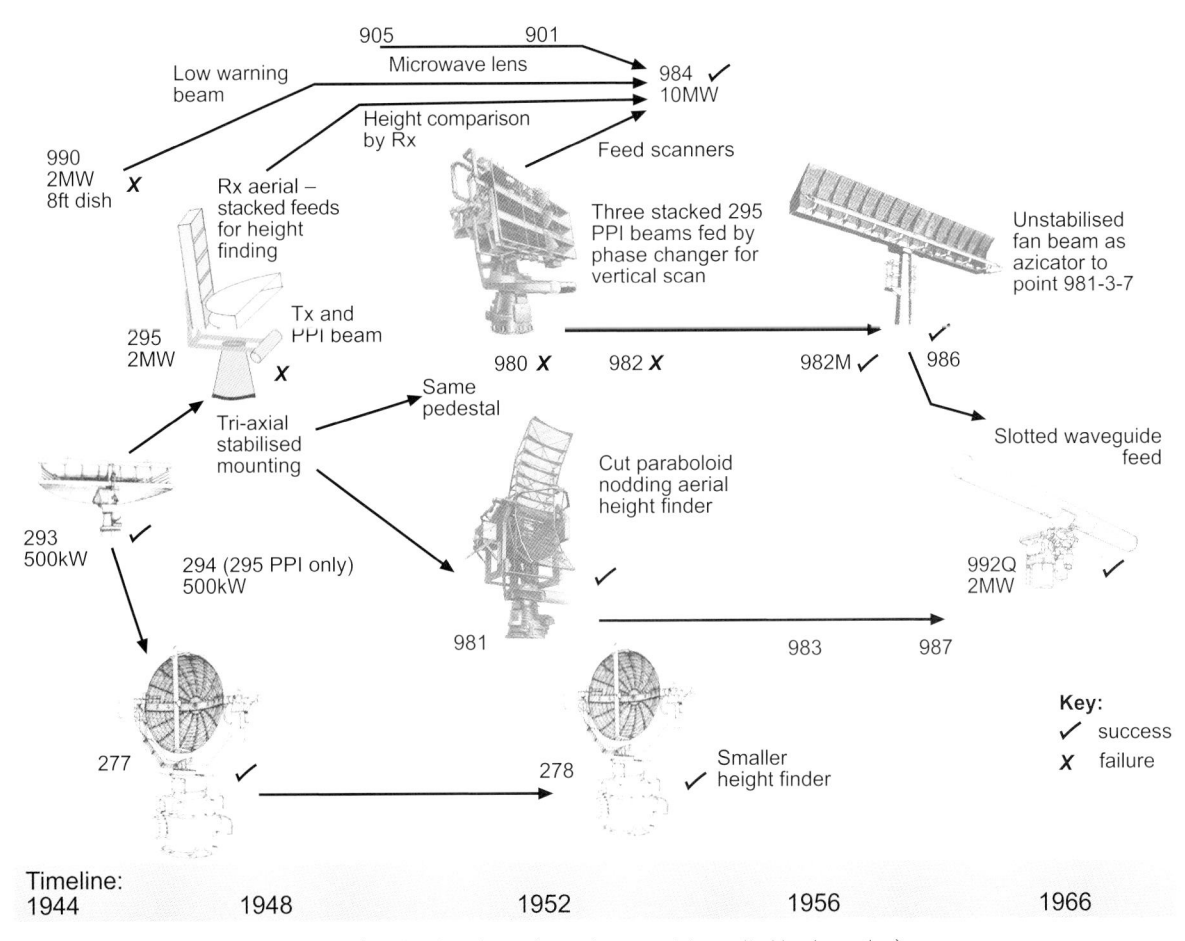

Timeline: 1944   1948   1952   1956   1966

Key:
✓ success
✗ failure

Fig 4: FD Radar Technology Exploitation (Graphic by John Jordan using material supplied by the author)

the bearing of the target, and the accompanying heightfinder searching in elevation by 'nodding', until the target was acquired and its height calculated.

Although there were disappointments with the FD series of radars, particularly Types 980–982 (until the latter 'came good' as the 982M azicator) there was a significant amount of 'carry forward', as features were exploited into later programmes such as Type 984. The failures were due to the combination of insufficient power (2MW magnetrons were not available until the later Types 984 and 992), and the ill-understood impact of fading and glint on Radar Cross Section (RCS)/Radar Echoing Area (REA). The work did contribute to the later Types 901 and 984 radars with RF scanner feeds.

CDS and Radar Type 984

The Comprehensive Display System (CDS) and the associated Type 984 three-dimensional radar were at sea operationally from early 1958 in the carriers *Victorious*, *Hermes* and *Eagle*. CDS (see the author's article in *Warship 2016*) was mostly valve-based analogue electrical computing technology and handled only air tracks. It should be emphasised that it was in advance of contemporary US systems: the US had initially fitted SP scanning pencil beam heightfinders for aircraft control during the Second World War, but then went onto the organ-pipe mechanical scanning SPS-8, and later to SPS-52 electronic frequency scanning systems.

Radar Type 984 (as Project 'Postal') was initiated in 1947 and designed during the early 1950s; a complete development model was installed at ASWE (Building Y East/West) in the autumn of 1955 and was set to work in 1956. This was prior to commissioning the first ship fit in HMS *Victorious* in January 1958, and its subsequent operational success in exercises against the US Navy in Exercise 'Riptide' in 1959 was reported by Benjamin (*Five Lives in One* – see Sources). Radar 984 was very tightly coupled to the Comprehensive Display System (CDS); it was far more successful for 3D volumetric cover than the contemporary US system, the large AN/SPS-2 radar, and CDS led the equivalent US AIO technology by about five years. The accompanying graphic (Fig 5) shows an exploded view of the Radar Type 984 aerial, highlighting the scanners, feed horns, and the stabilisation axes.

Type 984 had four separate 2.5MW 2.5μsec 400pps S-band transmitters, switched by a common mercury-pool modulator. The radar coverage was substantially solid from the horizon to 40,000 feet or more, out to the range limit of 180 nautical miles, shown in the vertical coverage diagram (Fig 6).

The lens had a diameter of about 14 feet, and comprised a large number of aluminium waveguides of different lengths and cross sections, all spot-welded together. This converted a spherical *wave* front into a plane *phase* front; like the Fresnel lens in a lighthouse, the 'stepped' profile limited excessive bulk at the centre of the array.

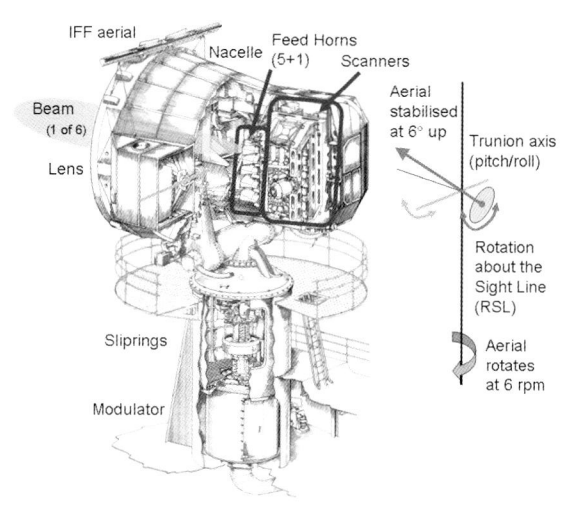

Fig 5: Type 984 aerial outfit AKY, with IFF Mk 10 Aerial Outfit AMD on Top (BR2307, HMS *Collingwood*, courtesy of Clive Kidd)

Table 4: **Type 984 Parameters (from BR1992B)**

Frequency:	S Band (2750–2900MHz); power 10MW
Pulse:	prf 391pps; varied to eliminate 2nd trace echoes; length 2.5μsec
Range:	discrimination 500yds; accuracy ±500yds; minimum range 1000yds
Bearing:	accuracy 1/2°; discrimination 1.8°
Beamwidth:	horizontal 1.7° for each beam; vertical 23°
Accuracy of height information:	±1000 feet at 60 miles, ±2000 feet at 100 miles
Rotational speed:	4.56/6rpm

Five pencil beams were generated, each scanning a sector of about 5.5 degrees vertically. Scans were slanted upwards and forwards, with instantaneous flyback after 25prfs. The rotation of the mounting at 6rpm contributed to the horizontal scan; the scanner itself was set at 55 degrees. Diagonal orientation placed the top of any scanner ahead of the bottom of the next higher one. This allowed them to be arranged for continuous vertical cover without gaps or unwanted overlaps. The scanners, feed horns and lens combined to give a complex beam pattern, illustrated in Figure 7.

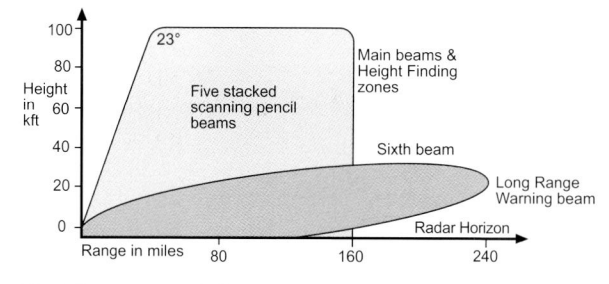

Fig 6: Type 984 Vertical Coverage Diagram and Cross-section of Beam Scanning (Drawn by John Jordan using material supplied by the author)

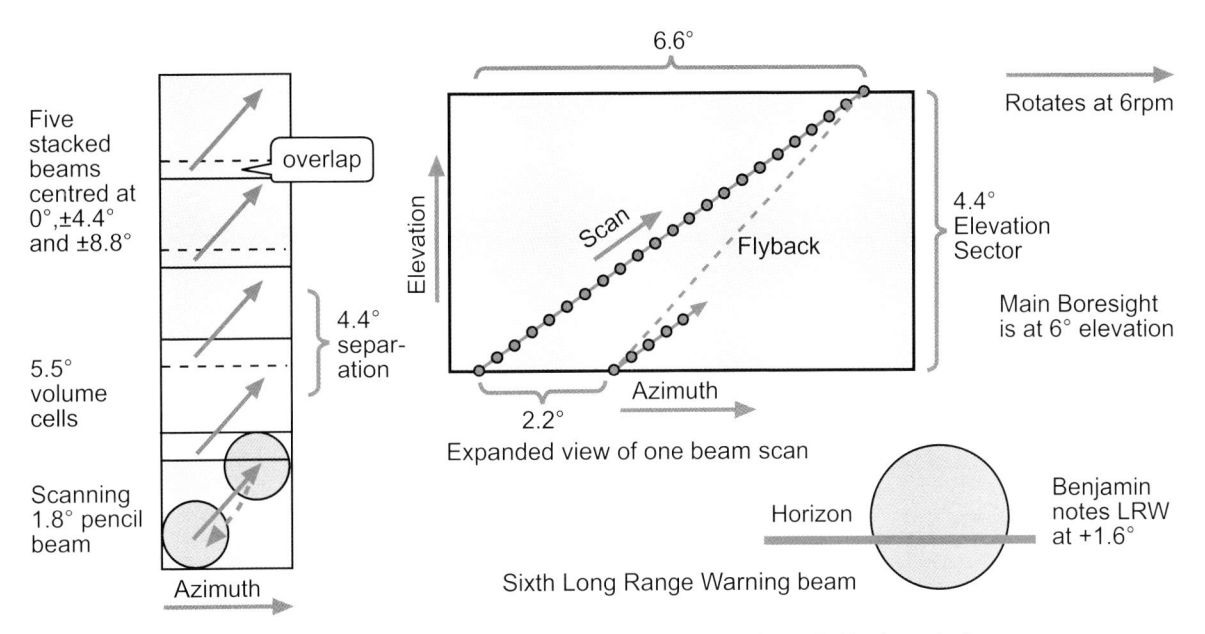

Fig 7: Type 984 Cross-section of Beam Scanning (Drawn by John Jordan using material supplied by the author)

The nacelle boresight axis was inclined up at 6 degrees, with the fixed beam 6 at 10.75 degrees down (appearing in the 'up' sense after the lens, and noted by Benjamin as being at +1.6°). Other beams were at ±4.4 or ±8.8 degrees relative to the axis. Beams 1–5 were scanned at a +55° slant, but fixed beam 6 was set at -45°. In order to improve the signal-to-jamming ratio for close-in targets, the power of various beams was tapered (see Table 5).

A log-law complex receiver had a gain with up to 120dB dynamic range. The IFF aerial was added on top of the nacelle, and a fourth transmitter with a fixed feedhorn behind the lens generated a non-scanning warning beam, at about 0.75 degrees elevation. The four magnetrons were on different S-band frequency subbands, but 'pulling' caused a degree of pulse-to-pulse frequency hopping to supplement the larger scanner-to-scanner frequency diversity as a further anti-jam (AJ) feature. Type 984 also introduced quasi-random pulse interval jitter, and the radar was designed around a rapidly-acting automatic frequency control (AFC), so that receivers were aligned to the new frequency as soon as the radar trace emerged from the close-in 'ground clutter'.

The antenna nacelle for the Type 984 radar was 14ft in diameter, 18ft long, and weighed approximately 27 tons. It had two-axis stabilisation plus Rotation about the Sight Line (RSL), via a Metadyne drive from a Stabiliser Type 2 (see BR2307 series).

Ship's staff from *Victorious* made it clear that Type 984 with CDS was a challenge (see 'CDS - A Maintainers View', NER Vol 13 No 4 dated Apr 1960, 124–140), but that the complexity was managed by parallel paths and redundancy. The overall component count for Type 984 with CDS was:

Valves: 8,912
Resistors: 47,400
Capacitors: 11,750
Relays: 2,115
Switches: 1,372

The system watchkeeper at the RME desk was able to 'dial up' the circuit to be monitored in the nacelle. Support documentation included full binary fault-finding logic to support 'divide-by-two' diagnostic testing. The NER article stressed lessons learned about the ageing of valves, monitoring video performance, and block component changes (relays and valves) prior to wearout. Benjamin also notes the significant advances made in terms of connector and PCB technology.

Table 5: Transmitter Power Allocation

Scanner	Beam	% Power			
		TX1	TX2	TX3	TX4
1 (lowest)	5 (high)	88.9%	–	–	–
2	4	–	66.6%	–	–
3	3	–	–	100%	–
4	2	–	33.3%	–	–
5 (highest)	1 (low)	11.1%	–	–	–
Low cover sixth beam (fixed)	–	–	–	–	100%

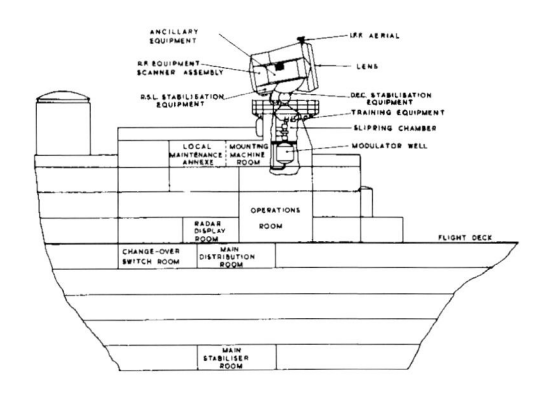

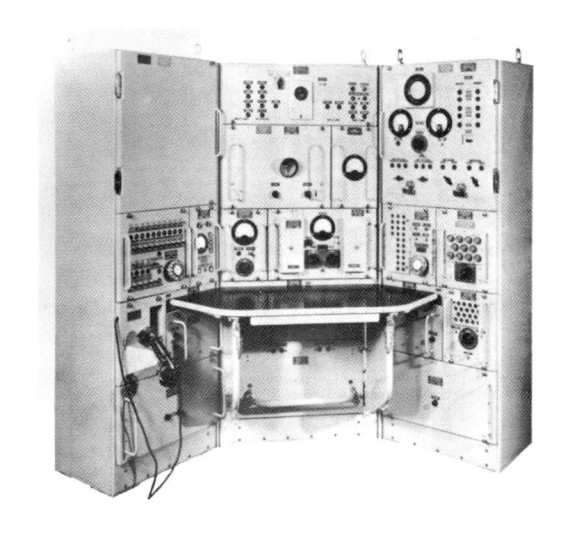

Fig 8: Radar Type 984 in Aircraft Carriers, and Diagnostic Desk Outfit RME (BR2307, HMS *Collingwood*, courtesy of Clive Kidd)

Aircraft Control Aids

The first aircraft control aids included the wartime YE beacon used by 'Pylon' units to marshal returning aircraft. This was superseded in postwar aircraft carriers by an early TACAN system (Type 957) for navigation back to the ship, and by carrier-controlled approach (CCA) radars (Types 961/962, then 963) for the final approach. In *Ark Royal* this was later augmented by a US SPN-35 radar to handle the F-4 Phantom. For images of some of these radars see Figure 9.

Gun Direction Radars

Radar Type 992 supported the complex GDS3 TI system in the cruisers of the *Tiger* class and scanned at up to 90rpm to give a rapid update of target position. It was succeeded by 992Q, which moved onto a 21ft slotted waveguide aerial that rotated at up to only 30rpm.

The 1962 'Notes on Above Water Weapons'[3] introduced Type 993 with AKD, to be fitted from HMS *Leander* (spring 1963 onwards). This was to feature a new office equipment using modern components and designs in lieu of the wartime 293 with its many modifications. The requirements were that the aerial (a quarter cheese 14ft x 7ft, tilted up at 15 degrees) should be able to work across a wider range of frequencies using plug-in transmitters, on one of three sub-bands (from a total choice of eight magnetrons), giving a time to change frequency and re-tune in the order of 15 minutes. This was not an ECM measure, but was intended to counter mutual interference within a force. Type 278 was the same modern office equipment, applied to the 277Q heightfinder in missile destroyers (DLGs) and Type 41 AD frigates, with a parametric amplifier in the radar receiver.

TNA records and ASWE annual reports also mention Type 976 (dual E/F- and I-band radars on a single pedestal) for light frigates such as Type 14 and 17, and the transportable Project 'Avocado' radar with complex mechanical scanning and photographic displays, for minewatching in defence of harbours and anchorages.

The Development of New Radars

Post-war, there was an emphasis on countering jamming; RN and RAF radars were vulnerable to Soviet jamming with the carcinotron (= backward wave oscillator), to be countered in the first generation complex set (Type 984)

Fig 9: Aircraft Control Systems (HMS *Collingwood* CHC website, courtesy of Clive Kidd)

[A] YE beacon.

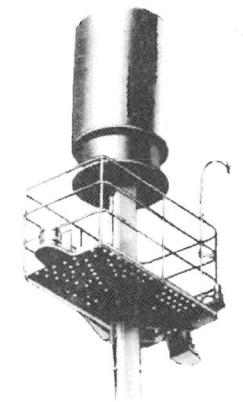

[B] TACAN: Type 957 in AMG at masthead.

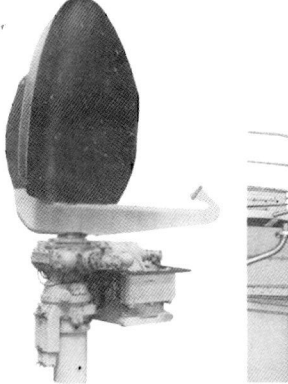

[C] and [D] Carrier-Controlled Approach (CCA) radar: Type 963 inside AKN, aft of carrier island.

Fig 10: Target Indication (TI) Radars

[A] Type 992 with AKC (12ft aerial); closed radome, pressurised; aerial rotation at 0–30rpm and 45/90rpm. Stabilised for pitch and roll.

[B] Type 993 with AKD quarter cheese (14ft); aerial rotation rate 24rpm. Not stabilised.

(HMS *Collingwood* CHC website, courtesy of Clive Kidd)

by using four separate transmitting frequencies, pencil beams and pulse-to-pulse jitter.

Dissatisfaction with Types 982/983 led Benjamin to examine Utility Fighter Direction radars in 1953 (for a mini-984, or V-beam solution). Experimental work 1956–1960 for a second-generation set (985) included frequency scanning, travelling wave tubes, phased arrays, multiple pulse-forming circuitry, and other techniques. Prior to project definition this was replaced by the Anglo-Dutch Project 'Broomstick' (988) intended for CVA-01 and the Type 82 destroyer; this was eventually cancelled in late 1968.

Cdr JAA McCoy was on the staff of the Controller of the Navy during the period 1978–80. Due to high-level frustration about the position with RN Air Defence radars, he documented the background to this long-running saga, and proposed a number of organisational and management changes.[4] Broadly, the dissatisfaction was that by 1960 the RN had a world-beating capability in Type 984, but almost two decades later had slipped backwards with only the WWII-era Type 965 and no new sets in prospect, despite significant research investment in cancelled projects.

The radar programme had included:

– NSR7932 for Type 965 improvements, a 'double bedspread' AKE aerial (965P), plus short-pulse MTI as 965Q and 965R; this became Project 'Reprise', accepted 1967.
– NSR7963 for Type 966, with long-pulse MTI and five selectable frequencies – side lobe suppression ECM aspects associated with this project 1958–1971 was deleted as a cost saving; accepted 1976.
– NSR7938 for a new long-range radar Type 1015, with dual-frequency UHF (430/600MHz) and back-to-back (or wrap-around) antennae, thereby avoiding the need for a mast-head site. Studies considered were 'Broomstick', the American Hughes SPS-52 and three UK industry offerings (by EMI, Plessey and Marconi). Later studies also considered further modernising Type 965, the US SPS-40B, the Dutch LW-08, and the Italian Selenia SPS-68. Type 1015 was cancelled late 1971, but led on to NSR7938 (Interim) that became 1022.
– NST7946 for a new TI radar, initially viewed as being a D-band pulse-doppler radar with a range of about 80nm (and up to 70° elevation), to support Seadart. With the cancellation of Type 1015 in 1971, the scope was widened to add surveillance, becoming STIR (1030). This included both low pulse repetition frequency (prf) for long-range surveillance, and high prf (with higher elevation) for TI, using back-to-back antennae. Later, an additional 'Half STIR' (1031) was proposed for air control in the new Support Carriers (CVS), which became the *Invincible* class. Studies considered were Marconi's S617N and S680N, the Selenia SPS-68 and RAN-3L, and LW-08, while a technology demonstrator with a stabilised back-to-back aerial was tested afloat in HMS *Grenville*; ASWE research also considered a lighter/neater TASP stabilisation mechanism. Both 1030 and 1031 were cancelled in 1979.

As a later enhancement, NSR7938 (Interim) long-range radar became Type 1022 based on Dutch (HSA) LW-08 office equipment (at D band, with a travelling wave tube and pulse compression, 150kW) plus a new unstabilised AZV aerial using a squintless linear feed that exploited MoD's prior investment in the STIR project. The original (1975) intention was that 1022 would be a stop-gap measure for *Invincible* alone, but Type 1022 was eventually fitted to eighteen units (all three CVS, HMS *Bristol* and the fourteen surviving Type 42s), plus training and shore reference sets.

The development sequence for this series of radars is shown in Figure 12. The reason for this convoluted picture was a long sequence of changes to projects, and some nugatory reversals of policy. After this period, the RN ended up with a wider fit of Type 1022, plus 967/968, and then 996.

McCoy's conclusions

McCoy's conclusion was that (by 1979) the RN had made little or no progress with air surveillance radars since Type 984, which had entered service in 1959; indeed it had gone backwards, now being reliant on the elderly Type 965.[5] After twenty years' work and many cancelled

Fig 11: Prototype 1030 Radar
Surveillance and Target Indication Radar (STIR), with two beams (one high, one low) and two modes (surveillance to 250nm and 70,000ft against a 4m2 target using low beam, and TI using both beams). The high beam had a 6–70° angle of sight and a range of 40nm to 80,000ft, the low beam 0–6° to 80nm. Initially envisaged as D band (early 1970s), later moving to E/F band; 5MW peak power, pulse repetition frequency 900pps in TI mode and 300pps in surveillance mode at pulse length 1μs, compressed to 50ns in receiver. Other features were side lobe cancellation, PRF jitter and high/low beam interleaving.
The trials aerial had a pivot between the antennae for simple stabilisation, but a novel TASP mount on skewed (A+M) axes was proposed for weight saving. Type 1030 was cancelled, but the array became AZV for 1022. (NER HMS *Collingwood*, courtesy of Clive Kidd)
Left: An early version of the 1030 radar trialled in HMS *Grenville* in 1973. This antenna had a single face and was not stabilised. (See also title pages.) (C & S Taylor)

projects the only new radar in prospect was Type 1022, which had not been selected for its excellence, nor on cost-effectiveness grounds, but merely to make up the foreign element in an industrial workshare agreement.

McCoy (para 58–88) suggested that this was due to problems in the following areas:

– Technical issues about the trade-offs between surveillance and TI functions were oversimplified for the approving committees, leading to wishful thinking. Staff requirement documents had grown to 108 pages even for relatively simple platforms.
– The Naval Staff Requirement was regularly 'detuned' to match the Procurement Executive's view on feasibility, leading to a dangerous understatement of the real requirement. Instead, the PE should have been required to present the technical, financial and timescale options that were available.
– There were simply too many stages in the project: Pre-Feasibility, Feasibility, Project Definition, Full Development, and Production, leading to slippage in time. The combination of multiple committee levels/stages and qualified approval or re-submission gave many opportunities for a change of direction or 'rabbit holes'.
– Key committees (ORC and DEPC) often deferred unpalatable decisions by calling for more studies. A further key factor was the turnover in key decision-making staff during the lifespan of the project. Committees were sometimes deferential to rank or to compulsive speaking rather than to reasoned argument.
– There was over-management due to the number of layers. McCoy touted the Special Project Executive (SPE) model used for Stingray, suggesting the same for Type 43 as a total project entity, including GWS31, STIR and the SM1A gas turbine. Such 'vertical' project teams would be similar to the much later Type 45 PCO/prime contract construct.

The slatted array of Type 1022 on a Type 42 destroyer. Above it, atop the foremast, the antenna for Type 1006 can be made out. (John Jordan)

McCoy's overriding recommendation was to reduce the layers and stages in the approval process: only Category B (>£500M) projects should go to the Operational

POSTWAR RADAR DEVELOPMENT IN THE ROYAL NAVY

Pre-1960 | **1960** | **1970** | **1980**

984 in carriers
965 elsewhere
992/993 for TI

exploited as: 968 E/F band / 967 D band Pulse Doppler

985, 2nd gen 1956–60 abandoned ✗

UK/NL 988 *Broomstick* 1960 to late 1968 ✗

Reduced version in RNIN *Tromp*

Built on 1953 Utility FD work (mini 984 or V beam), Frescan, TWT phased array, multi-pulse

NSR7932 965 MTI short pulse/ double AKE 1966 to 1967 ✓

NSR7963 966 MTI long pulse, 5 frequency but SLS ECM deleted 1972 to 1976 ✓

996 E/F stacked beam

982M/983 for FD

986/987
993

965 AD replacement

NSR7938 new LR radar 1015, with back-to-back dual-frequency UHF Aerial 1965 to end 1971 ✗

NSR7938 interim LR radar 1022 1973 to 1975 ✓

wider fit of 1022

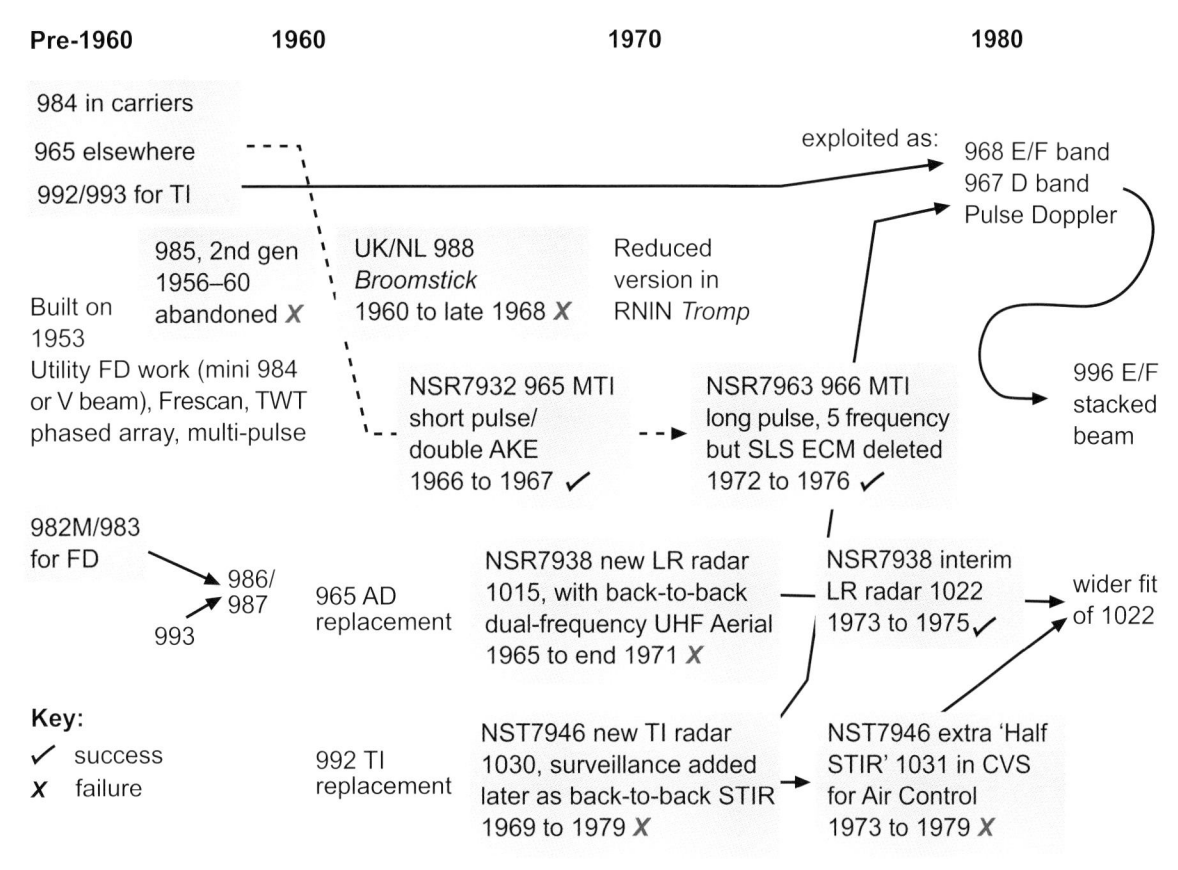

Key:
✓ success
✗ failure

992 TI replacement

NST7946 new TI radar 1030, surveillance added later as back-to-back STIR 1969 to 1979 ✗

NST7946 extra 'Half STIR' 1031 in CVS for Air Control 1973 to 1979 ✗

Fig 12: McCoy Radar Narrative (Graphic by John Jordan using material supplied by the author)

Requirements Committee. The principal checkpoint should be when PE were able to present real costed options to the Naval Staff and, once endorsed, implementation should then proceed without further check. A 75 per cent solution now was better than 100 per cent at some future point (or never). When McCoy returned to MoD five years later, he believed that some organisational improvements had been made.[6]

Other considerations

It is important to note the huge interplay between each radar project and:

- ships' programmes/fitting opportunities, size and weight, choice between Type 42 refit or new-build Type 43/44
- cost savings through deleting ECM/EPM features
- inputs from uncertain parts of the research programme (eg MRSL 'Spoof' for ECM)
- synchronisation with evolving programmes such as GWS31, 909M and Type 43/44
- a focus on low prf for air surveillance, and high prf TI (also at higher angles) for Seadart, as a back-to-back array
- nomenclature (Type 1015): the IFF block was Types 1010/11/12/13, with a gap before later IFF 1017/18. (Why was an air defence radar included in this number sequence? Logically it should have been either Type 989 [post-'Broomstick'], or a number below 1022>1030.)

There was, however, some later exploitation of the projects which failed to carry through:

- The STIR antenna was developed as Type 1022 aerial outfit AZY.
- The STIR D-band pulse-doppler radar became Type 967.
- The SLS was developed as 967M and 996.

Immediately before the Falklands, the research programme developed a HF surface wave radar that looked over the horizon in order to provide a 'guard ring' warning of impending low-level air raids.[7] This was demonstrated in the early 1980s from a Blockhouse trials site near Milford Haven, and subsequently using HMS *Londonderry* and a Royal Fleet Auxiliary (RFA). Despite showing some promise, this project was overtaken by the Falklands conflict, which triggered a crash programme to develop a Sea King AEW variant using a Searchwater radar, to give a degree of airborne early warning (later improved as the Mk 7 ASaC, and then the 'Crowsnest' programme to move to a Merlin airframe).

Type 993 office equipment was updated to 994, using the commercial AWS-4 radar, but retained the quarter-cheese AKD, except for RFA *Argus* and the Offshore

Patrol Vessel deployed as the Falklands guardship after the conflict, which had AKV using a Cosec^2 antenna.

Post-Falklands War

The 1982 Falklands War was fought with antique radars such as Types 965R and 992Q, plus a few 967/968 in Seawolf ships, and only a single Type 1022 in *Invincible*. This provided a major impetus to develop replacement radar, which led to Type 996 via a Cardinal Point Specification (CPS). The intent was to develop a capable surveillance and TI radar that could support both Seadart and Seawolf, with some height measurement capability to shorten the acquisition times of the 909 and 910 fire control radars. Figure 12 shows the resultant antenna and its coverage.

One option would have been to fit Type 967/968 more widely. However, this was a large, expensive system combining two radars, and was by then almost middle-aged. It was also heavy, and Type 996 offered the chance to explore new technology for a lighter system (with the goal of less than 1,000kg (0.98 tons) at the masthead (essential for Type 42 refits).[8]

There were a number of issues. Radar 996, and its associated plot & track extractor (Outfit LFA), were contracted from different suppliers against separate specifications, which raised a number of integration problems. Type 996 required enhanced spares support via the AVIMP programme, while LFA required a short-term improvement as LFAST, and then full replacement as Outfit LFE. Type 996's output as synthetic processed video was deemed not good enough for collision avoidance, and IMO/SOLAS rules required ships with 996 to adopt another E/F-band bridge radar, as Type 1008.

Phased Array Radar

There was early US work on planar naval radar arrays (AN/SPS-32 and -33) in USS *Long Beach* and *Enterprise*, fitted four-square around the superstructure. This was followed by a multi-function beam steering radar for the US Navy with the Typhon programme, which began with Johns Hopkins University Applied Physics Laboratory studies in May 1957 and ran through to June 1966. This developed a prototype AN/SPG-59 radar, operating in C band and fitted in the trials ship USS *Norton Sound*. This was based on a dielectric ball (a spherical luneborg lens) to produce focused RF beams. Technology limitations led to Typhon being terminated and replaced by the Aegis programme, which reached operational service in USS *Ticonderoga* (CG-47) in January 1983.

The Aegis radar (SPY-1A) had four flat faces, fixed to the ships superstructure, but moved to S band. It was initially based on high-power crossed-field amplifier transmitter tubes, using a waveguide to feed a large array of passive ferrite phase shifters (4,350 elements) that made up the 12ft square antenna, all under computer control to form detection and tracking beams.[9] SPY-1 has since been refined via component upgrades as far as model -1D, but remains a passive array, fed by a relatively small number of high-power transmitters.

The UK and NATO started somewhat later (1977), but concentrated on an active array radar, developing rela-

The Type 22 frigate HMS *Cumberland*, seen here in June 2005, shows the combined 967M/968 array at the masthead, just above the UAF ESM antenna. Type 1007 is fitted on a spur halfway down the mast. (John Jordan)

The Type 23 frigate HMS *Westminster*, seen here in March 2005, shows the 996 antenna (Outfit ADQ with IFF interrogator Outfit AMO) at the masthead, immediately above the UAT ESM array; the aerial was stabilised, and rotation was 30rpm. The model in the Type 23 frigate was Type 996(1), that in the Type 42 destroyer and the *Invincible*-class CVS 996(2). *Westminster*'s foremast also has 1007 on one spur, and the GSA8 electro-optic tracker ball for the 4.5in Mk 8 gun on another. The radar tracker above the bridge is Type 911 for Seawolf. (John Jordan)

POSTWAR RADAR DEVELOPMENT IN THE ROYAL NAVY

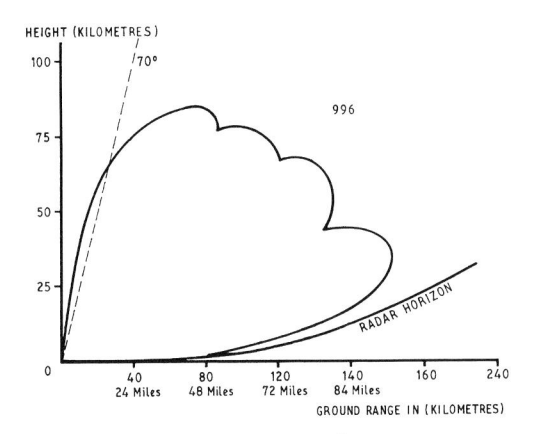

Fig. 13: Type 996 Vertical Coverage Diagrams
The radar operated in the E/F band, and used cyclic hopping or frequency agility via a travelling wave tube (TWT); peak power was 137kW peak power, and MTI employed complex PRFs at two pulse lengths (frequency modulated). There were three transmit beams (+ long range – see upper diagram right) with interleaved receive beams [lower diagram]. (BR1982d, HMS *Collingwood*, courtesy of Clive Kidd; graphic by John Jordan using material supplied by the author)

Table 6: **Weights of Principal Radar Equipments**

Weight (tons)	992Q/ADN	967M/968/AZT	996/ADQ	997
Antenna	0.65	2.20	1.07	1.31
Below-decks	1.95	9.51	2.68	2.76
Total weight	2.60	11.71	3.75	4.07

tively low power transmitter modules (announced in 1984) that formed the face of the array.[10] This had each element generating 2.5W peak power, under digital control of waveform and beam forming, with 1,500 modules per face. This led to a series of technology demonstrators: MESAR 1 and 2, followed by ARTIST. Both Aegis and Sampson work in the S band (now E/F band). The Aegis peak effective radiated power (ERP) is *ca* 4MW fed by 8 transmitters, whilst MESAR proposed 1,500 modules per face at 2.5W peak power, for a peak ERP of 3.75kW.

This technology now forms the basis of two in-service naval radars:

– 'Sampson' MFR (Radar Type 1045): two-faced array enclosed within a rotating spherical shield; fitted in Type 45 destroyers (isd 2009).
– 'Artisan' MRR (Radar Type 997): single-faced rotating array, replacing 996; fitted in Type 23 frigates and *Queen Elizabeth* class carriers (isd 2017).

The UK is the project leader with two manufacturers: Ferranti Edinburgh/Selex (Seaspray E-radar) and Siemens-Plessey (now BAe) at Cowes. The other new radars in the Type 45 destroyer are the Long Range Radar (Type 1046), based on the Dutch SMART-L radar, and a new navigation radar (Type 1047).

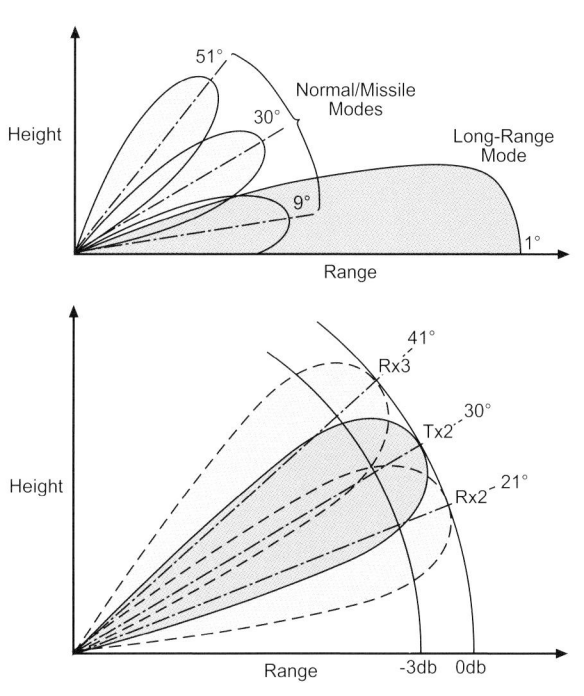

Identification Friend or Foe (IFF)

The wartime Mk II, II and IV IFF persisted through until the early 1950s, when they were superseded by Mk 10. Mk II worked on 22–38MHz, 39–51MHz and 54–82MHz, while Mk III worked on 157–187MHz and Mk IV worked on 400MHz, all using a variety of 'toasting fork'-style aerials. Mk V was developed as a contingency against growing enemy spoofing capabilities on 950 and 1,150MHz with some coding, but was not implemented. Mk X (later Mk 10) was introduced from the 1950s with the current 1,030 and 1,090MHz challenge and response, and featured several modes and growing complexity (4,096 codes).

In the 1970s Mk 10 IFF was updated as Type 1010, which added side lobe suppression for better bearing accuracy, and was fully integrated into the command system using computer control of the challenge and response sequence; the decoded ID was held with other main track table information. The most recent Mk 12 IFF added cryptography, and has become essential for Coalition operations involving the United States. The IFF modes now supported are SIF (Selective Identification Feature), Mode 1, 2, 3/A, Mode S, and encrypted Modes 4 and 5 SIFF (Successor IFF):

Fire Control

The wartime metric (*ca* 600MHz) fire control radars were replaced by centimetric sets (Types 262, 274 and 275) on I or E/F band. Post-war these were supplemented by Type 901 (the intended tracking radar for LRS1) that subsequently became the Seaslug guidance system; 902 would have been the associated gathering system but was subsumed within 901's RF scanning system.

Type 901 was the contemporary of the Type 984 FD

109

The Type 45 destroyer HMS *Defender*, seen here in June 2014, shows the spherical Sampson array (Type 1045) at the top of the foremast, plus the Long Range Radar (LRR – 1046) aft of the pole mast, above the hangar. (John Jordan)

radar, albeit about 4 years later, with similar technology in the lens aerial and mechanical scanning. This radar changed significantly between 1946 and 1951 from being two 6ft parabolic dishes to the final 9ft microwave lens on an unmanned director (as GMS1). The lens was a matrix of waveguides of different lengths to introduce phase delay and build a plane wavefront. The eventual director is shown in Figure 14, and the perspex model was a design technique to develop a stiff lightweight structure that could handle significant torques whilst maintaining overall RF alignment.

Type 901 was one of the most powerful fire control radars, with a 750kW I-band magnetron, and was also one of the most accurate. This derived from a conical scan, with simultaneous dual-lobes on receive, able to exclude target propeller effects, or jamming, plus a very agile sub-reflector (the beam deflection unit); it was able to fine-point the beam through the lens without the servo noise that came from the heavy radar dish structure coarse tracking its target. This was necessary to give an accurate beam for the Seaslug missile to 'beam ride' up to the target.

Type 903 was the anglicised version of the US Mk 35 radar for MRS3, while **Type 904** was the same radar in GWS22 directors.[11]

Type 905 was a research tool as the 'ultimate' gunnery radar. The staff requirement for Medium Range System Mk 5 (MRS5) was raised in 1944. Development ran from Oct 1946 to Mar 1950, and was followed by a contract to continue with development of Radar Type 905 that examined radar target glint in the 'Netting' trials, using Lancaster and Mosquito targets between October 1950 and November 1951.

The elevating structure included an inner gimbal, allowing three-axis servo drives (elevation, training, and traverse), the latter catering for high-angle tracking, where the two base axes provide only poor control authority near the zenith. Type 905 was a simultaneous phase comparison monopulse, using an 'eggbox' plate

HMS *Iron Duke*, seen here in June 2014, shows the new Artisan radar (Type 997) at the masthead, and 1007 on a sponson, just above the SCOT satcom dish. The ship still has the 911 tracker above the bridge, marking the ship as still being fitted with Seawolf. This tracker is replaced by an uplink aerial when Sea Ceptor/CAMM is fitted in lieu of Seawolf. (John Jordan)

aerial. The four 3ft antennae, which were combined in a square array with clipped corners, gave a 6ft overall diameter. The lens was illuminated by four horns, and the RF part of the radar was based on the X (now I) band CV368 magnetron (200–250kW),

Reports up to April 1950 mention a separate search radar, using scanning elements for acquisition. This would have involved two additional lenses 6ft x 3ft illuminated by a line source such as a Forster scanner or a slotted waveguide array, and fitted outside the main 905 radome; this was likely to have been Type 906 (thereby mirroring Types 901 and 902).

While Type 903 achieved a range of 30,000 yards in service and Type 901 up to 65.000 yards (both figures against aircraft of unspecified size), Type 905 (which was intermediate in both RF power and dish size) would probably have been capable of 40–50,000 yards.

Types 907 and **908** were radars for SGS1 and TOM directors.

Type 909 (initially codenamed 'Desertcar') was developed for Seadart, while **Types 910** and **911** were Seawolf trackers that added TV, then Thermal Imager (THIM) channels. In parallel, **Type 912** was adopted for both gun and Seacat control in the commercially-designed Type 21 frigate by Vosper. When Sea Ceptor/CAMM is fitted in lieu of Seawolf in Type 23 frigates, the 911 trackers have been replaced by uplink aerials.

Types 930–931–932 were added to low-angle director control towers, and to the US Mk 37 director for splash spotting, but were not carried through to the postwar systems with 275 or 903 radars. They used an extra display (part of the AFCB10 console) to correct for fall of shot during surface fire.

Conclusions

The RN was a leader in the development and application of radar during the Second World War. In the immediate aftermath of the war it consolidated, bringing those ships that were to be retained up to a common late-war standard. Progress then stalled, with a series of problem Fighter Direction sets that did not live up to expectations, largely because the development of the higher powered magnetron that were needed had not kept pace.

The RN then succeeded in developing a world-leading radar in Type 984 with its multiple scanning beams for height measurement in service by 1958. With 984, the associated CDS and DPT data link, the RN was then at least five years ahead of the US Navy. Although planned for cruisers and FADE escorts, it was fitted in full form only in aircraft carriers.

There was then a 20-year hiatus through to 1980, with senior staff frustration voiced by McCoy. This long saga included an offshore procurement of Type 988 from the Dutch (on workshare grounds driven by the need to offset purchase of UK equipment), cancelled with CVA-01, followed by the failure of the Stabilised Target Indication Radar (STIR), which had been tied to the large Type 44 destroyer cancelled by John Nott. There was some limited exploitation via 1022, but the Falklands war in 1982 was largely fought with antique radars (Types 965R and 992Q).

This led to Type 996, acquired via a CPS (then in vogue and championed by Dr Keily), but the stand-back

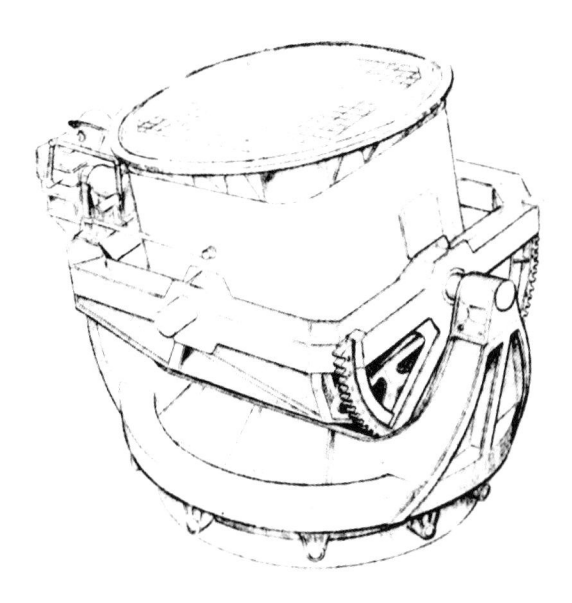

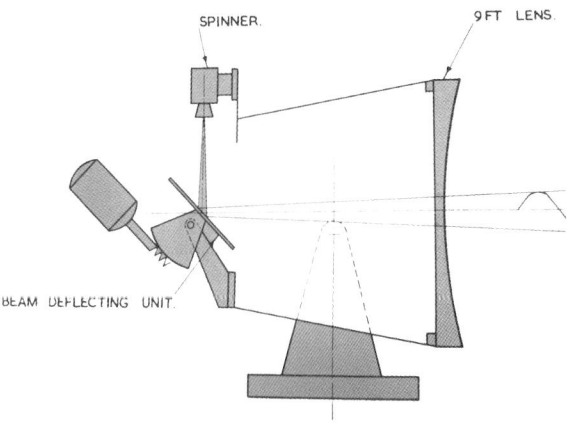

Fig 14: Seaslug Director Model and Layout.
(Author's collection)

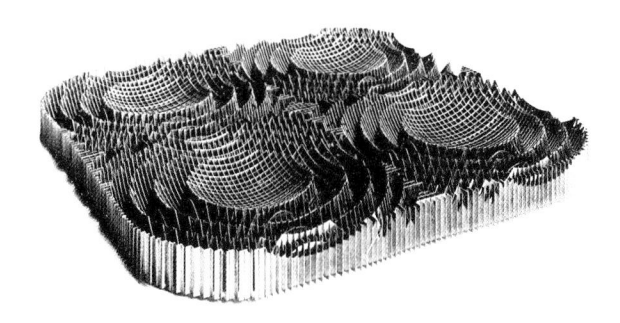

Fig 15: The Type 905 'Netting' Director and its 'Egg Box' Lens Antenna.
(Author's collection)

procurement method resulted in a string of flaws, plus a series of problems with spares and documentation.

Finally, UK work on active antennae, using low powered transmit modules, bore fruition with MESAR>ARTIST>'Sampson' and then Type 997, albeit after the USN had shown the way, employing a brute-force approach with the Aegis passive phased array radar, SPY-1A.

Across the period, there was generally good basic technology research. Some of the form-fit replacement programmes were successful, but there were repeated failures with more ambitious projects, aimed towards larger ships or complex refits, that made cancellation more likely. Each decision fully considered offshore options.

During the bulk of the postwar period, technical development was Government-led, via the combination of research scientists (ASWE), procurement & development engineers (DGSW), and serving RN staff embedded in the projects as Naval Applicators. This was a creative partnership, responsible for most of the major improvements. However, by the mid-1980s a more 'hands off' approach to contractors became the norm, and tensions between the MoD elements were amplified when the Procurement Executive moved to Abbeywood in 1994.

Most of McCoy's recommendations still hold true, though he also championed vertically-integrated project portfolios. The Special Project Executive did succeed with Polaris and with the heavyweight torpedo programmes; however, it has faced more recent criticism with the Type 45 Prime Contracting Organisation (PCO), which may have been unduly dominated by the industry prime, which an underpowered MoD side failed to hold to account.

Wider Lessons Learned

Engineering people matter. Running hot or 'sweating the assets' may look like macho management, but (like the NHS) your human capital simply walks, in an escalating retention vortex that becomes the limiting factor on sustaining future capability.

A view from outside is that where there were originally only a few project managers in the 1970s and 1980s who were not up to the job, nowadays the process and checklist/tickbox culture means that now only a team of 'all stars' can succeed.

With Cardinal Point Specification the onus is on the manufacturer's product to pass contractual milestones, and broadly 'what you see, is what you get'. This may be adequate to specify desired performance, but the more intangible qualities of being 'user friendly' or 'jackproof' cannot be included. These were the traditional business of naval applicators, but they are now excluded. Problems over interpretation often emerge long after the tender evaluation process; but once the contract has been placed, any change is likely to lead to cost escalation and to jeopardise timely completion.

Experience is that innovation is largely 'bottom-led', and is not engendered by top-down directive control. One problem is the senior wish to be seen 'to do something', overriding more pragmatic considerations.

Annex

Nomenclature

Individual RN equipments are described by a range of titles or different 'Nomenclature' schemes. Generally Type numbers cover radio, radar and sonar transmitters, while three-letter Outfits are aerials, receivers and all else. Generally, older equipment was identified by two letters and a number, but this evolved into the current three-letter Outfit title, with variants in parentheses, *eg* Outfit ABA(2) or, in the case of radio, radar and sonar sets, by a type number, *eg* Radar Type 996. Type suffixes M–P–Q–R refer to sequential modifications, while X–Y–Z are used for development models; numbers in parentheses, *eg* (2), identify variants in configuration. This is described in CB03329 (ADM 239/627), but has no easily decodable structure – unlike the US JAN system, where for AN/SPY-1 the SPY explains the system functionality. The most recent introductions have reversed the sequence (*eg* Outfit 4KMA). The radar numbering sequence is shown in Figure A1.

Systems may also be known by their project open name or colour codeword (*eg* 'Orange Crop'). Finally, there are a few instances, largely concerning communications equipment, where the UK numbering scheme mirrors the equivalent US structure (UK/URC-613 was the Code Division Multiple Access modem used by RN satcoms). Airborne systems tend to use both a codeword name and an ARI (airborne radio installation) number.

All these different systems coexist side-by-side in documentation; there are a few unexplained gaps where a nomenclature was issued for planning purposes but the project was subsequently cancelled before delivery.

Personnel

Across the postwar period, radar was operated by the Direction specialisation, part of the Seaman Branch of the Navy. RP ratings and direction officers were trained by HMS *Dryad*, later becoming the Warfare Branch. Radars were maintained by the Electrical Branch, with Radio Electrical Mechanics and Artificers (REM/REA) and Electrical Officers trained at HMS *Collingwood*, later becoming the Weapons Electrical branch, then the Weapon Engineering branch. Both establishments are now co-located at *Collingwood* as the Maritime Warfare School.

Examples of Radars

HMS *Collingwood*'s museum has a complete Type 271 wartime radar set, while postwar radars (262, 274, 275, 277, 293 and 960) can be seen aboard HMS *Belfast* and HMS *Cavalier*. Note that while the operations rooms are open, the radar offices are usually closed, and access would need to be negotiated with the respective curators.

Acknowledgements:

The author wishes to thank Lt-Cdr Clive Kidd, RN Rtd (HMS *Collingwood* Museum), Cdr James McCoy, RN Rtd, and Dr Alan Moore for their help in preparing this article.

Abbreviations

3D or 3d	Three-dimensional	GDS	Gun Direction System
AD	Air Defence (ship)	IFF	Identification Friend or Foe
AEW	Airborne Early Warning	IMO/SOLAS	International Maritime Organisation/Safety of Life at Sea
AFC	Automatic Frequency Control		
AIO	Action Information Organisation	ITU	International Telecommunications Union
AJ	Anti-Jam	JNS	Journal of Naval Science (originally Journal of the RN Scientific Service)
ARTISAN	Advanced Radar Target Indication Situational Awareness & Navigation		
		LPI	Low Probability of Interception
ARTIST	Advanced Radar Technology Integrated System Testbed	MESAR	Multi-function Electronically Scanned Adaptive Radar
ASWE	Admiralty Surface Weapons Establishment	MTI	Moving Target Indication
BR/CB	Book of Reference/Charge (or Confidential) Book	NAAW	Notes on Above-water Weapons
		ORC	Operational Requirements Committee
CCA	Carrier Controlled Approach (radar)	PCO	Prime Contracting Organisation
CDS	Comprehensive Display System	PE	(MoD) Procurement Executive
CPS	Cardinal Point Specification	RCS/REA	Radar Cross Section/Radar Echoing Area
CVA-01	New Aircraft Carrier design, cancelled in 1966	SIF	Selective Identification Feature (IFF)
CVS	*Invincible*-class aircraft carrier	SMART	Signal Multi-beam Acquisition Radar for Targeting
CW	Continuous Wave		
DEPC	Defence Equipment Priorities Committee	SPE	Special Project Executive
DGSW(N)	Director General of Surface Weapons (Navy)	STIR	Stabilised Target Indication Radar
DLG	Destroyer Light Guided Missile	TACAN	Tactical Air Control and Navigation
ECM	Electronic Counter Measures	TEWA	Threat Evaluation and Weapon Assignment
EPM	Electronic Protection Measures	TI	Target Indication
FADE	Fleet Air Defence Escort	TWT	Travelling Wave Tube
ERP	Effective Radiated Power	VCD	Vertical Coverage Diagram
FD	Fighter Direction (radar)		

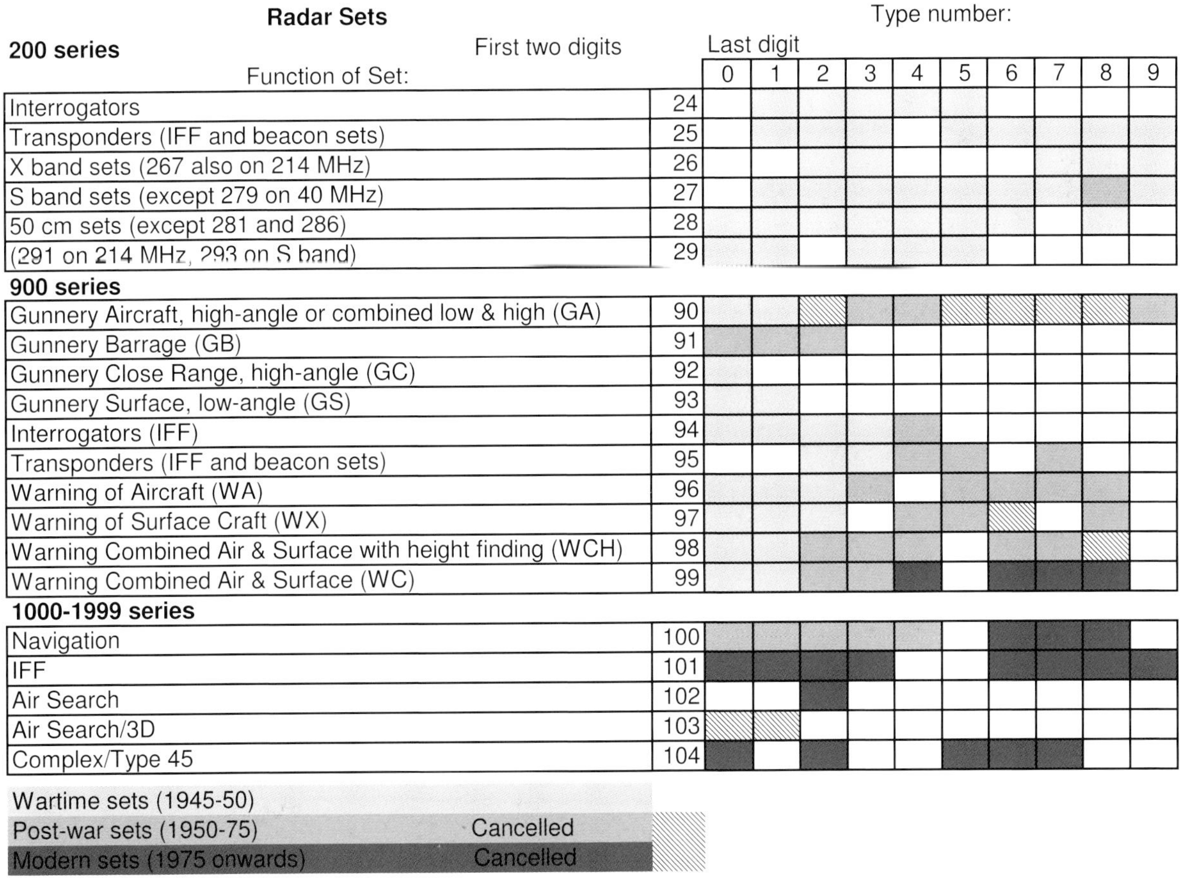

Radar Sets

Type number:

Function of Set:	First two digits	Last digit 0	1	2	3	4	5	6	7	8	9
200 series											
Interrogators	24										
Transponders (IFF and beacon sets)	25										
X band sets (267 also on 214 MHz)	26										
S band sets (except 279 on 40 MHz)	27										
50 cm sets (except 281 and 286)	28										
(291 on 214 MHz, 293 on S band)	29										
900 series											
Gunnery Aircraft, high-angle or combined low & high (GA)	90										
Gunnery Barrage (GB)	91										
Gunnery Close Range, high-angle (GC)	92										
Gunnery Surface, low-angle (GS)	93										
Interrogators (IFF)	94										
Transponders (IFF and beacon sets)	95										
Warning of Aircraft (WA)	96										
Warning of Surface Craft (WX)	97										
Warning Combined Air & Surface with height finding (WCH)	98										
Warning Combined Air & Surface (WC)	99										
1000-1999 series											
Navigation	100										
IFF	101										
Air Search	102										
Air Search/3D	103										
Complex/Type 45	104										

Wartime sets (1945-50)
Post-war sets (1950-75)   Cancelled
Modern sets (1975 onwards) Cancelled

Fig A1: Radar Type Numbers

RN Radar Development Roadmap

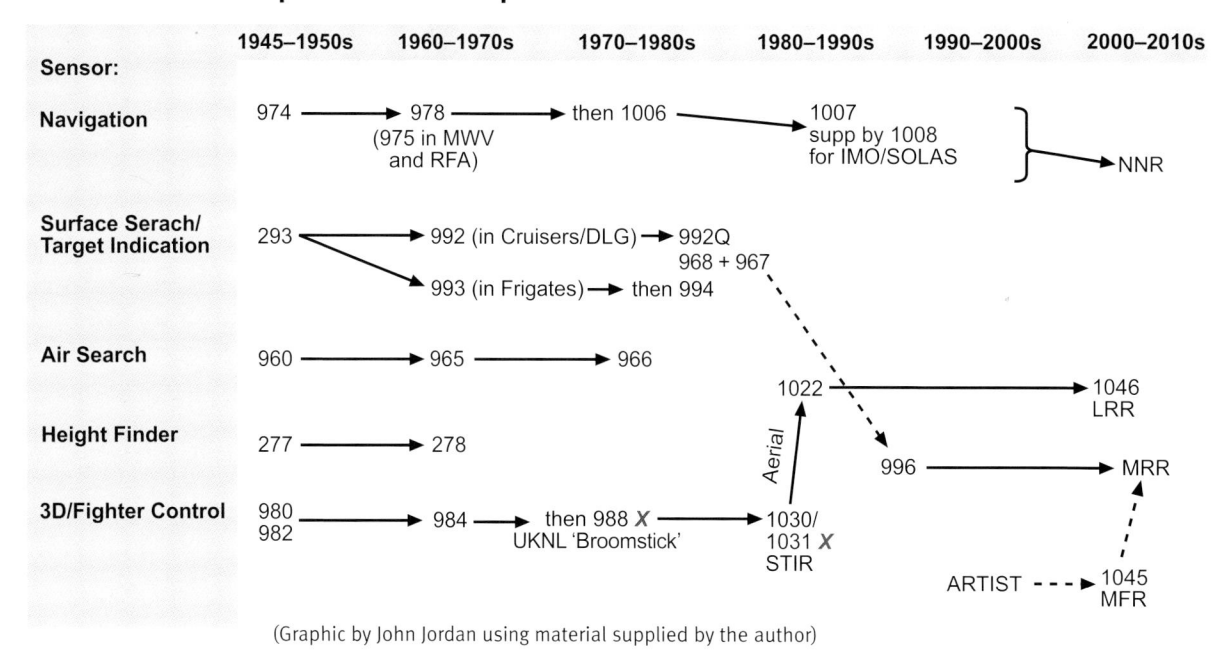

Sensor:	1945–1950s	1960–1970s	1970–1980s	1980–1990s	1990–2000s	2000–2010s
Navigation	974 →	978 → (975 in MWV and RFA)	then 1006 →	1007 supp by 1008 for IMO/SOLAS	} → NNR	
Surface Search/ Target Indication	293 →	992 (in Cruisers/DLG) → 993 (in Frigates) →	992Q 968 + 967 then 994			
Air Search	960 →	965 →	966	1022 →	→ 1046 LRR	
Height Finder	277 →	278			996 →	MRR
3D/Fighter Control	980 982 →	984 →	then 988 X UKNL 'Broomstick'	→ 1030/ 1031 X STIR (Aerial)	ARTIST - - →	1045 MFR

(Graphic by John Jordan using material supplied by the author)

Sources:

American Society of Naval Engineers (ASNE) Journal.

Benjamin R, *Five Lives in One*, Parapress (Tunbridge Wells, 1996).

Benjamin R, 'The Post-war Generation of Tactical Control Systems', *JNS* Vol 15 No 4, Nov 1989, 262–275.

HMS *Collingwood* Museum, via website www.rnmuseum-radarandcommunications2006.org

Howse D, *Radar at Sea*, All Naval Radar Trust with Macmillan (London, 1993).

Kingsley FA (Ed), *Radar: the Development of Equipment for the RN 1935–45*, and *Radar and the Application of Other Electronic Systems in the RN in WW2*, both Macmillan (London, 1995).

Johns Hopkins University: Applied Physics Laboratory (JHU-APL), via website.

Journal of Naval Engineering (JNE), via website https://jneweb.com.

Journal of Naval Science (JNS).

Lavington S, *Moving Targets: Elliott Automation 1947–67*, Springer (London, 2011).

Marland P, articles published in *Warship 2014, 2015, 2016* and *2021*.

McCoy J, 'The Provision of Air Defence Radars to the Fleet', 6 Dec 1979, private communication.

Moore A, 'MESAR (Multi-Function Electronically Scanned Adaptive Radar)', IET Conference Proceedings 1997, CP_19971631.pdf.

Naval Electrical Review (NER).

Navigating & Direction Officers Association, website ndassoc.co.uk.

Peters J, *Echoing Down The Years, ASRE Remembered*, Elgar Press (Alverstoke, 2002).

Ratsey OL, 'As We Were: 50 years of ASWE History 1896–1946', internal document, 1974.

Schofield VAdm RR, *The Story of HMS Dryad*, Keith Mason (Havant, 1978).

Vidler N, 'The Story of ASWE', internal document, June 1969.

Endnotes:

[1] Naval Radar Trust books (FA Kingsley). These build on a series of articles in the *Journal of Naval Science*, and are the main reference for radar, displays, fire control, and EW development through to *ca* 1950. In addition, R Benjamin's article covers postwar C2 and radar Type 984 up to 1958.

[2] This section is based on CA Cochrane and material from the *Collingwood* museum.

[3] NAAW was published from 1958–1971 on a semi-annual basis by the Weapons Department, and replaced Progress in Naval Gunnery (PING); most are held by TNA.

[4] 'The Provision of Air Defence Radars to the Fleet', 6 December 1979, private communication.

[5] Derived from the RAF WW2-era Type 15 GCI radar, working on 200MHz.

[6] Covering comments, email McCoy to Marland, 11 December 2014.

[7] Note that this is separate from the longer-range HF skywave system fitted ashore, for example used by the Canadian authorities to monitor their offshore fishing grounds, or the Australian 'Jindalee' strategic surveillance system.

[8] N Bailey, *Radar Type 996*, *JNE* Vol 32 book 3, December 1990, 543–552 (www.jneweb.com).

[9] ASNE *Naval Engineers Journal*: Bryce Inman, 'From Typhon to Aegis', May 1988, 62–72, and Milton Gussow, 'Typhon: A Weapon System Ahead of its Time', July 1997, 53–62.

[10] Eugene Billam, 'Phased Array Radar and the RN', *JNS* Vol 12 No 2, May 1986, 93–103.

[11] Type 904 is the only case of a nomenclature being re-used, since the number was also associated with the proposed MRS4 gunfire control system in the 1950s, but was later reallocated to an entirely different radar for GWS22 Seacat.

AFTER THE *SOVETSKII SOIUZ*:
SOVIET BATTLESHIP DESIGNS 1939–1941

Soon after the construction of the *Sovetskii Soiuz* class was begun, the Soviets embarked on preliminary design studies for follow-up battleships. Although the work never progressed beyond the initial stages, there were several interesting developments before the German invasion of 22 June 1941 put an end to the process. **Stephen McLaughlin** describes the course of the work and the resulting designs.

By the late 1930s the Soviet warship design process had been formalised into an elaborate, multi-step negotiation between the People's Commissariat of the Navy (hereafter simply the Navy) and the People's Commissariat of the Shipbuilding Industry (hereafter the Shipbuilding Industry). The Navy's Main Naval Staff (*Glavnyi morskoi shtab*, or GMSh) began the process; assisted by its technical advisory body on warships, the Scientific-Technical Committee (*Nauchno-technicheskii komitet*, or NTK), it would draw up the Operational-Tactical Requirements (*Operativno-takticheskie zadanie* or OTZ, equivalent to the Royal Navy's 'staff requirements'). These spelled out the roles a ship was intended to fulfil and the basic characteristics – armament, protection, speed and range – it would require to carry out those roles.

The OTZ would be submitted to the Shipbuilding Industry, where they would be studied by TsNII-45 (*Tsentral'nyi nauchno-isledovatel'skii institut* 45 = Central Scientific-Experimental Institute 45), which was in effect a warship research group; the relevant central construction bureau (*Tsentral'nyi konstruktorskii biuro*, or TsKB), might also examine the OTZ. One or both would establish the feasibility of the OTZ by producing 'pre-sketch designs' (*predeskiznye proekty*), basically outline or preliminary designs showing various combinations of armament, protection and speed.

The pre-sketch designs, accompanied by suggested Tactical-Technical Requirements (*Taktiko-tekhnicheskoe zadanie*, or TTZ) would be sent to the Navy where, if they met with approval, they would be taken as the basis for the final TTZ, which enumerated in some detail the ship's characteristics – approximate displacement, maximum draught, armament, armour thicknesses, speed, range, and so forth. These amounted to the contract specifications, and would be handed over to the appropriate design bureau – for battleships, this was TsKB-4 – which would develop a series of increasingly detailed designs, culminating in the technical design, which would form the basis for the thousands of construction drawings needed to build a warship. None of the designs discussed in this article got past the pre-sketch phase.

It was all very neat in theory, but in practice it could get quite messy. Since two separate commissariats were involved, disputes could only be resolved at the highest level of the government. Two examples may be cited from the design of the Project 23 (*Sovetskii Soiuz* class) battleships: in November 1936, during a meeting of the Council of People's Commissars, the Navy had complained that the ships designed by the Shipbuilding Industry were larger than necessary, while at a later session the arguments between the Navy's representative, Fleet Flagman 2nd Rank I S Isakov, and the head of the Shipbuilding Industry, I F Tevosian, became so vehement that they refused even to face one another.[1] The government generally came down on Industry's side, so the Navy wound up with ships less capable than it had expected. This had certainly been the case in the *Sovetskii Soiuz* class, where the speed was only 28 knots instead of the specified 30, and the anti-aircraft armament comprised eight 100mm guns (vs twelve) and thirty-two 37mm (vs forty); there were also several unsatisfactory features in the protection scheme.[2] For its own part, the Shipbuilding Industry thought the Navy's requirements were unrealistic. Both viewpoints had an element of truth: the designers were inexperienced, especially when it came to large warships, and the Navy tended to be overly optimistic regarding what could be achieved on a given displacement. But the upshot was that there was a good deal of distrust between the two commissariats; in particular, the Navy regarded the compromises in the 60,000-tonne Project 23 as a reflection of the shortcomings of the Shipbuilding Industry.

It is therefore hardly surprising that once the Project 23 design had been finally approved by the government on 13 July 1939, Engineer-Flagman 3rd Rank A A Frolov, chairman of the NTK, began thinking about a better battleship. On 27 August 1939 he sent a memo to Isakov, who was responsible for shipbuilding, pointing out that:

> At the present time Project 23 battleships are under construction in Leningrad and Nikolaev, with launch dates in May–June 1941. That is, in one year and ten months, after the launch of *Sovetskii Soiuz* and *Sovetskaia Ukraina*, the question of the future utilsation of the slipways will arise

Table 1: Post-*Sovetskii Soiuz* (Project 23) Designs 1939–1940

	A Project 23[1] Pre-sketch Design	B Project 23*bis* Pre-sketch Design (9-gun)	C Project 23*bis* Pre-sketch Design (12-gun)	D Project 24
Date	13 Jul 1939	Dec 1939	Dec 1939	Oct 1940
Displacement:				
standard	59,150 tonnes	60,800 tonnes	?	59,882–60,146 tonnes
full load	65,150 tonnes	66,800 tonnes	74,700 tonnes	65,535–65,800 tonnes
Length:				
overall	269.4m	285.9m	?	c270m
waterline	260m	278m	287.7m	260m
Beam:				
maximum	38.9m	39.9m	?	38.9–39.9m*
waterline	36.4m	37.1m	38m	36.4m*
Draft (design)	10.1m	10.27m	?	10.1m*
Armament:	9 – 406mm (3xIII)	9 – 406mm (3xIII)	12 – 406mm (4xIII)	9 – 406mm (3xIII)
	12 – 152mm (6xII)	12 – 152mm (4xIII)	12 – 152mm (4xIII)	12 – 152mm (6xII)
	8 – 100mm (4xII)	12 – 100mm (6xII)	12 – 100mm (6xII)	16 – 100mm (8xII)
	32 – 37mm (8xIV)	32 – 37mm (8xIV)	32 – 37mm (8xIV)	44 – 37mm (11xIV)
Aviation:				
catapults:	1	1	2	1
aircraft:	4	4	?	3
Protection:				
citadel length	148.4m / 57% wl	201.0m / 72.5% wl	?	as Project 23?
upper belt	25mm	25mm	25mm	20mm
main belt	375–420mm @ 5°	380mm @ 8°	375–420mm @ 5°	375–420mm @ 5°
longitudinal b/hd	—	—	—	50mm
fwd belt	220mm	20mm	20mm?	220mm
fwd transverse b/hd	285mm @ 30° + 230mm	425+50mm	230mm	250mm @ 30° + ?
aft transverse b/hd	365–75mm	420+50mm	365–75mm	380mm*
foc's'le deck	25mm	25mm	25mm	20mm
upper deck	155mm	155mm	155mm	155mm
middle deck	50mm	50mm	50mm	50mm
Underwater protection:				
system	Pugliese for 123m 'American' for 33m	'American' for 164m	'American'?	'American'
depth amidships	8.2m	8.0m	8.0m?	6.5–7.0m
designed resistance	750kg TNT	750kg TNT	750kg TNT?	700kg TNT
Machinery:				
engines	3 sets turbines	3 sets turbines	3 sets turbines	3 sets turbines
boilers	6 x 3-drum	6 x 3-drum	6 x 3-drum	6 *or* 12 x 3-drum
horsepower (natural)	201,000shp	201,000shp	201,000shp	201,000shp
horsepower (forced)	231,000shp	231,000shp	231,000shp	231,000shp
Speed:				
natural draught	28 knots	30 knots	28 knots	28 knots
forced draught	29 knots	31 knots	?	29 knots
Range at 14.5 knots:				
full load of fuel	5,960nm	5,770nm	?	6,000nm*
max load of fuel	7,250nm	N/S	?	?
Hull form:				
block coefficient	0.657	0.615	?	?
length:beam ratio	7.15	7.50	7.57	?

Notes:

[1] Project 23 (*Sovetskii Soiuz*) is included for comparison.

[2] Starred figures (*) have been measured from drawings or estimated based on Project 23.

Sources:

Vasil'ev, *Lineinye korabli tipa 'Sovetskii Soiuz'*, 129, 131, 133; Vasil'ev, 'Lineinye korabli VMF SSSR predvoennykh proektov', 59, 64; Gribovskii, 'Pervaia poslvoennaia korablestroitel'naia programma', 7.

.... We have every opportunity over the approximately two years remaining to modify Project 23 .... It is necessary to begin the development of the TTZ for the ship quickly, all the more so, ... since the Shipbuilding Industry (TsKB-4 and TsNII-45) will take the initiative and present their own [fully] developed design for a battleship ....[3]

The last point is particularly interesting. Frolov was in effect saying: 'unless we act first, we're going to have to settle for another unsatisfactory battleship designed by the Shipbuilding Industry'. He enumerated the areas of Project 23 that were in need of improvement:

- range to be increased to 10,000nm
- maximum speed to be not less than 30 knots
- the number of 100mm anti-aircraft guns to be increased to twelve
- the machinery to be improved
- protection against torpedoes to be improved
- the armour protection scheme to be simplified
- the possibility of increasing the number of main battery turrets to be explored.

But when Isakov discussed Frolov's memorandum with the Navy commissar, Fleet Flagman 2nd Rank N G Kuznetsov, they agreed that it would be 'inopportune' to initiate a new battleship design – this would have been tantamount to admitting that the Project 23 design was flawed, a potentially dangerous move in a nation still recovering from the purges of the late 1930s.

Project 23*bis*

Nevertheless, Frolov's concerns about the Shipbuilding Industry stealing a march on the Navy were well-founded, as on 19 September 1939 – just three weeks after his memo – the deputy commissar of the Shipbuilding Industry ordered TsNII-45 to study new battleship designs. The memo instructed TsNII-45 to work up materials establishing 'the tactical-technical requirements for the battleships of the second series' – the assumption seems to have been that the original OTZ for Project 23 were still valid. The designers were to base their work on the experience gained in designing Projects 23 (*Sovietskii Soiuz*) and 69 (the *Kronshtadt*-class battlecruisers), as well as ongoing experimental work. The goals were:

- an increase in speed to 30 knots
- an increase in the 100mm battery from eight guns to twelve
- improvements to the anti-torpedo protection system
- a simplification of the armour protection scheme without reducing its resistance
- the elimination of various other defects that had been noted while developing Project 23.

These aims were similar to those Frolov had enumerated, with several exceptions – most notably the absence of any reference to improving the range, revisiting the propulsion plant, or increasing the number of main-battery turrets. What the Shipbuilding Industry was aiming for was not a completely new design, but what in the US Navy of today would be termed a 'Flight II' version of Project 23, retaining its main features (basic structural configuration, propulsion plant, armament) while making incremental improvements in other areas. Given the relatively slow Soviet process, an interim project of this type was the only way to ensure that the design would be ready by the time *Sovetskii Soiuz* and *Sovetskaia Ukraina* cleared the slipways. This relatively modest goal was reflected in the designation given to the work as Project 23*bis*, the '*bis*' being a borrowed French term for 'repeat' or 'encore'.

Within TsNII-45 the work was assigned to a team headed by naval constructor L A Gordon; the most difficult task his group faced was increasing the speed from 28 to 30 knots. In Project 23 a sub-optimal hull form had been deliberately accepted, primarily to maintain a deep torpedo protection system abreast the forward main-battery turrets. Since Project 23*bis* had to retain Project 23's propulsion plant, the only way to increase the speed was to improve the hull form. TsNII-45 therefore conducted a series of model tests at its experimental basin, and as a result the pre-sketch design showed a hull 16.5m longer than that of Project 23. This increased the length-to-beam ratio from 7.14:1 in Project 23 to 7.49:1, and allowed a finer hull form with a block coefficient of 0.615 instead of 0.658. This of course meant a greater displacement, but the tests indicated that 23*bis* could still achieve 30 knots at the machinery's normal output (201,000shp), and 31 knots with forced draught (231,000shp), compared with 28 and 29 knots respectively for Project 23. The improved hull form also gave a slight improvement in fuel economy at higher speeds, although it was equivalent to that of Project 23, or even slightly less, at lower speeds due to the increased frictional resistance of the greater immersed area of the longer hull.

The next item on TsNII-45's agenda was increasing the 100mm battery from four twin turrets to six, as originally specified for Project 23. In that design the assumed blast effects of the 406mm guns at extreme angles of training, combined with strict guidelines on locating magazines, had forced the designers into a very cramped arrangement of the secondary and heavy AA batteries, leaving space for only eight 100mm guns in four twin turrets. Gordon's team, contending with the same blast criteria, gained some space by slightly increasing the distance between the second and third main battery turrets, but more importantly they replaced the six twin 152mm turrets of Project 23 with four triple turrets of the type designed for the contemporary Project 68 (*Chapaev*) class light cruisers. This made it possible to insert two additional 100mm turrets, but required a 0.7m increase in the waterline beam. The hull volume gained was used to reduce the height of the superstructures by one deck level and arrange the magazines of the auxiliary batteries more conveniently.

WARSHIP 2022

Project 23*bis* 9-Gun Variant

Displacement: 66,800 tonnes full load
Length: 278m wl
Beam: 37.1m wl
Horsepower: 201,000shp = 30kts
231,000shp = 31kts
Armament: 9 × 406mm (3 × III)
12 × 152mm (4 × III)
12 × 100mm (6 × II)
32 × 37mm (8 × IV)

© John Jordan 2020

118

AFTER THE *SOVETSKII SOIUZ*: SOVIET BATTLESHIP DESIGNS 1939–1941

Project 23*bis*, December 1939. This design was drawn up by TsNII-45, a warship research bureau. Using the same machinery as *Sovetskii Soiuz*, the hull was lengthened by 16.5m to increase speed from 28 to 30 knots. Note the extremely long runs of the wing propeller shafts (about 113m) due to the placement of the engine rooms forward of the boiler rooms. Externally the most obvious change from *Sovetskii Soiuz* was the concentration of the secondary battery in four triple turrets instead of six twins, which made it possible to boost the number of 100mm HA guns from eight to twelve. The aircraft hangar was beneath the quarterdeck, covered by a sliding hatch, and the aircraft would have been lifted out by the qua'terdeck cranes. Note that some of the discrepancies in published drawings, including the placement of major bulkheads, have been reconciled.

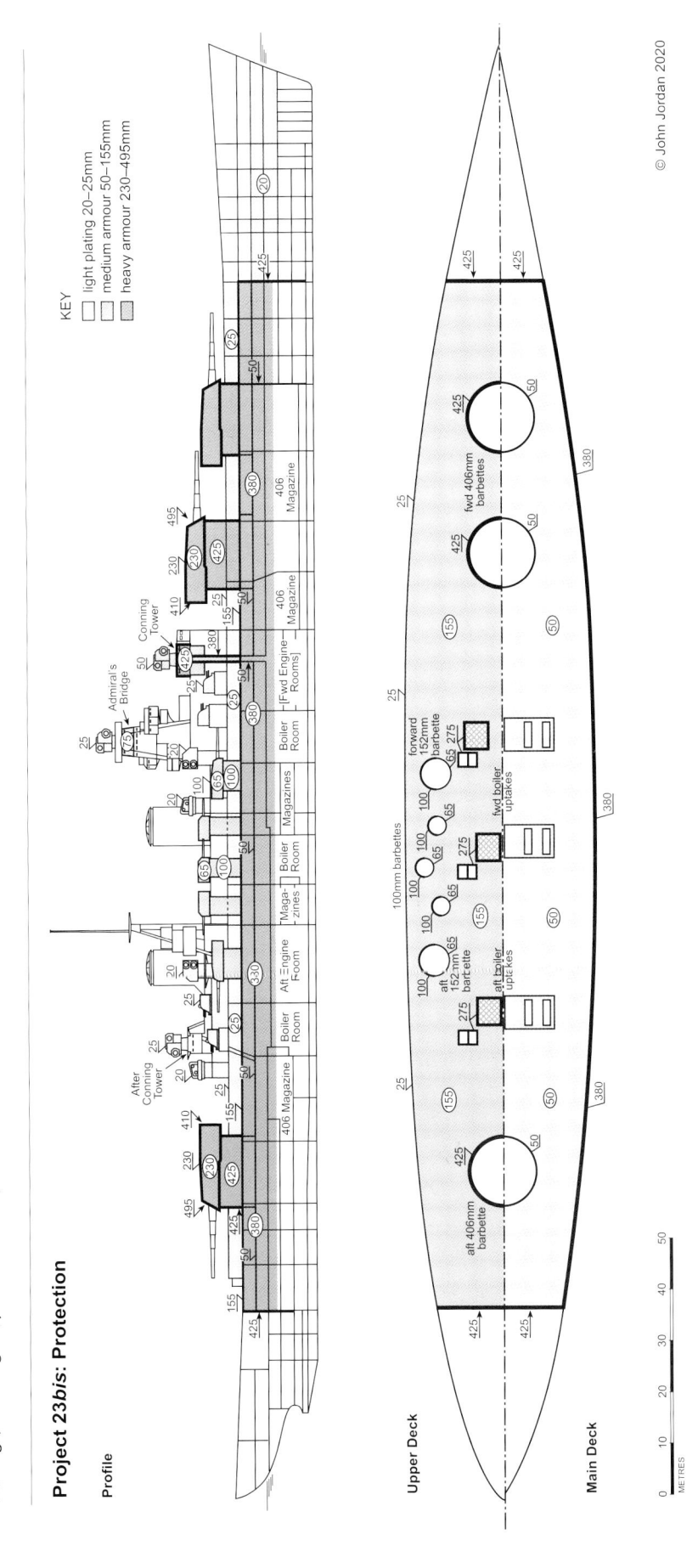

Project 23*bis*: Protection

Profile

Upper Deck

Main Deck

Project 23*bis*: armour scheme; thicknesses are in millimetres. The armoured citadel comprised a uniform 380mm belt inclined at 8 degrees covered by a 155mm upper deck, with a 50mm 'splinter' deck at main deck level. It was closed off fore and aft by 425mm transverse bulkheads. The citadel protected 72.5 percent of the waterline, extending forward of turret No 1 to cover the diesel generators and aft of turret No 3 over the steering gear, making it much longer (and heavier) than in contemporary foreign battleships. There are some inconsistencies in published drawings regarding the protection of the main battery directors; the thicknesses shown here seem the most plausible.

WARSHIP 2022

Project 23*bis* 12-Gun Variant

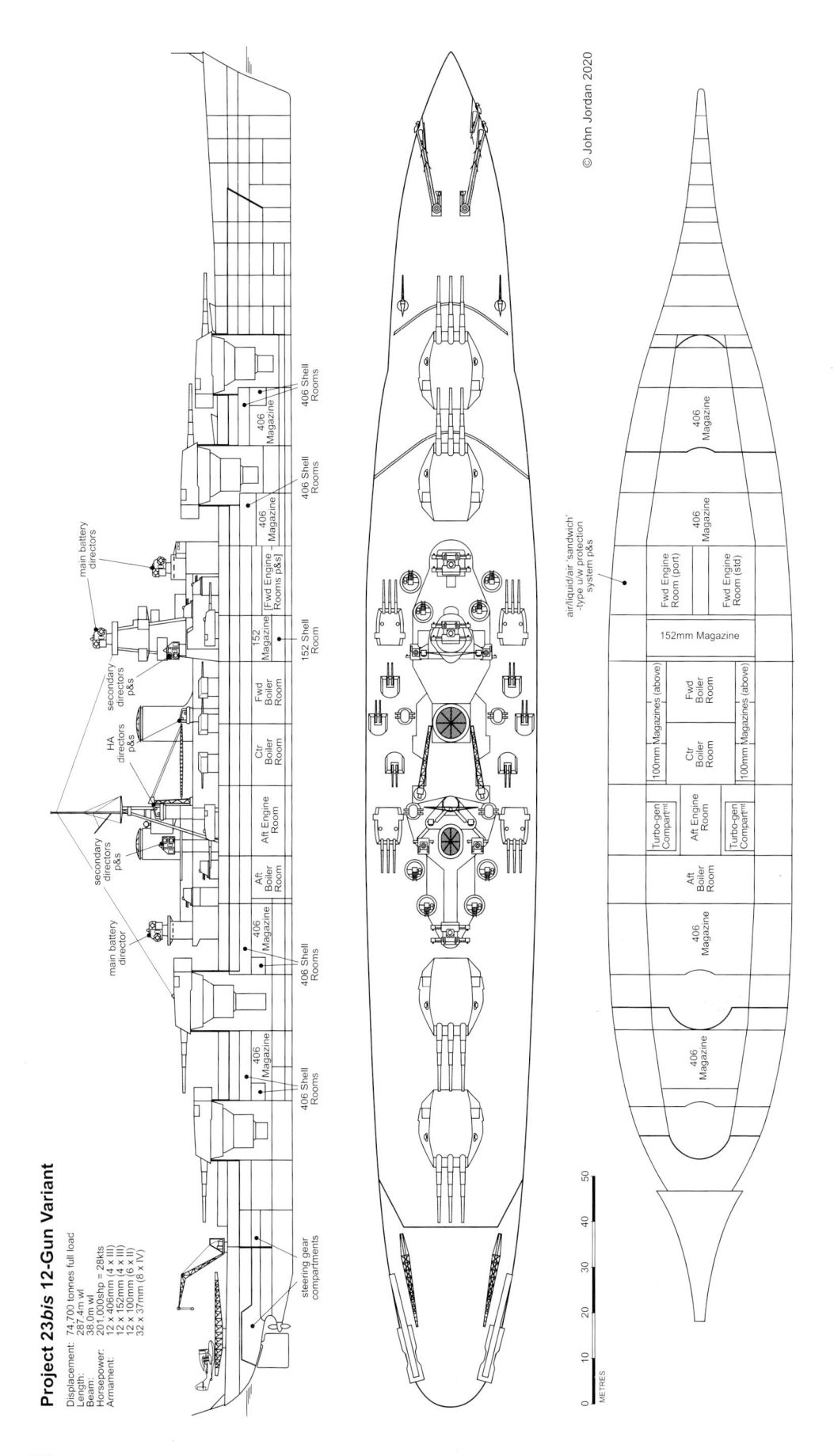

Displacement: 74,700 tonnes full load
Length: 287.4m wl
Beam: 38.0m wl
Horsepower: 201,000shp = 28kts
Armament: 12 x 406mm (4 x III)
12 x 152mm (4 x III)
12 x 100mm (6 x II)
32 x 37mm (8 x IV)

© John Jordan 2020

Project 23*bis*, 12-gun variant. This differed from the 9-gun version in retaining several features of the *Sovetskii Soiuz* design, including an armour belt that varied in thickness from 420mm to 375mm over the citadel, an inclined forward transverse bulkhead, and a centreline rudder. Its great size and relatively low speed (28 knots) led to its rejection at an early stage.

The third major issue was simplifying the protection scheme. In Project 23 there were no fewer than twenty-five different plate thicknesses; the main belt alone employed 420mm, 406mm, 390mm, 380mm and 375mm plates. This complex arrangement had been devised to meet the requirement of providing uniform protection when fighting an enemy 35–50 degrees off the bow; as the hull grew broader the thickness of the plates could be reduced because a shell coming from forward bearings would strike at more and more oblique angles. However, the fabrication of so many different types of extremely thick plate posed technological problems for Soviet armour plants still struggling with basic manufacturing processes. Another defect of Project 23 was an odd 'step down' in the 380mm main belt abaft the third 406mm turret; above this was 180mm side armour. This clumsy arrangement had been adopted to reduce weights aft because trim problems had arisen as the design progressed; despite this, in its final version the ship still had a 0.6m trim by the stern, which was considered excessive.

In Project 23*bis* these problems were resolved by having a uniform belt of 380mm inclined at 8 degrees instead of 5 degrees as in Project 23. The increased inclination improved the belt's resistance to penetration while saving sufficient weight to eliminate the 'step down' in the belt aft. Calculations indicated that this would make the belt impenetrable by a nominal 1,000kg armour-piercing 406mm shell with an initial velocity of 880m/sec striking from angles of 35–50 degrees off the bow at ranges greater than 80 cables (16,000 yards) instead of 84–88 cables (16,800–17,600 yards) in Project 23. The horizontal protection was identical in both designs, the decks becoming vulnerable to the same 1,000kg shell at ranges greater than 155 cables (31,000 yards). In Project 23*bis* the armour arrangements forward were also simplified, the main belt being extended towards the bow, eliminating the need for Project 23's 220mm water-line belt forward; a single 425mm transverse bulkhead closed off the citadel, providing better protection against end-on fire than the two separate bulkheads of Project 23.

Underwater protection was another feature that required improvement. In Project 23 the Italian Pugliese system had been adopted for most of the length of the citadel because of its comparative lightness, but towards the stern the hull lines and propeller runs had ruled out this system, so a multi-bulkhead system on the American pattern had been employed for the aftermost section. However, trials with two full-size and fifteen 1:5 scale caissons, completed in the autumn of 1939, demonstrated that a liquid-loaded multi-bulkhead system offered superior resistance to torpedo explosions. It was therefore adopted in Project 23*bis* throughout the length of the citadel. This form of side protection was also considerably easier to build and repair than the Pugliese. Calculations indicated that the new design should have been able to survive two torpedo hits (750kg TNT warheads or their equivalent) on the same side of the ship even with the unarmoured above-water hull destroyed.

A final improvement in Project 23*bis* worth noting concerned the rudder arrangements. In Project 23's original design there was a large rudder abaft the central propeller with smaller side rudders in the wake of the propellers for the wing shafts. However, tests with a large self-propelled model had shown that if the big rudder became jammed due to accident or battle damage, the smaller rudders could not overcome its effects and the ship would turn in circles. This was especially worrying since the steering gear of the big rudder was outside the citadel and had only splinter protection. In Project 23 the decision had reluctantly been taken to delete the central rudder entirely, which increased the ship's tactical diameter from 3.18 ship-lengths to 4.7. In Project 23*bis* the central rudder was eliminated from the start; presumably the surface area of the side-rudders was increased to compensate. This had the added benefit of freeing up space below the quarterdeck, so the aircraft hangar could be located there instead of being placed under the guns of the after 406mm turret. The hangar would have accommodated two aircraft, with a third on the catapult.

TsNII-45 also briefly studied a twelve-gun version of 23*bis* with a fourth main battery turret superimposed above the after turret. Not surprisingly, it was almost 8,000 tonnes heavier than the nine-gun design. It employed the same graduated scheme of belt armour as

Project 23*bis*: Midship Section

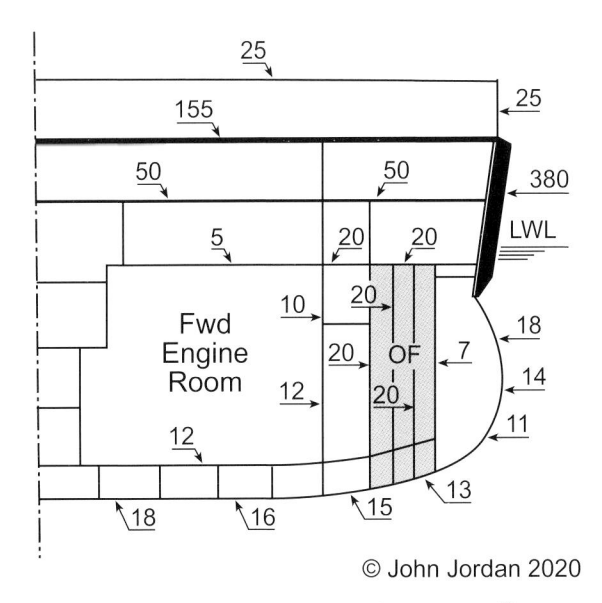

Project 23*bis*: midship cross-section; thicknesses in millimetres. The 380mm belt was thinner than in the *Sovetskii Soiuz* class, so the angle of inclination was increased from 5 degrees to 8 degrees to compensate. The clumsy mixture of Pugliese and 'American' multi-bulkhead systems used in the underwater protection of the earlier design was replaced by a full-length liquid-loaded multi-bulkhead system.

Project 23, suggesting that the design was abandoned before work on the nine-gun version was completed; it was considered too large and, at 28 knots, too slow to meet the Navy's requirements.

The Next Step: Project 23NU

Gordon's team at TsNII-45 had managed to accomplish virtually everything they had been asked to do, all with only a modest increase in displacement of 1,650 tonnes (see Table 1 column B, and Table 5). The design work was completed in December 1939 and presented to the Shipbuilding Industry and the Navy in January 1940. The Navy seemed to consider the results of TsNII-45's studies promising, and on 3 March 1940 Fleet Flagman 2nd Rank L M Galler, chief of the Main Naval Staff, approved the 'OTZ for the Improvement of Design no 23'.[4] Taking Project 23*bis* as a basis, it outlined the following changes from Project 23:

> **Armament:**
> a) The number of 152mm guns should remain unchanged, but the possibility of using four triple turrets instead of six twin turrets should be studied, maintaining the ability to bring six guns to bear on forward bearings.
> b) The number of 100mm guns should be increased from four twin mountings to eight; fire control should allow the engagement four targets simultaneously.
> **Speed:**
> Maximum speed is to be increased to 30 knots.
> **Protection:**
> a) The requirements for armour protection remain unchanged, but a more 'rational' arrangement of the armour should be studied.
> b) Underwater protection should be able to withstand a charge of 500kg instead of the 750kg warhead assumed in Project 23 [see Table 2 column A].

The reasoning behind the last point is not stated, but it may be that the 750kg requirement was considered excessive, given that the largest Soviet torpedo warhead in service was 400kg, and those of other nations were of similar size.[5] Even the most powerful torpedo in the world, the Japanese 24in Type 93, of which the Soviets were almost certainly unaware, had a warhead of only 490kg.

On 5 March the Navy's Shipbuilding Administration forwarded Galler's OTZ to TsKB-4, which had been responsible for Project 23. The new design was designated Project 23NU (for *novyi ulushennyi* = 'new improved'), but unfortunately information on the design is fragmentary and no drawings are available. It is possible that no complete pre-sketch designs were drawn up, since the bureau seems to have submitted only a draft of the TTZ for the Navy's approval.

At this point the Navy clearly expected a more-or-less direct development of the Project 23*bis* design. This is confirmed by a memorandum from Navy Commissar Kuznetsov to the Shipbuilding Industry dated 25 July 1940; he wrote that '… the Navy's OTZ for a new battleship design in 1940 will not be issued … and the construction bureau should work on Project 23*bis* (the modernisation of Project 23), taking into account German, American, and domestic experience and the experimental work already carried out'.[6] In other words, the Navy believed that Project 23*bis* could serve as the basis for a satisfactory design, so new staff requirements were unnecessary. The German experience Kuznetsov mentioned was obtained through the Soviet delegation sent to Germany in the wake of the August 1939 Molotov-Ribbentrop Non-aggression Pact; headed by shipbuilding commissar Tevosian, its naval experts and constructors had inspected *Scharnhorst* and *Bismarck* as well as other warships. As one historian has observed, to Soviet naval constructors still lacking in practical experience it was 'the cumulative mass of small observations they had made of German practice' that proved to be of the most valuable result of the visit to Germany.[7] The delivery to Leningrad of the incomplete heavy cruiser *Lützow* in April 1940 provided even more opportunities to study the technical features of German ship construction. As for American experience, the Soviets had studied the Gibbs & Cox design for a giant hybrid carrier-battleship in November 1938, and while they did not purchase it, they had probably picked up some useful details; moreover, there were ongoing espionage efforts in the United States to obtain information about warships.[8] For example, as early as 1931 Soviet agents had been directed to obtain information about the American 'stable vertical control apparatus for fixing the aim of a gun on target regardless of the motion of a ship', and indeed a 'gyrovertical' (*giro-vertikal*) was incorporated in post-Project 23 designs.[9]

The Shipbuilding Industry was only too happy to agree that a new design from scratch was unnecessary, and in September Kuznetsov reported to A A Zhdanov, secretary to the Central Committee of the Communist Party, who was responsible for naval affairs and shipbuilding:

> At the present time neither the NKSP [People's Commissariat of the Shipbuilding Industry], nor the NK VMF [People's Commissariat of the Navy] is ready for full-scale design work on a new battleship.[10]

Meanwhile TsKB-4 was making slow but steady progress on the new design, despite being hampered by the fact that it was still heavily involved with the working drawings needed for the construction of the Project 23 ships. The most striking element of Project 23NU was a radical change in the belt armour. In the *Sovetskii Soiuz* class it had been inclined at 5 degrees, in Project 23*bis* it was 8 degrees, but in 23NU the angle of inclination jumped to 15 degrees. In its cover memorandum TsKB-4 noted that this secured the maximum weight savings while allowing a 40mm reduction in the thickness of the belt plates; however, 'with the goal of improving the ship's protection it was decided to limit the reduction to only 30mm'.[11] As a result, the belt, which in Project 23 had

Table 2: Battleship Design Requirements 1940

	A OTZ Main Naval Staff	B Draft TTZ TsKB-4	C TTZ Main Naval Staff
Date	3 Mar 1940	Sep 1940	27 Sep 1940
Displacement:			
standard	N/S	N/S	59,650 tonnes
full load	N/S	up to 70,000 tonnes	N/S
Length (wl)	N/S	less than 300m	as Project 23
Beam (wl)	N/S	less than 40m	as Project 23
Draft (max)	N/S	10.1m	N/S
Armament:	9 – 406mm (3xIII)	9 – 406mm (3xIII)	9 – 406mm (3xIII)
	12 – 152mm (6xII or 4xIII)	12 – 152mm (6xII)	12 – 152mm (6xII)
	16 –100mm (8xII)	16 –100mm (8xII)	16 –100mm (8xII)
	?? – 37mm	44 – 37mm (11xIV)	32 – 37mm (8xIV)
		20 – 12.7mm	
Aviation:			
catapults	N/S	1	N/S
aircraft	N/S	3	N/S
Protection:			
main belt	as Project 23 with improvements	220–390mm inclined 15°	as Project 23 with improvements
decks (total)	N/S	230mm	N/S
Underwater protection:			
system	N/S	improved 'American'	N/S
depth amidships	N/S	6.5–7.5m	6.5–7.5m
designed resistance	500kg TNT	750kg TNT	N/S
Machinery:			
horsepower	201,000shp	201,000shp	201,000shp
speed	30 knots	28–29 knots	28 knots
range at 14.5kts	N/S	'about' 8,000nm	5,960nm

Notes:
OTZ = *Operativno-takticheskoe zadanie* (Operational-Tactical Requirements)
TTZ = *Taktiko-tekhnicheskoe zadanie* (Tactical-Technical Requirements)
N/S = not specified.

Source: Vasil'ev, *Lineinye korabli tipa 'Sovetskii Soiuz'*, 131.

---

varied from 420mm to 375mm, ranged from 390mm to 345mm in Project 23NU. Part of the weight saved was used to make the plates larger, allowing the side armor to extend 100mm farther below the waterline – although this might have been less a matter of increasing the extent of the armour than a geometrical necessity due to the steeper inclination of the plates. The reduced thickness of the belt apparently saved enough weight to accommodate the doubling of the 100mm battery and, as specified by the Navy, TsKB-4 also produced variants with twin and triple 152mm turrets. The light AA battery was also boosted by twenty 12.7mm machine guns.

For the underwater protection system TsKB-4 opted to go rogue, completely ignoring the Navy's requirements; as it explained in its cover memorandum:

> With regard to the PMZ [*protivominnaia zashchita* = anti-torpedo protection], TsKB-4 decided not to fulfil the requirement of basing it on a 500kg explosive charge …, since TsKB-4 knows that future torpedoes will have a greater power than is indicated by the GMSh [Main Naval Staff]. The side protection in the design should have a guaranteed resistance to an explosion of 700 kilograms of trotyl [TNT].[12]

This was all well and good, but it came at a considerable cost in increased displacement and reduced speed. Of four hull forms studied, only one met the 30-knot speed requirement, and it tipped the scales at over 70,000 tons and was almost 300m long (see Table 2 column B, and Table 3). This came as something of a shock to the Navy when the draft TTZ compiled by TsKB-4 were presented in September 1940, especially given that TsNII-45 had (optimistically) indicated that the much smaller Project 23*bis* could meet the requirements. As a result the head of the Navy's Shipbuilding Administration, Engineer-Rear Admiral A A Zhukov, rejected TsKB-4's proposed TTZ.[13]

Table 3: Project 23NU Variant Hull Forms circa September 1940

	Project 23*bis*[1]	1-609	1-615	1-632	1-650
Full load displacement	66,800 tonnes	70,600 tonnes	70,300 tonnes	69,500 tonnes	68,000 tonnes
Waterline length	278m	298m	295m	284m	268m
Beam:					
waterline	37.1m	38.5m	38.4m	38.3m	38.6m
over bulges	39.9m	40m	40m	39.9m	39.8m
Draft	10.27m	10.1m	10.1m	10.1m	10.1m
Speed with 200,000shp	30 knots	31.1 knots	29.7 knots	29.0 knots	27.9 knots
Froude number	0.296	0.295	0.284	0.282	0.279
Admiralty coefficient	217	256	225	205	183

Note:
[1] Project 23*bis* is included for comparison.

Source: Vasil'ev, *Lineinye korabli tipa 'Sovetskii Soiuz'*, 135.

Back to the Drawing Board: Project 24

The Main Naval Staff, having finally reconciled itself to the fact that it could gain only minor improvements over Project 23, quickly composed new tactical-technical requirements. These were completed on 27 September 1940 and were issued to TsKB-4 within a few days (see Table 2 column C, and Table 4). TsKB-4 referred to the new design as Project 24, but this seems to have been an 'in house' designation and caused some confusion, the bureau noting that some documents still referred to it as Project 23NU.[14] TsKB-4 worked quickly: it had taken more than six months to produce Project 23NU, but the new design series was ready after only a month, at the end of October 1940. This speedy result may have been because the designs were 'supplementary variants' (*dopolnitel'nye varianty*), suggesting that these were developments of earlier work – indeed, in many ways they appear to be modifications of Project 23 (see Table 1 column D).

The four pre-sketch designs were designated D-6-6.5, D-6-7, D-12-6.5, and D-12-7, with the first digit indicating the number of boilers and the second the depth of the underwater protection system in metres. Externally, they were very similar, differing primarily in the locations of the 152mm and 100mm turrets and other details. Their internal differences were more significant, and are noted in the captions to the drawings.

In the accompanying explanatory materials TsKB-4 pointed out that the new TTZ 'establish the permissibility of several serious departures' from those allowed in previous designs.[15] Most crucial was the elimination of the 30-knot speed requirement (Table 4 point 2), which meant that Project 23's hull form and basic structural arrangements could be retained, thereby considerably easing the task the design bureau faced.

TsKB-4 also noted that due to the changed international situation – that is, the outbreak of war in western Europe and the consequent irrelevance of treaty restrictions – standard displacement had lost its meaning. So despite being limited to the same displacement as Project 23 (point 1), the bureau assumed that this meant that the full load displacement was the significant criterion. This allowed a little leeway in meeting the size requirement by tinkering with fuel loads, as specified in points 12 and 13. By making more use of fuel-efficient diesel generators, the load on the main machinery and turbo-generators could be reduced, making it possible to decrease the fuel bunkerage by 350 tonnes. Combined with 500 tonnes or so saved in hull weight, this helped to compensate for the doubling of the 100mm battery (point 5), which was responsible for a weight increase of 454 tonnes for the additional four twin gun mountings plus a further 823 tonnes for the barbette protection, ammunition hoists, ammunition, and so forth. Thanks to the reduced fuel load, while standard displacement rose by about 730 tonnes, at full load it increased by only about 385 tonnes (see Table 5).

Point 7 of the new TTZ also resolved another problem caused by increasing the 100mm battery from four twin turrets to eight: finding above-deck space for the additional turrets. In Project 23 the 152mm guns had be located well clear of the blast from the 406mm guns when they were trained at extreme angles, but due to a lack of data very generous allowances for these blast zones had been adopted. This squeezed both the secondary and heavy AA batteries into a very cramped space amidships. However, among the data obtained from the visit to Germany was the 'Maximum [blast] pressure according to the materials from the report on the L/K [*lineinyi korabl'* = battleship] *Bismarck*', which suggested that the blast zone could be substantially reduced; instead of the 12m x 14m zone in Project 23, it was only 4m x 8m in Project 24.[16] So whereas in Project 23 the axis of the forward 152mm turret on each beam was about 27m from the muzzles of the 406mm guns of turret no 2 when the latter was trained to 150 degrees off the bow, in Project 24 they were only 12m away; similarly, the after 152mm turrets in Project 23 were 26m from the muzzles of turret no 3 when trained to 30 degrees off the bow, but in Project 24 they were only 20m away. This opened up space amidships for the doubled 100mm battery. However, TsKB-4 also hedged its bets, noting that it still did 'not possess reliable data on the effect of a gas cone [blast] on turrets and equipment...'[17] This caveat probably reflects uncertainties about the applicability of the German data to the Soviet 406mm guns.

Finding above-deck space for the additional 100mm turrets was one thing, but internal volume was also needed for the expanded magazine capacity they required. Here again the revised TTZ came to the rescue by relaxing the restrictions on the placement of magazines (point 8). Despite the higher temperatures associated with machinery spaces, magazines could be located adjacent to them so long as they were provided with 'partitions' (*ie* insulated by having double walls with an air space between). The magazines for the secondary and high-angle guns were in some cases also closer to the sides or the splinter deck than had been allowed in Project 23.

The only changes to the fire-control system were the addition of a second central gunnery post and the provision of a fourth stabilised high-angle director (SPN) for the 100mm battery; the latter reflected the OTZ approved back on 3 March 1940, which had specified the ability to engage four aerial targets simultaneously.

Point 6 specified that the armour protection scheme should repeat that of Project 23; as in that design, the inclination of the belt was 5 degrees and it employed the same thicknesses, varying from 420mm to 375mm along the ship's side. One major change, however, was the addition of a longitudinal 50mm splinter bulkhead approximately 1.75m inboard of the main belt, similar to that used in Russia's dreadnoughts of the First World War. This was the result of concerns that had recently arisen over the accuracy of the armour penetration formulae used in determining the plate thicknesses in Project 23. The Soviet designers had used the Jacob de Marre formula, which had originally been developed in the early 1890s, with a special coefficient ('K') to adjust the formula for face-hardened plate.[18] However, the value used for K in Project 23 had been called into question, recent experiments having shown that the interaction between shell and armour was much more complex than had been previously believed. This meant that Project 23's level of protection might be less than had been assumed, but the manufacture of thicker plates was ruled out by technical considerations. So the splinter bulkhead was the only solution the designers could find to improve the belt protection. In proposing this, however, TsKB-4 admitted that 'the thickness of this barrier is a doubtful quantity', so 50mm was little more than a guess.[19] To find the weight for this new bulkhead TsKB-4 eliminated Project 23's 50mm bulkhead between the magazines of turrets 1 and 2 and reduced the thickness of the forecastle deck from 25mm to 20mm. The horizontal protection was otherwise identical to that of Project 23, with a 155mm upper deck and a 50mm main deck.

The Navy also made a major concession in point 6; throughout the post-*Sovetskii Soiuz* design process it had hoped to replace the 180mm armour above the sides of the 'step-down' in the belt abaft the after main battery turret by extending the 380mm belt up to the upper deck; now the Navy agreed not only to the retention of the 'step-down', but also permitted the elimination of the 180mm armour. This saved a good deal of weight that was needed for improvements elsewhere.

In accordance with the requirements there were two variants of the side-protection system, one 6.5m deep and the other 7m deep; based on measurements of the available drawings, it appears that, while the waterline beam remained the same in all the designs, in those with 7m-deep bulges the maximum beam was 1m greater, which would probably have reduced speed somewhat. Using data from recent experiments with caissons, TsKB-4 developed 'box-shaped' bulges based on those of the American aircraft carrier *Saratoga*.[20] But the heavy bulk-

---

Table 4: **The Main Naval Staff's TTZ, 27 September 1940**

1. Displacement not to exceed that of Project 23.
2. Maximum speed to be the same as that of Project 23 (that is, 28 knots, 29 with forced draught).
3. The length of the citadel to be no greater than that of Project 23.
4. The depth of the underwater protection system to be 6.5–7.0m.
5. The secondary battery to comprise twelve 152mm guns in six twin turrets; the long-range high-angle battery to be sixteen 100mm guns in eight twin turrets.
6. The arrangement and thicknesses of the armour to be as in Project 23, except that the 180mm vertical armour aft is to be deleted.
7. The secondary battery could be located within the main battery's blast zones as used in Project 23.
8. Magazines for all calibres could be located in spaces contiguous to boiler and engine rooms as long as they had partition bulkheads with an air space.
9. In addition to a variant with six boilers, one should be developed with twelve boilers.
10. Reduce the lengths of the propeller shafts as much as possible.
11. Locate the boilers as low as possible.
12. A portion of the turbine-driven auxiliary machinery could be replaced with electrically-driven auxiliaries.
13. At economical speed the diesel generators could be run alongside the turbo-generators.
14. In other respects the machinery to remain unchanged from that of Project 23.

Source: RGA VMF, F r-441, op 3, d 443.

WARSHIP 2022

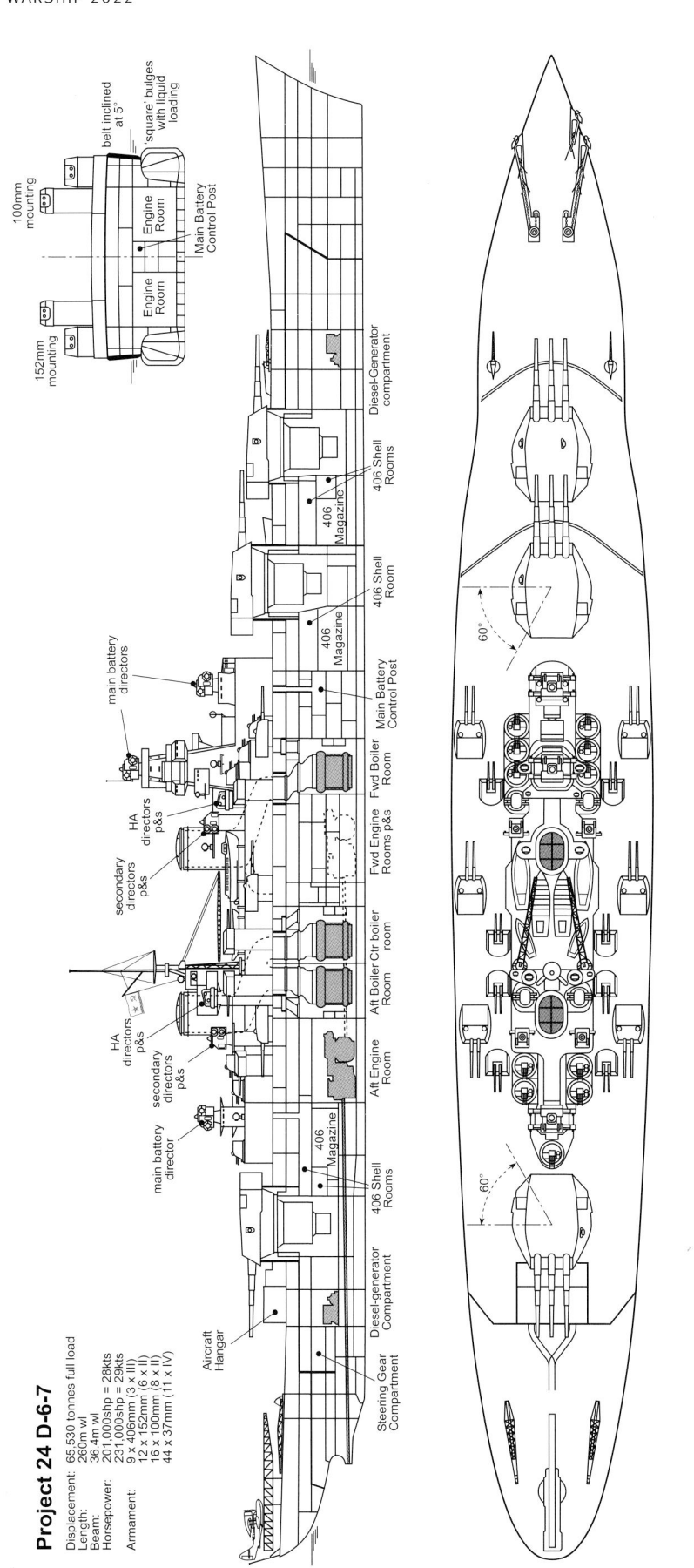

Project 24 D-6-7

Displacement: 65,530 tonnes full load
Length: 260m wl
Beam: 36.4m wl
Horsepower: 201,000shp = 28kts
231,000shp = 29kts
Armament: 9 × 406mm (3 × III)
12 × 152mm (6 × II)
16 × 100mm (8 × II)
44 × 37mm (11 × IV)

© John Jordan 2020

Project 24, variant D-6-7, October 1940. Drawings of three of the four variants in this series have come to light. In these designs, developed by TsKB-4, the battleship design bureau, the first digit indicated the number of boilers, the second the depth of the torpedo protection system in metres. All the Project 24 designs shared a hull form similar to that of the *Sovetskii Soiuz* class, but with 'square' rather than semi-circular bulges. The armour scheme was likewise based on that of the earlier design, including a belt that varied in thickness from 375mm to 420mm, an inclined forward transverse bulkhead, and a 'step down' in the height of the belt abaft turret No 3. However, the machinery layout was revised, with the boiler rooms now forward of the engine rooms; this served to shorten the shaft runs, one of the Navy's requirements. D-6-7 differed from other designs in the series in having two equal-sized funnels, the result of trunking the uptakes from the forward boiler room and auxiliary boilers (located on the upper deck inboard of the centre 152mm turrets) to the forward funnel, while those of the two after boiler rooms vented through the second funnel.

AFTER THE *SOVETSKII SOIUZ*: SOVIET BATTLESHIP DESIGNS 1939–1941

Project 24 D-6-6.5

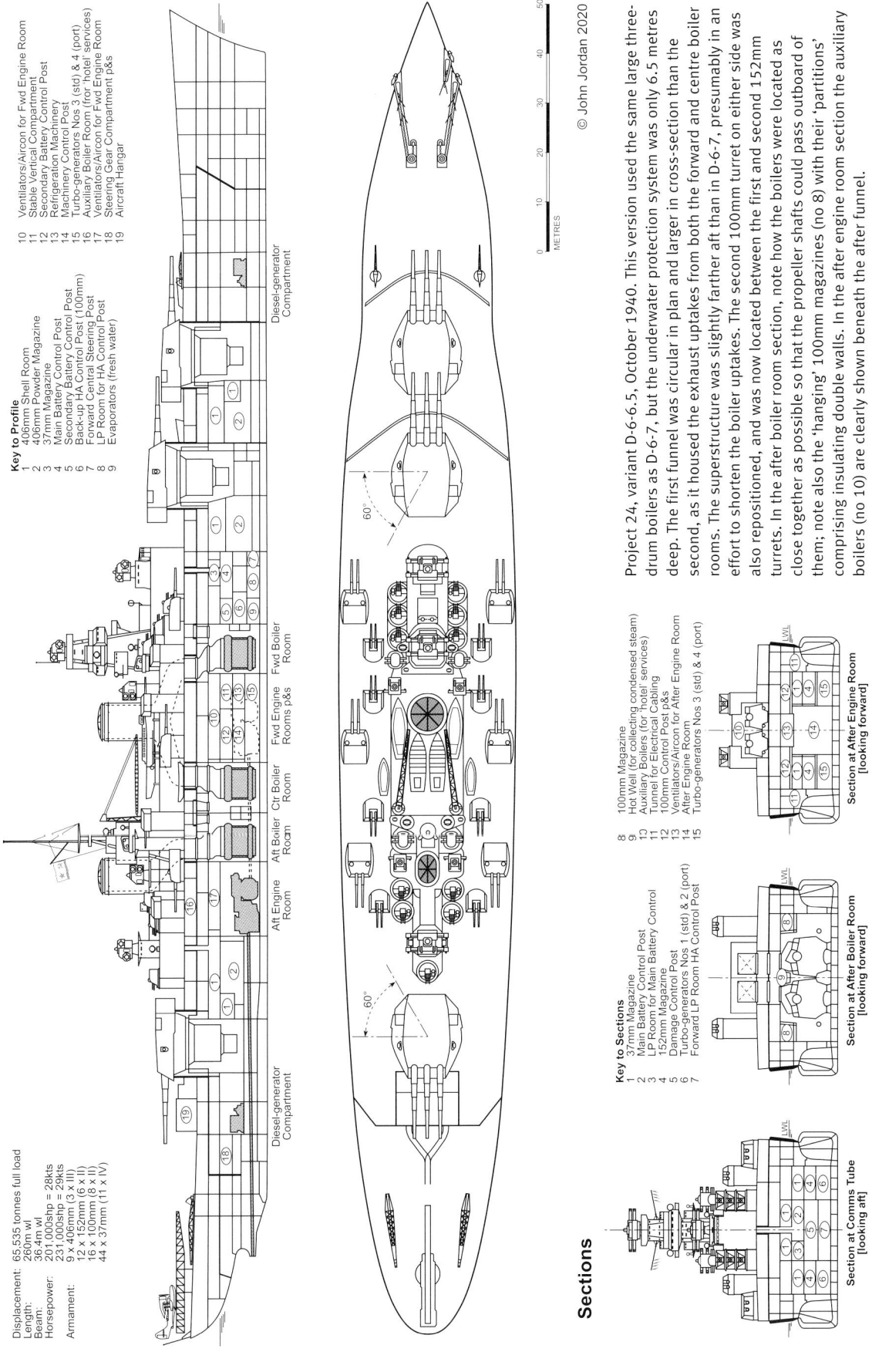

Displacement: 65,535 tonnes full load
Length: 260m wl
Beam: 36.4m wl
Horsepower: 201,000shp = 28kts
231,000shp = 29kts
Armament: 9 x 406mm (3 x III)
12 x 152mm (6 x II)
16 x 100mm (8 x II)
44 x 37mm (11 x IV)

Key to Profile
1 406mm Shell Room
2 406mm Powder Magazine
3 37mm Magazine
4 Main Battery Control Post
5 Secondary Battery Control Post
6 Back-up HA Control Post (100mm)
7 Forward Central Steering Post
8 LP Room for HA Control Post
9 Evaporators (fresh water)
10 Ventilators/Aircon for Fwd Engine Room
11 Stable Vertical Compartment
12 Secondary Battery Control Post
13 Refrigeration Machinery
14 Machinery Control Post
15 Turbo-generators Nos 3 (std) & 4 (port)
16 Auxiliary Boiler Room (for 'hotel' services)
17 Ventilators/Aircon for Aft Engine Room
18 Steering Gear Compartment p&s
19 Aircraft Hangar

© John Jordan 2020

Project 24, variant D-6-6.5, October 1940. This version used the same large three-drum boilers as D-6-7, but the underwater protection system was only 6.5 metres deep. The first funnel was circular in plan and larger in cross-section than the second, as it housed the exhaust uptakes from both the forward and centre boiler rooms. The superstructure was slightly farther aft than in D-6-7, presumably in an effort to shorten the boiler uptakes. The second 100mm turret on either side was also repositioned, and was now located between the first and second 152mm turrets. In the after boiler room section, note how the boilers were located as close together as possible so that the propeller shafts could pass outboard of them; note also the 'hanging' 100mm magazines (no 8) with their 'partitions' comprising insulating double walls. In the after engine room section the auxiliary boilers (no 10) are clearly shown beneath the after funnel.

Sections

Key to Sections
1 37mm Magazine
2 Main Battery Control Post
3 LP Room for Main Battery Control
4 152mm Magazine
5 Damage Control Post
6 Turbo-generators Nos 1 (std) & 2 (port)
7 Forward LP Room HA Control Post
8 100mm Magazine
9 Hot Well (for collecting condensed steam)
10 Auxiliary Boilers (for 'hotel' services)
11 Tunnel for Electrical Cabling
12 100mm Control Post p&s
13 Ventilators/Aircon for After Engine Room
14 After Engine Room
15 Turbo-generators Nos 3 (std) & 4 (port)

Section at After Engine Room [looking forward]

Section at After Boiler Room [looking forward]

Section at Comms Tube [looking aft]

127

Project 24 D-12-7

Displacement: 65,800 tonnes full load
Length: 260m wl
Beam: 36.4m wl
Horsepower: 201,000shp = 28kts
231,000shp = 29kts
Armament: 9 × 406mm (3 × III)
12 × 152mm (6 × II)
16 × 100mm (8 × II)
44 × 37mm (11 × IV)

Key
1 406mm Shell Room
2 406mm Powder Magazine
3 LP Room for HA Control Post (100mm)
4 Stable Vertical Compartment
5 Main Battery Control Post
6 37mm Magazine
7 LP Room for Main Battery Control
8 Damage Control Post
9 Forward Central Steering Post
10 Evaporators (fresh water)
11 Turbo-generators Nos 3 (std) & 4 (port)
12 100mm Magazine
13 152mm Shell Room
14 152mm Powder Magazine
15 Secondary Battery Control Post
16 Machinery Control Post
17 Ventilators/Aircon for After Engine Room
18 Steering Gear Compartment p&s
19 Aircraft Hangar

© John Jordan 2020

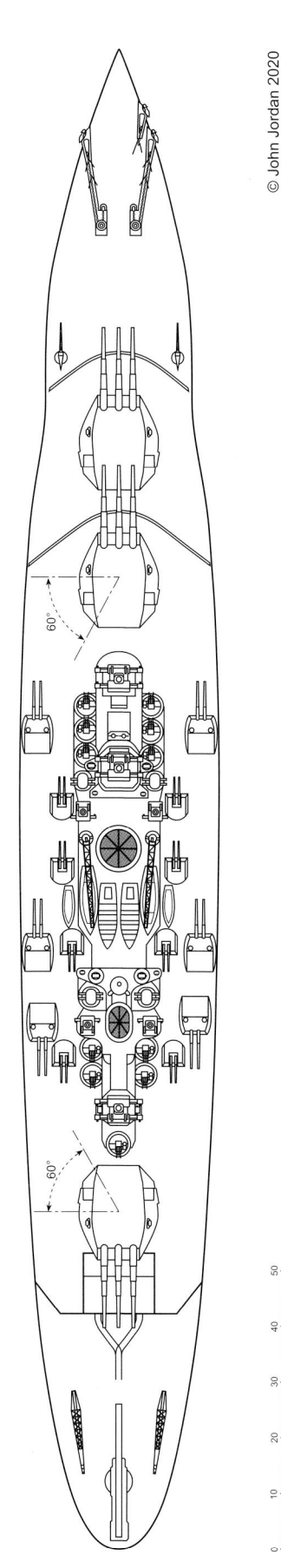

Project 24, variant D-12-7, October 1940. This design employed twelve of the smaller boilers developed for the *Kronshtadt*-class battlecruisers, arranged four athwartships in three boiler rooms. A short transverse compartment was inserted between the two after boiler rooms, and the centre 152mm turrets were moved aft so that they were directly above their magazines. The two forward 100mm turrets on either side were also shifted aft to clear the 37mm mounts. As a result the boat deck had to be rearranged, with the cranes facing aft instead of forward. In the original plans, the bow shows greater sheer than in the other variants.

head was similar to that of the Pugliese system, running first inward and then curving downward to the ship's bottom. There were two bulkheads within the bulge, and apparently the double bottom continued around the curve of the bilges parallel to the outer hull plating, acting as a third bulkhead. It is not clear if the system included liquid-loading of some of the compartments, but this seems probable.

As for the machinery plant, point 9 of the TTZ had specified the development of two options: one with six boilers of the same three-drum type used in Project 23, and one with twelve boilers of the smaller model used in the Project 69 (*Kronshtadt* class) battlecruisers. Except for their size, these had generally similar parameters to those of Project 23: the working pressure of 37kg/cm^2 was the same in both types, while their working temperature was slightly greater (380° vs 370°C), and their steam production was a little more than half that of the bigger boilers (90 tonnes/hour vs 162 tonnes/hour). TsKB-4 noted the combined steam production of the twelve smaller boilers was greater, and that they provided some advantages in terms of survivability. Nevertheless, the bureau recommended the six-boiler plant, noting that it took up less space, allowed a simpler arrangement of the machinery and pipework, and was 137 tonnes lighter.

Project 24 also satisfied the Navy's requirement for shorter propeller shafts than those of Project 23 (point 10), where their great length was considered a vulnerability. The earlier design had featured alternating engine and boiler rooms, as was standard practice in many navies of the era, but with the engine rooms forward of the boiler rooms – hence the centre shaft was 79m long, while the outboard shafts were 106m. The same layout had been adopted in Project 23*bis*, but in Project 24 the boiler rooms were forward of the engine rooms, which reduced the total lengths of the shafts to 65–81 metres (65–77m in variant D-6-6.5).

Point 11 of the requirements, that the boilers should be placed as low as possible, seems to have been a reference to the fact that in Project 23 all of the boilers had to be raised well above the inner bottom so that the wing propeller shafts could pass underneath them. In Project 24 these boilers were grouped closer to the centreline so that the wing shafts could run outboard of them, allowing them to be positioned closer to the inner bottom.

On the Eve of War

In Project 24 TsKB-4 had managed to meet all the requirements set out in the Navy's 27 September 1940 TTZ, but only because the Navy once again had to forego the long-desired 30-knot speed. This showed more clearly than anything else that Project 24 was still a compromise – with the launchings of *Sovetskii Soiuz* and *Sovetskaia Ukraina* originally planned for May–June 1941, there was insufficient time to produce a design from scratch that could incorporate all of the Navy's requirements. But while work had been proceeding on the various design studies, those launch dates had begun to slip. In January 1940 they were postponed to October 1941 – still reasonably close to the original dates – but by October 1940 it had become clear that work on the ships was progressing much more slowly than anticipated, due in large measure to late or incomplete deliveries of steel and components. So by the end of 1940 the launches were expected no earlier than June 1943. With the lead time now extended to almost three years an interim design such as Project 24 was no longer necessary – there was now sufficient time to develop a completely new and (it was hoped) far more satisfactory battleship design. As a result, on 27 December 1940 Engineer-Rear Admiral N V Isachenkov, head of the Navy's Shipbuilding Administration, signed off on the 'Plan of Design Work for the NK [People's Commissariat] of the Navy for 1941', which stated that the technical requirements for a new battleship design should be developed by July 1941, and a sketch design by March 1942. The plan included the note:

> Design work has demonstrated that the Project 23 battleship is not optimal for its displacement in terms of its combat power. Designing battleships is a very lengthy process and requires uninterrupted experimentation over the course of several years by means of developing sketch designs.[21]

The Main Naval Staff began developing completely new requirements in January 1941, which were approved by Navy Commissar Kuznetsov on 8 May. This 'preliminary OTZ' for a battleship provides some insight into the lessons the Navy had gleaned from earlier design work and reports from the war raging in western Europe. It stated that:

> … the basic purpose of the battleship is fighting the enemy's battleships, both in seas bordering our own territory and on the oceans. The battleship's secondary roles include suppressing large-calibre shore batteries, destroying positions on the enemy's flanks and bombarding targets in [the enemy's] rear.[22]

The heavy emphasis on supporting the army's seaward flank may be noted in passing; this was a regular feature of almost all Soviet warship requirements and reflected the prevailing trend in the Red Army's military theories. The OTZ went on to specify the following:

- The main and secondary batteries were to be as in Project 23.
- The heavy AA battery was to consist of sixteen 100mm guns in stabilised mountings.
- The light AA battery was to be 44 x 37mm.
- A variant of the design with the 152mm and 100mm guns replaced by twenty-four 130mm dual-purpose guns was to be studied.
- There were to be six aircraft, capable of being recovered while the ship was steaming at 18 knots (presumably by using something akin to the Hein mat).

Protection Systems

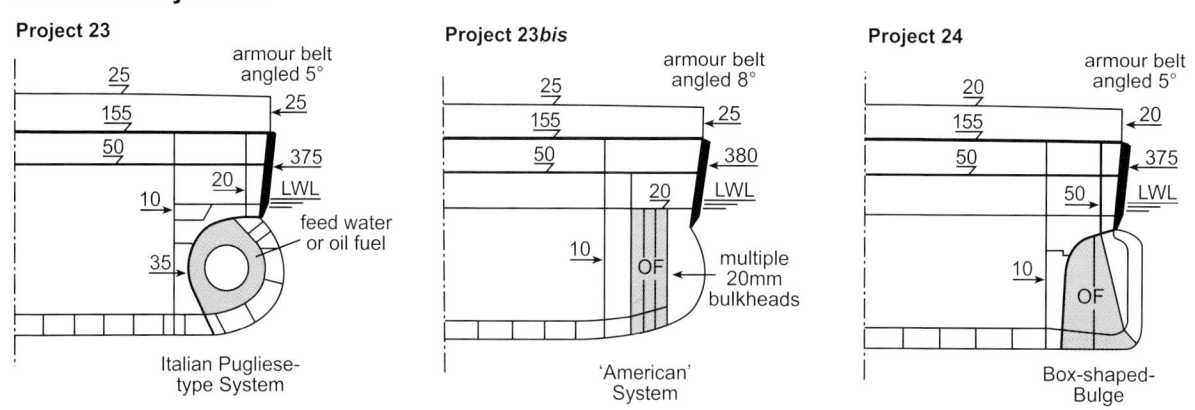

Comparative midship half-sections of Projects 23, 23*bis* and 24; armour thicknesses are in millimetres. The so-called 'American' liquid-loaded side protection system had proven superior to the Pugliese system in tests, so it was adopted in Projects 23*bis* and 24. In the latter, however, it took an unusual form, with one bulkhead sloping outward to the double bottom. Although none of the available drawings show its compartments filled with oil fuel, it is assumed that the system did indeed feature liquid loading.
(Drawn by John Jordan)

- The armour was to provide an immunity zone from 80 to 200 cables (16,000 to 40,000 yards) against 406mm shells, as well as protecting against 1,000kg armour-piercing bombs dropped from an altitude of 5,000m; also to be studied was the possibility of protecting against 2,000kg AP bombs.
- The torpedo protection was to resist warheads of 500kg; there should also be under-bottom protection against non-contact weapons.
- Speed was to be not less than 30 knots, cruising speed 24 knots, economical speed 18 knots.
- Range was to be 10,000nm at 18 knots.[23]

The speed and range requirements led the Shipbuilding Industry to classify this as a fundamentally different design from the preceding work, so it now officially received the designation Project 24 that TsKB-4 had been using unofficially for the previous series of pre-sketch designs.[24] The most stringent – not to say unrealistic – specifications concerned the increased protection, with the outer edge of the immune zone moved out to 200 cables (40,000 yards), as opposed to 155 cables (31,000 yards) as in earlier designs. As for protecting against 2,000kg AP bombs, that would have been almost impossible – Project 23's deck protection had been intended to withstand only 500kg HE bombs.

The stabilised 100mm mountings mentioned may have been a domestic version of the German 10.5cm twin AA mountings, examples of which the Soviets expected for the ex-*Lützow*. The interest expressed in substituting a single battery of 130mm dual purpose guns for separate low-angle and high-angle batteries may have been inspired by information about similar British (5.25in/133mm), French (130mm) and American (5in/127mm) guns; it harked back to early design studies in 1935–1936, which had considered similar uniform DP batteries for the secondary armament of battleships.[25] By 1940 a prototype of just such a 'universal' mounting –

Table 5: Post-*Sovetskii Soiuz* (Project 23) Designs 1939–1940: Weights (metric tons)

	Project 23[1]	Project 23*bis*	Project 24 Pre-Sketch Design Variants			
	Tech Design	Pre-Sketch Design	D-6-6.5	D-6-7	D-12-6.5	D-12-7
Date	Jul 1939	Dec 1939	Oct 1940	Oct 1940	Oct 1940	Oct 1940
Hull	20,188	20,280	19,659	19,658	19,678	19,728
Armour	23,306	24,500	23,670	23,666	23,706	23,845
Armament	8,547	8,720	9,021	9,021	9,021	9,021
Ammunition	1,920	1,990	2,041	2,041	2,041	2,041
Machinery	3,727	3,740	3,984	3,984	4,004	4,004
Crew & supplies	642	640	687	687	687	687
Margin	820	870	820	820	820	820
Standard displacement	59,150	60,800	59,882	59,877	59,957	60,146
Fuel, feed water, lube oil	6,000	6,000	5,653	5,653	5,653	5,654
Full load displacement	65,150	66,800	65,535	65,530	65,610	65,800

Note:
[1] Project 23*bis* is included for comparison.

Source: Vasil'ev, *Lineinye korabli tipa 'Sovetskii Soiuz'*, 131.

meaning it could engage surface, aerial, and shore targets – was under construction. This B-2-U twin turret used a new 130mm/55 gun and would have had a maximum elevation of 85 degrees instead of the 45 degrees of previous 130mm mountings. Plans called for it to be installed in the Project 35 destroyers, which were scheduled to be laid down in 1941.[26]

However the Germans attack on 22 June 1941, just six weeks after the new OTZ were issued, soon found the Soviet Union fighting for its existence; future battleship plans were largely irrelevant. Nevertheless work continued on some elements of Project 24, particularly underwater and under-bottom protection, after TsKB-4 had been evacuated to Kazan, more than 800km east of Moscow. Some of the concepts developed there would be incorporated into postwar battleship designs that differed fundamentally from their prewar predecessors. However, that is a story for another day.

Acknowledgements:
I owe a deep debt of gratitude to Sergei Vinogradov, who guided me through my visit to the Naval Archives in St Petersburg in 2002 and has provided much valuable material in the years since. John Jordan once again worked his magic, creating a comprehensive series of drawings from a difficult batch of originals and offering insights gleaned from his work. And as always, my wife Jan Torbet provided invaluable service as an editor.

Sources:
The major published sources on these designs are:

Vasil'ev, A M, 'Lineinye korabli VMF SSSR predvoennykh proektov', *Gangut* No 16 (1998), 53–66.

Vasil'ev, A M, *Lineinye korabli tipa 'Sovetskii Soiuz'*, Galeia Print (St Petersburg, 2006). [Note that the relevant portions of this book are almost word-for-word repeats of the earlier article, which has somewhat more detailed data tables.]

Gribovskii, V Iu 'Pervaia poslevoennaia korablestroitel'naia programma VMF SSSR (1946–1955 gody)', *Gangut* no 12 (1997), 3–24.

Markelov, V P, 'Sozdanie zashchity sovetskikh linkorov v predvoennye gody', *Tsitadel'* No 11 (2004), 103–29.

In addition I have used the notes I took from the following documents during a visit to the Russian State Archives of the Navy (RGA VMF) in July 2002. Note that in the following citations, 'F' stands for *fond* (collection), 'op' for *opis* (list), 'd' for *delo* (file):

F r-441, op 3, d 421: TsKB-4, 'Predeskiznyi proekt. Ob'iasnitel'naia zapiska po bronirovaniiu', *circa* September 1940.

F r-441, op 3, d 443: TsKB-4, proekt 24, 'Dokladnaia zapiska po dopolnitel'nym variantam pr. 24', *circa* October 1940.

F r-411, op 3, d 444: TsKB-4, proekt 24, 'Predeskiznyi proekt. Dopolnitel'nye varianty: D-6-6.5, D-6-7, D-12-6.5, D-12-7. Izmenenie nagruzki proekta 23 dlia proekta 24', *circa* October 1940.

F r-441, op 3, d 445: TsKB-4, drawing 24-A96-53/16107ss, 'Predeskiznyi proekt, Dopolnitel'nyi variant D-12-7. Obshchee raspolozhenie. Prodol'nyi razrez i Verkhnii vid', scale 1:200, dated 31 October 1940.

Endnotes:
[1] S V Molodtsov, 'Stalinskie linkory', *Briz*, 1996, no 9, 17, 20. At this time the 'bourgeois' term 'admiral' was not used; fleet flagman 2nd rank (*flagman flota 2-ranga*) was equivalent to vice admiral.

[2] See Stephen McLaughlin, 'Stalin's Super-Battleships: The *Sovetskii Soiuz* Class', *Warship 2021*, 8–28.

[3] Vasil'ev *op cit* 2006, 126. Note that Vasil'ev identifies the writer as A S Frolov, but several reliable sources identify the chairman of the NTK as A A Frolov. Engineer-flagman 3rd rank (*inzhener-flagman 3-ranga*) was equivalent to an engineer-rear admiral.

[4] I P Spasskii (ed), *Istoriia otechestvennogo sudostroeniia*, Sudostroenie (St Petersburg, 1994–1996), 4:295, says Galler approved the OTZ on 3 April 1940, but this appears to be incorrect.

[5] Iu L Korshunov and A A Strokov, *Torpedy VMF SSSR*, Gangut (St Petersburg, 1994), 8; John Campbell, *Naval Weapons of World War Two*, US Naval Institute Press (Annapolis, 1985).

[6] Vasil'ev *op cit* 1998, 61.

[7] J N Westwood, *Russian Naval Construction, 1905–45*, Macmillan (London, 1994), 210.

[8] Allen Weinstein and Alexander Vassiliev, *The Haunted Wood: Soviet Espionage in America – The Stalin Era*, Random House (New York, 1999), 26–27.

[9] Katherine A S Sibley, *Red Spies in America: Stolen Secrets and the Dawn of the Cold War*, University Press of Kansas (Lawrence, KS, 2004), 21.

[10] Vasil'ev *op cit* 1998, 61.

[11] RGA VMF, F r-441, op 3, d 443.

[12] RGA VMF, F r-441, op 3, d 443. Published sources generally state that the system was designed to resist 750kg, but this document indicates 700kg as the correct figure.

[13] The earlier 'flagman' ranks had been replaced by admiral, vice admiral and rear admiral on 7 May 1940.

[14] This confusion still persists; Vasil'ev *op cit* 2006, 134, and Gribovskii, 7, have drawings of the same design, but the former labels it Project 23NU and the latter Project 24.

[15] RGA VMF, F r-441, op 3, d 443.

[16] RGA VMF, F r-441, op 3, d 445.

[17] RGA VMF, F r-441, op 3, d 443.

[18] Nathan Okun, 'Armor and its Application to Warships', Part 2, *Warship International*, Vol XIV, No 2 (1977), 98–99.

[19] RGA VMF, F r-441, op 3, d 443.

[20] For a diagram of *Saratoga*'s bulges, see Norman Friedman, *U.S. Aircraft Carriers: An Illustrated Design History*, US Naval Institute Press (Annapolis, 1983), 42.

[21] Vasil'ev *op cit* 1998, 66.

[22] Vasil'ev *op cit* 2006, 136.

[23] Vasil'ev *op cit* 2006, 136, 138.

[24] E A Shitikov, V N Krasnov and V V Balabin, *Korablestroenie v SSSR v gody Velikoi Otechestvennoi voiny*, Nauka (Moscow, 1995), 43–44.

[25] See Stephen McLaughlin, *Russian and Soviet Battleships*, US Naval Institute Press (Annapolis, 2003), 373, 380–1.

[26] A V Platonov, 'Otechestvennye pribory upravlenniia artilleriiskoi strel'boi', *Tsitadel'* No 6 (no 1, 1998), 92–115, at 113; Russian Wikipedia article on the B-2-U mounting, accessed 23 October 2020.

THE GENESIS OF YOKOSUKA NAVY YARD

In this first part of a two-part study, **Hans Lengerer** puts the development of the major IJN dockyard at Yokosuka into its historical context.

The construction of the Yokohama Iron Mill and the Yokosuka Shipyard was begun in the middle of the 1860s and was the largest industrial enterprise undertaken by the Tokugawa Shogunate in the last years of the Bakumatsu period. It was implemented despite the opposition of the finance commissioner Oguri Tadamasa and his supporters, with a view to strengthening the position of the central government (the *Bakufu*) against some of the largest feudal domains (*Daimyate*), which openly opposed the *Bakufu*. Due to the preferential treatment accorded by Britain to the *Daimyate*, the *Bakufu* turned to the new French representative, the envoy Léon Roches, who responded readily to the invitation to assist the *Bakufu*. The French hoped to establish closer mutual relations between France and Japan, obtaining privileges and reducing British influence in the process. During a period marked by conflicts in domestic and foreign policy in Japan, political considerations were a more decisive factor than national defence and industrialisation, which would later take centre stage.

Both facilities were planned by the *Ingénieur du Génie maritime* François Léonce Verny, a graduate of the Ecole Polytechnique, who came to Japan in 1865, directed the works and remained there as director of Yokosuka Shipyard until 1875. He was supported by a French mission composed of selected specialists of all professions and a Japanese committee that directed the native Japanese involved in the project. The influence of the French mission was not limited to warship construction, but also included matters of organisation, administration and finance. In addition, Japan's future skilled workers were educated in a school specially created for this purpose and managed by members of the French mission.

The work started in 1865 and progressed comparatively quickly. One part was completed prior to the Meiji Restoration. After that the work slowed and ceased altogether during the turbulent period of the Boshin civil war, but in 1871 further progress was realised and the first graving dock was completed. At that time the Yokosuka Shipyard was administered by the newly-established Industry Ministry, but it was transferred to the Navy Ministry in 1872. Yokohama Iron Mill lost its *raison d'être* but Yokosuka, often called 'the cradle of the Imperial Japanese Navy', continued to progress, first under French direction and then, from 1876, under Japanese management. The creation of the Yokosuka Navy Yard (as it was called from 1903) marked the beginning of a new epoch in the modern history of Japan with the introduction of western-type heavy industry. It still exists today and is used by the JMSDF and also the US Navy. In this respect the history of the Yokosuka NY is a microcosm of the evolution of Japan from a medieval feudal state to the modern state of the postwar era.

The foundation and early build-up phases of the then-Yokosuka Shipyard can be seen as a unique product of early Franco-Japanese relations, and was important with respect to:

- the foundation of an industrial base by cooperation between states
- technology transfer (metal fabrication, propulsion by propeller, steam engine and boiler), and the acquisition of skills by the application of theory into practice
- the adaptation of foreign technology to meet particular Japanese needs, with a final goal of independence of foreign aid.

Background

As a consequence of the lessons the Japanese mission involved in the American-Japanese Treaty of Commerce of 1858 had obtained in the USA, the *Bakufu* took a number of decisions for the modernisation of the organisation of the ground and naval forces, with the chief emphasis on the latter. These decisions, taken in early 1861, resulted in the order of warships abroad; it also aimed to promote the domestic building of western-type warships, and planned the combination of the naval forces of the Shogunate and the 29 *Daimyate* under a unified command, to be stationed at six bases. The realisation of this ambition would prove difficult, as those tasked with the study recognised that it could not be executed in conformity with the feudal state system: the power of the central government would need to be strengthened, that of the *Daimyate* considerable reduced.

Finance commissioner Oguri Tadamasa (1827–68), who had been in the USA in 1860 as a member of the diplomatic mission for the exchange of the treaty ratification documents, concurred with this assessment. He had returned with the conviction that Japan under the

THE GENESIS OF YOKOSUKA NAVY YARD

A Japanese artist's conception of Yokosuka shipyard and port in 1879. The long three-row building at bottom right is the officers' living quarters. To the left of it is the Machinery School and above it, at the extreme right (four-row building) is the Shipbuilding School. In the centre of the map is the anchorage. The boat moored to the jetty on the right is the 10hp ferry, which served the population as well as the workforce. At the time the map was drawn Japan's first armourclad *Fusō*(i) was under repair (left). To her right the Imperial Yacht *Jingei* is anchoring. The second of the three docks is drawn as completed despite the fact that it was under construction at that time. Beyond the docks are Verny's residence (left), the Main Office for the shipyard (centre) and the church. To the left of the docks two ships under repair can be seen. (US Navy)

rule of the Tokugawa Shogunate was in need of a thorough modernisation. The differences between the Shogunate and the Imperial Court deepened, resulting in an open rebellion of the Chōshū domain against the Shogunate. This led the central government to conclude that, calling on 'support from outside', it should annihilate its enemies and thereby create the basis for similar actions against other potential resistance movements directed against the authority of the *Shōgun*. 'Support from outside' in this context did not mean recourse to foreign troops, but to the execution of a modernisation programme in the course of which feudal decentralisation was to be abandoned in favour of a united state under the Tokugawa Shogunate. However, there was a perceived need for Japan to be developed not only politically, but also militarily and technically into a progressive state. One problem was the selection of a foreign power to support this programme,[1] another the need to obtain the agreement of the Elder Privy Councillors.

French Policy Towards Japan

The political influence of France was relatively small at the beginning, and trade was also insignificant. The first representative of France in Japan, the consul Gustave Duchesne de Bellecourt (1817–81), had been instructed to direct his policy in line with the actions of the British representative, and he acted according to his instructions.[2] On the other hand, an influential group of politicians and industrialists requested a stronger engagement with Japan and China. Napoleon III, whose free trade policy promoted the opening of markets to promote the growth of French industry, was particularly receptive such demands. The expansion of trade was expected to result in conflict with Britain,[3] so the French envoy would need to show considerable determination to promote French interests. However, following a change of Foreign Minister and a consequent revision of French policy in the Far East, the current envoy submitted his resignation. The former consul general at Tunis, Léon

133

Roches, had more than 30 years experience as a diplomat in North Africa and as an officer in the *Marine Nationale*. Appointed on 8 October 1863, he left France on 19 January 1864 and arrived at Yokohama on 27 April; he handed over his credentials, dated 27 October 1863 and signed by Napoleon III, the following day. His predecessor, Duchesne de Bellecourt, departed on 30 May.

When Roches arrived in Japan the internal political situation was turbulent but favourable, and soon offered him the opportunity to demonstrate an independent policy and to reinforce French influence over the *Bakufu* government. Since the outbreak of the Civil War the USA no longer had any diplomatic influence, and Roches' American colleague, Robert Hewson Pruyn (1815–82), who departed Japan in May 1865, was no longer trusted because of his treatment of the order for three warships and the handling of the $600,000 prepaid by the Shogunate. Relations with Britain had rapidly deteriorated after 1862, and the two countries seemed to be moving in the direction of an open war. When the British envoy, Rutherford Alcock, returned from an extended holiday[4] on 2 March 1864 and attempted to resolve the problem of the closing of Shimonoseki Strait by the Chōshū domain with an Allied 'punishment' action, Roches failed to give his backing and abandoned his resistance only when the ratification of the Paris Convention was rejected and there was no longer any chance of a compromise. The bombardment of the batteries in the Shimonoseki Strait took place from 5 to 8 September 1864, and ended with the conclusion of a convention with Chōshū on the 14th and another with the Shogunate on the 22nd. On 30 September Roches suggested to the French Foreign Minister:

> We ought to persuade the Shogun that we are his friends and natural allies .... We must avoid doing anything that may become a cause for the Shogun's government to be unpopular. We must aid and encourage him so that not only will he be motivated by friendship toward us, but that he understand that his real interest is at one with ours ....[5]

The Ikeda Mission

The beginning of this new policy line had been made already by the Ikeda Mission that had negotiated the Paris Convention. With the dispatch of the mission, headed by Ikeda Nagaaki, and proposed by de Bellecourt, the Shogunate government asked the treaty powers to agree to the closure of the treaty port of Yokohama and the designation of Nagasaki and Hakodate as free ports. In addition, it wanted the compensation claims resulting from the bombardment of the foreign vessels during their passage through the Shimonoseki Strait reduced and its own inactivity excused.

The mission, comprising 37 individuals, departed Yokohama on 6 February 1864 on board the French warship *Monge* and arrived in Paris on 18 April. The envoys were received by Napoleon III on 3 May and the negotiations with the French Foreign Minister Drouyn de Lhuys began on the 7th. De Lhuys offered military support to the Shogunate against internal resistance. The purchase of a number of elderly warships was offered, and Ikeda ordered a single frigate; however, this ship was never delivered and no other ships were purchased.

Political reform was also addressed in the negotiations with officials of the foreign ministry. The advice of the Belgian aristocrat Comte Descantons de Montblanc is of particular interest. He compared Japan with 16th Century France and proposed the establishment of a strong central government and the crushing of the *Daimyate*. This view chimed with that of Oguri, and was also close to that of Roches, who was now offered the opportunity to support the Shogunate government with a project that combined both military and industrial elements.

Planning for the Yokohama and Yokosuka Iron Mills

According to the *History of Yokosuka Navy Arsenal* (*Yokosuka Kaigun Senshō-shi*) the origin of the construction of the Yokosuka Shipyard was the delivery on 8 December 1864 of a letter signed by the elder statesmen Mizuno Tadakiyo, Abe Masatō and Suwa Tadamasa to the French envoy Léon Roches, formally requesting French personnel and equipment to establish a modern shipyard and factories. The driving force behind this request was the Commissioner Oguri Tadamasa, who had addressed his request to Roches via the Inspector (or 'Censor') Kurimoto Sebei.[6] The decision of Mizuno to ask France for support was based on the suggestion of these two influential men.

The friendly relations between Kurimoto and the secretary to the French legation, Abbé Eugène-Emmanuel Mermet Cachon, was well known, and at the beginning of December 1864 Kurimoto was permitted by Assistant Councillor to the *Shōgun* Sakai Tadamasu to visit the French legation unaccompanied. This was contrary to standard protocol, and was permitted so that Kurimoto could better evaluate the true intentions of the French and obtain further concessions.

Repair of the *Shōkaku Maru*

Sakai had been made aware of the good relations between Kurimoto, Cachon and Roches by the mediation of the former for the repair of *Shōkaku Maru*. This small steamer (350 tons) had been completed in New York in 1857 and purchased from a British trader in 1863 for US $35,000. It was a typical example of the poor judgment of the Japanese in commercial transactions with the West. Only a year after the purchase, urgent repairs were necessary.[7] However, Japan had no specialists capable of making the necessary repairs and lacked the money to have the repair undertaken at Shanghai. Instead Kurimoto was asked to use his good offices with Roches and to ask for support. The first talks ended with Roches' declaration that the captain of a warship had the sole right to decide, but on board the frigate *La Guerrière*, anchored in Yokohama port, was not only the ship's commanding officer but also the C-in-C of the French Naval Force for the China Sea and Japan (Asia Fleet), Rear Admiral Jean-Louis-Charles Jaurès, who was due to visit the legation that same day at 18.00. Jaurès agreed to the request, and following a meeting with Commodore Kinoshita Kingo and the consent of the Shogunate, French specialists[8] repaired the ship starting on 16 November 1864, the work taking less than two months.

The progress of the repair work was monitored by Kurimoto and Kinoshita and the latter's companions, who admired the technical capabilities of the French seamen. When Jaurès rejected the offer of money, remarking that the technicians 'are already paid by the French government and, in addition, they are glad to have something to do during their time in the anchorage in the port', the Japanese side was even happier.[9]

Oguri Tadamasa's Idea for a Shipyard

Oguri[10] had a good working relationship with Kurimoto, and appears to have been convinced of French technical capabilities and the goodwill of the French government towards the *Bakufu* government. The satisfactory course of events leading to the repair of the *Shōkaku Maru* strengthened the Oguri–Kurimoto–Roches axis and served to promote the policy of the French envoy towards Japan.

Oguri had recognised the importance of the navy for the US government, and emphasised its significance after returning to Japan. He was an advocate of naval expansion and the establishment of building facilities near Edo to provide the necessary support infrastructure. Oguri was aware of the advice brought back from France by the Ikeda Mission as regards political reform and support (with which he could identify), and he was also informed about the revision of the French East Asia policy. Both aimed at the support of the Shogunate government in conflicts with the rebellious *Daimyate*. A second campaign against Chōshū was about to begin, and in this sense the repair of the *Shōkaku Maru* was also a kind of support. The main problem, however, could not be resolved by such a temporary measure. Oguri believed that the *Bakufu* would obtain the power necessary to represent the Japanese state by building a navy, and favoured a programme of military and technical modernisation along French lines and with French support.[11] After the defeat of Chōshū, this would prepare the way for armed actions against other possible centres of resistance in order to achieve a unified state under the *Shōgun*.

The creation of a navy needed to be underpinned by the construction of a shipyard capable of building and repairing ships,[12] and the need for self-administration required the location of the shipyard to be close to the capital, because Nagasaki was too distant from Edo. A meeting was held at Oguri's house to which Kurimoto was invited to discuss the building of a shipyard with French support. Oguri gave the following explanation for the lack of progress thus far:

- When the *Bakufu* planned the building of the Nagasaki Iron Mill, the *Daimyō* of Saga, Nabeshima Naomasa, bought a steam engine and various machine tools from the Netherlands for the repair of the corvette that had been ordered, and wanted to build a shipyard within the border of the *Daimyate*. However, there were no engineers capable of planning and guiding its execution and, in addition, the *Bakufu* lacked the money for such a major investment. In view of this situation the *Daimyō* presented the machinery to the Shogunate.
- The *Bakufu* planned to use these machines as the basis for the building of a shipyard and an iron mill in Nagaura Bay (Sagami domain, Miura peninsula, near Uraga). A committee was duly established and the bay was surveyed. After that, planning had to be halted because the Shogunate experienced the same problem as Saga: there were no engineers in Japan who could plan and build an industrial plant of this nature.
- The proposals for the use of the machines in the Nagasaki Iron Mill or the 'expansion plants' were rejected. The machines had not been operated since their delivery and were rusting away.
- It was wrong to ignore the generosity of the *Daimyō* of Saga and to allow his gift to rust. The realisation of this project with the technical support of the French should be seriously considered.

Oguri ended his explanations with the instruction to Kurimoto to visit Nagaura Bay with Thibaudier and other Frenchmen to investigate the proposed location. But Kurimoto's knowledge of shipyards and iron mills was limited, so he proposed a meeting with Roches to discuss the issue. A preliminary meeting could be arranged immediately (Oguri hired a hotel at Kanagawa), and Roches, who himself was likewise no expert in these matters, asked for Rear Admiral Jaurès to attend the following meetings.

Selection of a Location

Roches saw this request as the perfect opportunity to offer economic and military assistance to the *Bakufu* in line with the new French policy towards Japan. In this project, which was to become the *Bakufu*'s most important industrial enterprise, both of these elements were present. In addition, it was undertaken with a view to stabilising the political situation by suppressing the anti-*Bakufu* forces.[13] After obtaining the agreement of Jaurès the envoy decided to back the project, and elected to contact his government and to work up a draft proposal. On the Japanese side Oguri and Kurimoto succeeded in convincing Privy Councillor Mizuno of the necessity of the project. Roches proposed to hire the naval engineer Verny[14] for the planning and execution and for a survey of the proposed location. On 24 December 1864 Oguri,

Kurimoto, Kinoshita and the Kanagawa commissioner, Asano Ujisuke, welcomed Roches, Jaurès and the captain of the *Semiramis* aboard the *Jundo Maru* and showed them Nagaura Bay. The investigation by the French revealed insufficient water depth. In contrast Yokosuka Bay not only possessed the necessary depth of water but was also well protected – the shape of the bay resembled the French naval port of Toulon. It was therefore Yokosuka that was selected for development.

The Shogunate began its own survey by a Japanese 'warship group', headed by Nagata Seizō and composed of nine men, on 31 December 1864. Their task was to verify the French survey results and the geological condition of the beach, and to investigate the surrounding area. The group was also instructed to select suitable positions for docks, slipways, and defensive positions.[15]

The Meeting on 6 January 1865

On 6 January 1865 the decisive meeting between Roches and Privy Councillors Mizuno, Abe and Suwa took place. The main items on the agenda were the hiring of Verny as director of the whole project, the dispatch of a mission to Europe for the purchase of machines and materials for the shipyard, and the purchase of guns for the second campaign against Chōshū. But the primary issue was finance. Roches proposed a monopoly of the silk trade by the Shogunate, with marketing exclusively via French traders. The proposal was controversial, as it would have brought the Shogunate into conflict with Britain, and was not agreed.[16]

The Japanese side agreed to the hiring of Verny and also accepted the broad outline of his planning proposal. In reports dated 16 and 17 January 1865 Roches informed the French foreign minister that, following negotiations with the Privy Councillors over the past month, he had received the instruction to build a shipyard in the vicinity of Yokohama at a cost of 25–30m French francs (equivalent to £1–1.2m sterling). He also reported further consultations with Rear Admiral Jaurès about this matter and referred to the building of the shipyard as 'necessary for the conquest of Chōshū', thus underlying the foremost purpose of the project: to strengthen the power of the central government (*Bakufu*). Roches also informed the foreign minister about the decision of the *Shōgun* to despatch a mission to Europe to request personnel and materiel. He recommended permitting the delivery of the guns that had been ordered because this had been agreed at the Paris Convention, and pointed out that other powers would furnish them if this request was rejected. The French foreign ministry gave their full agreement on 18 March 1865; the decision marked the beginning of total support by France of the Shogunate government.

Verny's Basic Concept

François Léonce Verny (1837–1909) entered the French Navy as an engineer after graduating from his studies at the Polytechnique and first worked in the naval arsenal at Brest. In 1862 he was ordered by the Navy Ministry to go to China and to establish a small shipyard at Ningpo for the construction of small river gunboats, which were urgently needed by the French East Asia Fleet (then under the command of Jaurès) for operations against Chinese pirates operating in the mouth of the Yangtze, who were becoming more and more audacious. Verny had only a few skilled workers and a small amount of usable materiel at his disposal and was forced to improvise. Despite this, Jaurès was able to report to the French Minister of Marine Chasseloup Laubat on 18 April 1864 that the quality of the river gunboats was 'good'. In the same report the high speed and the low fuel consumption of the ships was praised, and Jaurès also highlighted the establishment of the repair dock of the shipyard with 'few personnel and little materiel', and added that this dock was also used by British ships. Verny's combination of technical knowledge and organisational ability – qualities that would become even more apparent during the construction of Yokosuka Shipyard – also attracted the attention of the French Foreign Ministry, which appointed him Vice Consul.

Verny arrived at Yokohama from Shanghai on 6 February 1865 on the instructions of Jaurès, and was instructed to work out a proposal on the basis of a meeting previously held between the French and Japanese sides. He began this work at once and completed the draft within the very short period of two weeks, even though the content was comparable to a book. During this period he also visited the intended building site and checked the data already collected. His draft was composed of two parts: the Yokohama Iron Mill and the Yokosuka Naval Arsenal.[17] The difference between these industrial plants was made very clear: Yokohama was to be nothing more than the preparatory facility for the operation of the Yokosuka arsenal and was to serve two principal functions: the repair of warships, and the training of Japanese personnel by working alongside their French counterparts.

For the Iron Mill, Yokohama-Motomura (the later Muromachi) was proposed. In contrast, the Yokosuka plant was a large shipyard with docks, slipways and every shop necessary for the building of new ships and weapon storage. Thus the two plants had a fundamentally different character.

The Conclusion of the Building Contract

Roches informed the French foreign ministry in a report dated 28 February 1865 of Verny's stay in Japan, the investigation of the building locations, the drawing up of the proposal, its submission to the Privy Councillors and their agreement. He announced that Verny would submit his proposal to the Navy Construction Bureau after completing his work at Ningpo on 21 April.

Verny's proposal was formally endorsed by a building contract dated 24 February 1865 and a set of rules for the execution of the project bearing the same date. Verny was given total responsibility for the project and was also invested with extensive powers.

The principal content of the building contract (summarised), handed to Roches by the Japanese, was as follows:

– An iron mill, one large and one small dock, three slipways, a weapon storage depot and several buildings for officers and workmen; all to be completed in four years.
– As Yokosuka is similar to Toulon on the Mediterranean Sea coast in its physical characteristics, the iron works to be erected at Yokosuka shall be like that at Toulon, and shall cover two-thirds of the ground surface of the latter, namely length 818 metres (450 *ken*) and width 364 metres (200 *ken*).
– The total expenses of constructing the iron works, the docks and the slipways[18] shall be US $2,400,000, payable in four annual instalments of approximately $600,000 each. An initial payment of $600,000 will be made to France on exchange of contracts to cover the documentation for the first year.
– Both countries have agreed the aforementioned details.
– The envoy, acting with the agreement of the Admiral in charge, has ordered engineer[19] Verny to take on the post of director. On the Japanese side the finance commissioner Matsudaira, Commodore Kinoshita Kingo, Inspector Yamaguchi, Kurimoto Sebei and Asano will be responsible for the project.
– Both sides are to make every effort for close cooperation to guarantee the successful realisation of the project.

The contract was signed by Privy Councillor Mizuno Tadakiyo and the (younger) Privy Councillor Sakai Tadamasu.

The 'rules for execution' may be summarised as follows:

– Before the construction of the shipyard (at Yokosuka), a factory should be located at Yokohama and fitted with the machines bought from the Netherlands [the machines bought by the Saga *Daimyate* and presented to the *Bakufu*] and the USA [the machines ordered in 1860] to make Japanese workers skilled in the operation of the machines under the guidance of French officers, and in ship repair.
– The area of the shipyard should be sufficient to provide jobs for about 2,000 Japanese workers and 40 Frenchmen, and should have at least two graving docks and three berths. For this purpose an area of 18 hectares will be necessary.
– The total expenditure for the construction of the living quarters, workshops, docks and berths is estimated to be $2,400,000, to be spent over four years.
– The French personnel to be hired for the establishment of the shipyard should comprise the director and heads of the shipbuilding, construction and the finance departments plus 33 workmen of all professions (see Table 1). They are to be selected and hired in France by the director.[20]

Table 1: French Naval Personnel

Position/Profession	Team Leader	Worker
Director (overall Head)	1	
Head of Shipbuilding	1	
Head Architect	1	
Head of Finances	1	
Clerk	1	–
Joiner	1	2
Caulker	–	1
Iron worker	1	3
Cast iron worker	1	3
Boiler maker	1	4
Lathe operators, etc	3	8
Locksmith	1	1
Material administrator	1	1
Architect	1	3
Total	11	26

Notes:
One of the iron workers was in charge of interior decoration; one of the lathe operators was in charge of machine production

Sources: *History of the Machinery of the Imperial Navy* (*Teikoku Kaigun Kikan-shi*) Vol 1, 232–33; *History of Yokosuka Navy Arsenal* (*Yokosuka Kaigun Senshō-shi*) Vol 1, 9–10.

– The annual salary of Verny will be $10,000; that of the director of the Yokohama Iron Mill (Ferdinand N Gautrin) £4,300.[21] The monthly salary of a head of department will be $400, of a foreman $150 and a skilled worker $75.
– The standard working day will be ten hours, and may be extended if required for the execution of urgent work. Although the French normally have Sunday off, one third of the French contingent should be involved in teaching and supervising the Japanese workers on Sundays.
– On the establishment of the shipyard the Japanese side will appoint a general director,[22] a head accountant, a head of materiel administration (warehouses) and a head interpreter. Other personnel will be appointed to perform roles under them.
– The Japanese government will establish a school within the shipyard to educate the engineers and technicians to be in charge of its operation in the future. Young people should be selected as students, and the lessons are to be translated into Japanese by the head interpreter. The students should study French alongside their technical lessons, and the foremen are to study in their spare time.
– Young workers are to be selected by the French director as students for the career of junior engineer. They are to be placed in the factory in the morning to be trained in the work and in the afternoon they will have the necessary lessons in theory. The curriculum (theory and praxis) should reflect that of the French Navy.
– The Japanese director will be dispatched to France to

negotiate an agreement covering the following three items:
(a) The preliminary planning of the shipyard should be submitted to the French Minister of Marine and his agreement obtained.
(b) A technical officer of the French Navy suitably qualified to be head of the shipbuilding department is to be appointed.
(c) This head of shipbuilding is to select junior engineers and workers from the among the personnel of French shipyards, to be hired by the Japanese government (via the shipyard).

– The machines and materials purchased from France are to be delivered by 31 March 1866. The director of the shipyard is to begin construction of the shipyard on 1 January 1867, and after the completion of the factories and berths the building of ships should be begun. By the beginning of 1869 the construction of the repair dock and the education of the Japanese workers should be completed.

Both documents were approved without dissent, and the *Bakufu* government handed the contract to envoy Roches as stated above; the contract document was dispatched to the French government by Jaurès.

Even though Yokohama Iron Mill and Yokosuka Shipyard were built under the same programme, the respective locations, sizes and functions were very different, so it is appropriate to describe them separately.

Yokohama Iron Mill

The purpose of the Yokohama Iron Mill was (i) to train Japanese workers to obtain a nucleus of skilled personal prior to the completion of Yokosuka Shipyard and (ii) to repair steam-powered ships, particularly their engines and boilers,[23] to apply the latest technology and obtain practical experience. The teaching of the techniques depended on French technicians and both parties agreed that one officer, with a good general knowledge of the whole industrial plant, should be made the technical chief and that two technicians should be placed under him, specialising in engineering and shipbuilding respectively, to supervise the Frenchmen chosen from a French naval dockyard and from French ships at anchor in Yokohama port. They were to train 100 selected Japanese wood and metal workers in Western-type industrial production processes and also to supervise the practical application of these skills. If one worker was skilled in his profession he was to pass on his knowledge and capabilities to another suitable individual.

In contrast to the technical side of the project, the administrative management would be in the hands of Japanese bureaucrats, who were also charged with financial operations including purchase and storage. The technical chief needed to secure the permission of the Japanese side for general expenditure and the execution of complex repairs.

The building of the Yokohama Iron Works was to be completed within one year and had a total budget of $70,000,[24] which was divided into $20,000 for the land and buildings and $50,000 for the purchase of materials and equipment beyond the machines already in hand, and wages for one year.

The work started in February 1865 with the laying of the foundation stone at Yokohama-Motomura and the buildings were completed on 13 October of the same year. The workshops were situated in the southwest of the town near the European settlement on an island formed by the various passages of the Homura Canal. Verny's plans show a machine (and assembly) shop, a copper smithy, a smithy and an iron foundry. The first three shops were located in a large building whose rear wing was for assembly work; the foundry was separate, in a second building situated near the first. The installation of the machinery began in November and the plant was working early in 1866. This impressive achievement was due to the good management of Verny and the leading French officers, and also to the efforts of the Japanese commission.

The machines and tools were those purchased in the USA in 1860 by the *Bakufu*'s First Embassy to the USA, and also some of those purchased by the Saga *Daimyate* in the Netherlands in 1856, donated to the *Bakufu* government in 1859, and transported to Yokohama during April/May 1865. Only a small amount of additional machinery was needed.

All the machinery was powered by a single 15hp steam engine located in a separate building half-way between the assembly shop and the copper smithy. The engine also drove two large ventilators, one of which served the forging and the casting shops while the other was solely for the foundry. As back-up there was an auxiliary steam engine of 6hp. Both these engines were of Dutch manufacture.

The construction of the Iron Mill was supervised by the French naval officer FN Gautrin, who was also responsible for the technical management. The chief administrator was Kurimoto.

The Yokohama Iron Mill had an extensive programme of work, and its products made a major contribution to the construction of the Yokosuka Shipyard. Aside from the repair of ships and other works – farm and household items were made from iron and sold for commercial purposes – a small steam tug for the Mill's own use and the 10hp engines for two ferries between Yokohama and Yokosuka were produced. In addition, army field guns were repaired.

The workload of the Yokosuka Iron Mill diminished greatly after the completion of Yokosuka Shipyard, but it fulfilled all the goals set for it from the outset by both the French and Japanese sides.

At the end of 1871 there were about 180 Japanese workers and five Frenchmen: the technical manager and four foremen each in charge of one of the workshops. However, the drafting of workers to Yokosuka was constant, thereby further reducing the importance of the Mill.

Yokosuka Shipyard

The construction of the Yokosuka Shipyard was carried out in three phases:

1 preparation of the ground (begun May 1865)
2 building of the factories and accommodation (begun April 1866)
3 construction of the docks (begun April 1867).

It was decided to build Yokosuka Shipyard over an area of 245,807m² (75,359 *tsubo*) facing three bays, and the land was subject to a compulsory purchase order from local farmers and fishermen. The preparatory work consisted of levelling the high ground and reclamation, and was undertaken with a workforce of 300 men in May 1865, shortly after the start of work on Yokohama Iron Mill. Prison labour was used to break up the hills surrounding the bay, and the stones were transported in American-made transport wagons supplied by Edō Kaiseisho (the predecessor of Tōkyō University). However, the work of the prisoners did not proceed well, mainly because of sickness; a contract was therefore awarded to a civilian entrepreneur, who also undertook the reclamation work.

While the levelling of the ground was still underway, the foundation stone was laid on 15 November 1865, and in April 1866, following the arrival of the first French architects and engineers from France, the construction of the factories and accommodation was begun. The Japanese, who lacked the knowledge to build Western-style housing, had to wait for their arrival before beginning this phase under their guidance.[25] After that, however, the tempo of construction accelerated, and by 9 November 1866 43 hired Frenchmen had arrived. The second phase was carried out in the order given to the chief of architecture by Verny: (i) accommodation for the French nationals, (ii) workshops and offices, (iii) ship building/repair slipways, (iv) rope and sail workshops and (v) school house. By the summer most of the residences and offices were in use and the building of the workshops was in progress. In the autumn a repair slip and some workshops were ready.

The first building sites were the wood working shops and the hauling-up slips on the inclined beach; the boiler shop (part of the copper smithy), the iron foundry and the machine shop followed. The shops for wood working were most urgent because the machine shop and the smithy of the Yokohama Iron Mill had been in operation for more than a year. In parallel with these works the harbour basin was dredged and the wharves built.

The hauling-up slips were necessary until such time as the graving docks could be completed. They ranked among the first installations because their construction was straightforward and required minimal time: Pebbles were used as the foundation on the beach and rails with a toothed wheel between them and safety barriers were mounted on the foundations. Vessels were hauled up by two winches located respectively on the beach and the slip. Their use decreased following the opening of the first dock. It was anticipated that they would be dismantled in 1874, following the completion of the second dock.

The construction of the building ways resembled that of the hauling-up slips, which was simple and inexpensive; they were used only for the building of small, wooden vessels. However, the use of wooden piles without masonry meant that maintenance costs were high.

The wood working shops included a saw mill,

An unidentified ship ready for launch at the Yokosuka shipyard. (US Navy)

Taken in 1867, this photo shows Yokosuka's first floating crane nearing completion. (US Navy)

carpentry, joinery, rigging shop and loft, a model joinery and a shop for ship's boats, and were fitted with machines of all descriptions. Wood was the principal building material at that time and various types of wood were used, each for a particular purpose: elm (*keyaki*), small oak (*nara*), Mongolian oak (*kashiwa*), oak (*kashi*), pine (*matsu*), Japanese cedar (*sugi*), spruce (*momi*), yew (*shioi*), chestnut (*kuri*), and also camphor (*kusunoki*).[26] Elm was used for frames, ship's sides and bottom, small oak for the frames of small ships' sides, oak for parts which were required in large numbers, pine for ships' sides and beams but also rigs, cedar for the customary joinery works and decorations. The use of spruce was abandoned in favour of other woods such as beech (*buna*) and mulberry (*kuwa*). The sum of 23,749 Ryō had to be spent on the building of the shops, 26,357 Ryō for the machines and tools, whose fitting required another 787 Ryō, making a grand total of 50,892 Ryō.

The ropery was a one-storied building 270 metres long equipped with machines bought from the French arsenal at Cherbourg. However, hemp was very expensive, and as long as the IJN had only a few ships, manually-produced ropes were purchased. The spending of 26,543 Ryō for this building therefore appears superfluous, but it was compensated in part by the production of sails.

The smithy comprised the large smithy, the locksmith's workshop and several working places for blacksmiths divided into two buildings and fitted with the most modern equipment. A sum of 45,419 Ryō was spent on machines and tools; fitting required a further 8,567 Ryō and the cost of the buildings amounted to 29,368 Ryō, making for a total of 83,354 Ryō.

The copper smithy cost roughly half as much, 40,814 Ryō, of which the greater part (28,843 Ryō) was again for machines and tools, for whose fitting an additional 1,749 Ryō were required. The copper smithy was housed in a building with a functional layout. Boilers were assembled in the central section, and to the left and right were the smithy and machine shop respectively. The original work of the copper smithy was of secondary importance, because all work requiring sheet metal was carried out here. The relationship between the costs of the smithy

Before the later Imperial Yacht *Jingei*, the small paddle-wheel steamer *Yokohama Maru* had been built to link Yokosuka and Yokohama. This photo shows her under construction in 1869. (US Navy)

This photo, taken in 1870, shows dry dock No 1 under construction. (US Navy)

and copper smithy was also reflected in the number of workers – c130 vs c60 in mid-1872.

The capacity of the iron foundry was comparatively small, but it was equipped with two blast furnaces and several cupola ovens. In 1872 an average of 120 tons of pig iron were produced monthly and were used within the shipyard. For the production of large parts a pit in the bottom of the main building was used, for smaller parts the usual moulds. The costs of the building amounted to 6,478 Ryō; the machines and tools, which included a 1-ton cupola oven and several tools for the Yokohama Iron Mill, cost 25,819 Ryō, and fitting expenses were 11,086 Ryō.

The assembly area and shop were among the last facilities completed; most of the machines had been installed by 1873.

The principal workshops were connected by a railway. Hydraulic power was widely used and was served by an extensive system of pipework.

In August 1866 Verny proposed the building of steamers to accelerate communication and transportation between Yokohama and Yokosuka, and also to employ one of them as a tugboat for the shipyard. However, six months were necessary for the preparation, which would cost $200 per month, so a small steamer was bought for $3,037 from an American merchant in Yokohama to bridge the gap before a 30hp and two 10hp vessels could be completed. The former was later named *Yokosuka Maru* and its engine was imported from France; the latter were No 1 and No 2 small steamers (ferries), and their engines were built at Yokohama Iron Mill in November/December 1866. These wooden vessels were the first craft built in Yokosuka Shipyard, and the engines were the first propulsion units to be manufactured at Yokohama Iron Mill; the project should therefore be regarded as a prime example of the cooperation of these two facilities.

In April 1867 the excavation of dry dock No 1 to the plans of Verny was ordered by the minister in charge of the building of the shipyard.[27] The principal dimensions were as follows: total length 114.56 metres (later lengthened by 5 *ken* = 9.09m to 123.65m); width at the entrance 30.93m above and 29.09m below; and depth 10.91m. The area was calculated as 22,009m²; the costs of the excavation were 16,914 Ryō.[28] The contract for the works was awarded to the entrepreneur Hashimoto Chōzaemon, and the excavation progressed rapidly.[29] In late October the first slabs of the base were laid, but a few weeks later the Tokugawa Shogunate fell and the Boshin civil war broke out.

Owing to the turbulence of the civil war the dock was not completed until 1871. The ceremony took place on 28 March in the presence of Prince Arisugawa Taruhito. The length from head to head was 124.30m, the width at the entrance 25m, the water depth at high tide 7.1m and at low tide 6.5m. The iron dock was manufactured in France at a cost of 29,568 Ryō, transported to Yokosuka and assembled. Four centrifugal pumps (Kent & Dumont), driven by a 24hp steam engine, could pump out the water within eight hours. The total expenditure amounted to 167,078 Ryō of which the main part, 101,411 Ryō, was spent on the masonry work, 41,560 Ryō for other materials (including the dock frame), and 23,607 Ryō for the excavation works.

The structure of the second dry dock was similar to the first, the only difference being the increased depth near the entrance for the dismounting of the rudder of a ship when the dock was empty. The principal dimensions were smaller: length 85.00m, width 12.50m, depth at high tide 5.60m and at low tide 5.00m. The pumps were of the same type as those used for No 1 dock.

Building Expenses

Table 2 shows that the initial calculation of the building expenses ($2,400,000) was sufficient and proves that this sum was not spent within the five-year period from March 1866 until February 1871.

Generally speaking, domestic materials and products were favoured, and imports were restricted to items not available in Japan. Iron, wrought iron and steel were purchased in Britain, other metals in France. The main supplier of leather, linen, coke and cement was also France, with coke transported to complete the ship's load. Tin, copper and medium-quality iron, iron sand, charcoal, tar, hemp, sesame oil, tallow, bricks, tiles and also stones were acquired locally, as were domestic products such as wood.

Table 2: **Building Costs**

Purchase of raw materials, semi-products and products	509,897 Ryō
Purchase of machines and tools	215,816 Ryō
Transport, travel and various freight expenses	140,793 Ryō
Salaries (paid monthly)	333,039 Ryō
Wages (paid daily)	188,263 Ryō
Building expenses	321,391 Ryō
Expenses for Yokohama Iron Mill 1869 & 1870	59,827 Ryō
Total	769,026 Ryō

Selection of Personnel and Purchase of Machines in France and Britain

All the machines, tools and other articles necessary to equip Yokosuka Shipyard were to be purchased in France according to paragraph 10 of the 'execution regulations'. However, before the departure of the Japanese delegation, Roches advised that they visit London and order machines in British factories to avoid arousing Britain's enmity. The *Bakufu* accepted the advice and decided to also dispatch the foreign commissioner (*gaikoku bugyō*) Shibata Takenaka and his suite to Britain.

The delegation departed Edo on 25 June 1865 on board a British ship and arrived in Paris on 16 September. After meetings with the ministers for Foreign Affairs, the Navy and the Army, the hiring of Verny as overall director of the project was finally decided. The delegation also secured the purchase of materials and machines and confirmed Verny's selection of personnel (experts, specialists and skilled workers). Verny had arrived in France earlier after finishing his work at Ningpo (see Roches' report), had received the Japanese delegation in Paris, and had negotiated beforehand with the government offices concerned. The selection of suitable individuals and the purchase of machines was therefore a comparatively simple matter for the delegation, which also received the full support of the French government. Verny's 'house' banker, Paul Flury-Hérard, who was already involved in trade between Japan and France, was charged with all financial transactions of the *Bakufu* and was appointed Japanese consul general.

The stay in London lasted about one month from the middle of December, and the scale of purchase was very small compared to that in France. The delegation then returned home via Paris, arriving at Edo on 12 March 1866. Prior to this, the delegation had been reinforced by Hida Hamagoro who had been ordered by the *Bakufu* on 24 February 1865 (date of the building contract!) to go to Paris as a member of the Shibata delegation after finalising the purchase of machines in the Netherlands for which he and two other technicians had been dispatched in September 1864.[30] Hida went to Paris in July and carried out his orders both in France and Britain.

The purchase of machines proved to be straightforward, but their acceptance by Verny was a different matter. A total of 974 machines of all descriptions to equip the future Yokosuka shipyard were purchased in the Netherlands, France and Britain (in the case of Britain mainly by companies represented in the Netherlands). When Verny was handed the list he approved most of them as necessary and suitable, but declared that those for the working of iron and steel should be accepted only after further investigation. However, most of them were already on board freighters on their way to Japan. Inspection by Verny would have required unloading, and possibly assembly and testing, resulting in further expense and loss of time. In view of this situation Shibata consulted Verny and decided that the machinery already embarked should be transported to Japan and

Table 3: Characteristics of Early IJN Warships

	Seiki	Amagi	Jingei	Kaimon	Tenryū
Completed	21 June 1876	4 April 1878	5 August 1881	3 March 1884	5 March 1885
Displacement	898 tons	926 tons	1,464 tons	1,375 tons	1525 tons
Dimensions	62m x 9.3m x 4m	62m x 10.9m x 4.4m	76m x 9.5m x 4.3m	64.4m x 9.9m x 5m	64.7m x 10.8m x 5.2m
Propulsion	2 loco boilers	2 loco boilers	4 loco boilers	4 loco boilers	4 loco boilers
	1 HCRC engine	1 HC engine	1 DA engine	1 HCRC engine	1 HCRC engine
	1 4-bladed prop	1 x 4-bladed prop	2 paddle wheels	1 x 4-bladed prop	1 x 2-bladed prop
	9.5 knots	11 knots	12.5 knots	12 knots	12 knots
Coal	130 tons	150 tons	208 tons	182 tons	256 tons
Armament	1x 15cm Krupp	1 x 17cm Krupp	4 x 12cm Krupp	1 x 17cm Krupp	1 x 17cm Krupp
	6 x 12cm Krupp	5 x 12cm Krupp		6 x 12cm Krupp	1 x 15cm Krupp
	4 x 25.4mm MG	3 x 8cm Krupp		1 x 7.5cm Krupp	4 x 12cm Krupp
		3 x 25.4mm MG		4 x 25.4mm MG	4 x 25.4mm MG
Complement	143	164	170	226	342

DA direct action
HC horizontal compound
HCRC hoizontal compound return connecting rod

mounted in the various shops after inspection by Verny.

Whereas Hida was of the view that Japan needed to learn to build iron ships, Verny objected on the grounds that before embarking on this difficult process wooden ships should be built to learn European shipbuilding techniques. Enomoto Kamajirō, who would later command the *Bakufu* Navy and become president of the short-lived Republic of Hokkaidō, was studying in the Netherlands at that time and came to mediate; Shibata also intervened. The disagreement was settled finally in Verny's favour.

Training Program at Yokosuka

In parallel with the building of Yokohama Iron Mill and Yokosuka Shipyard, Roches proposed that a programme of language and technical training be established in conjunction with both works. The object was stated as follows: '… in order that the Japanese government may in future years replace the Frenchmen in charge of shipbuilding with Japanese nationals, a school will be established at the shipyard to train persons of talent as engineers and technicians …'. The acceptance of this proposal (later underlined and reinforced by Verny) by the *Bakufu* reflected a key change in industrial policy

Seiki was the first warship built in Japan after the Meiji Restoration; she was laid down on 20 November 1873, launched on 5 March 1875 and completed on 21 Jun 1876. The French Director of the Yokosuka Shipyard was instructed to design a 2,600-ton flagship, but the building of such a large ship was impossible at that time, and he opted instead for a revised version of the French Navy's *Talisman*. *Seiki* was built of Keyaki wood. With her three masts and barquentine-type rig, a single slim funnel amidships, and a clipper bow she had an elegant appearance and became the model for a series of wooden sailing warships. The Meiji Tennō attended her launching ceremony (he arrived from Yokohama on board *Ryūjō*).

After participating in the Seinan (South-West) War of 1877, *Seiki* sortied from Shinagawa on 17 January 1878 under the command of Lt-Cdr Inoue Yoshika and embarked on the first cruise of an IJN warship to Europe, visiting numerous countries and ports and returning to Yokohama on 18 April 1879. During a passage to Owashi Bay with machinery training students on board she struck a rock in Tsugaru Bay on 7 December 1888 and sank. (Author's collection)

Laid down on 9 September 1875, launched on 13 March 1877 and completed on 4 April 1878, *Amagi* was a half-sister of *Seiki*. She was used mainly for surveying and monitoring the coast of Korea, but also surveyed the Ryūkyū Islands in 1887 before continuing her watch on Korea based on Jinsen (Chemulpo) in 1888. On 13 August 1890 she was reclassified as a training ship. She rescued people from the Ogasawara Islands in a heavy storm in 1892 and participated in both the Sino-Japanese and the Russo-Japanese Wars before being stricken on 14 June 1905. (Author's collection)

Designed by François Léonce Verny, then head of Yokosuka Shipyard, and his deputy Thibaudier, *Jingei* was the first ship to be laid down there – *Seiki* followed two months later – and was the only paddle-wheeler built for the IJN after the establishment of the Navy Ministry. It was a surprising choice, as in 1853 the *Bakufu* had been advised by the Dutch that the paddle wheel had been superseded in western construction by the screw propeller.

Jingei had two masts with a schooner rig. She was laid down on 26 September 1873, launched 4 September 1876 and completed on 5 August 1881. Completion was delayed by problems with the propulsion system – the principal cause was found to be the weakness of the structure around the paddle-wheels – and the fitting of many decorations. Following brief service as Imperial Yacht (July 1881–February 1886) she became a hulk for the Torpedo Training Station until stricken on 2 December 1893. She was employed mainly for training with the German Schwartzkopff-type torpedo but also for trials with the French Canet type. Her building costs of Yen 563,976 (¥388 per ton) were high and spent largely in vain. (Author's collection)

The sloop *Kaimon* was built under the 1877 building programme. She was laid down on 1 September 1877, launched on 28 August 1882 and was completed on 3 March 1884. She took 6 years 5 months to complete because construction often had to be suspended due to a shortage of manpower and funding (building costs amounted to ¥615,756). Designed for coast defence under the direction of Rear Admiral (naval constructor) Akamatsu Noriyoshi, she was rigged as a three-masted barque, but had also a horizontal, surface condensing, return-connecting, double expansion engine which developed 1,267ihp. After the Sino-Japanese War the rig was much simplified: fighting tops were fitted and Gatling guns mounted, sails removed. (Author's collection)

since the 1850s, when the Dutch naval missions took place. The new policy of theoretical education first, followed by practical application, was continued by the Meiji government without substantial revision.

Strictly speaking this programme covered the language and shipbuilding schools only. However, it extended to the informal training of the crew of the corvette *Fujiyama* by crew members of the French frigate *La Guerrière*.

Language School

Even before the conclusion of the building contract for the Yokohama Iron Mill and Yokosuka Shipyard, Roches proposed the establishment of a language school to train Japanese translators in various fields, particularly the specialised technical branches necessary for the two projects. Remembering the problems at Nagasaki created by the rejection of the proposal of Gerhardus Fabius for the workers to learn Dutch before undertaking that project, the Privy Councillors agreed,[31] and the (younger) Privy Councillor Sakai Tadasuke was ordered to select a suitable site at Yokohama. Instruction began on 1 April 1865 with 23 students, all sons of 'high-ranking' families, ranging in age from 14 to 20. Mermet de Cachon was appointed director and Roches 'president'. The envoy was also asked to examine the students' progress. There is not the slightest doubt that the language school was useful for both sides: it contributed to the strengthening of the ties between Roches and the *Bakufu*'s Privy Councillors, increased support of the *Bakufu* by France in educational matters, and gave the students an insight into French civilisation – the curriculum was not restricted to purely technical matters.

Shipbuilding School

Known as *Yokosuka Seitetsusho Kosha*, this was the first scientific and technical school in Japan, and the only one in which naval architecture was taught as a special subject. It was established following a proposal Verny submitted to the *Bakufu* on 25 March 1867. He stated that:

> The organisation of the shipbuilding school is most important and should be implemented urgently. The purpose is to educate the young people to be capable to deal with the problems which the Japanese Government will face in the future.

The students, aged between 17 to 21 years, were to be examined in the French language before entry to ensure that they were able to follow the lectures. The course lasted three years. Beside the main subjects of algebra, geometry, and mechanics the students would also be instructed about finance. Two students at the top of each class were to be sent to Europe to continue their study for an additional two years.

These and other details of Verny's proposal indicate that he had in mind a syllabus similar to that of the French Naval Architecture School at Cherbourg, and reflect his vision of the 'Yokosuka Iron Mill' as a classical western-style shipyard operated exclusively by Japanese nationals.

Tenryu was the last of the series of five wooden sloops and was originally to have been a sister of *Kaimon*. However, errors were made in the stability calculations, and when she completed trials on 26 February 1885 her centre of gravity was found to be significantly higher than that of *Kaimon*, due in all probability to changes in fittings and equipment, together with a 40 per cent increase in coal bunkerage, on similar dimensions. When she was handed over to Yokosuka Naval Station on 5 March 1885 it was decided to fit wooden bulges (fabricated from full timbers) to improve stability, and *Tenryu*'s entry into service was delayed until the spring of 1886. After the Sino-Japanese War she was modified in similar fashion to her half-sister *Kaimon*. (Author's collection)

The first students were accepted in May 1867. They were divided into two classes:

1 the educated sons of *Samurai* families who were to be educated to technician (*gite*) and engineer (*gishi*) rank
2 the sons of farmers, merchants, etc, who were to be educated as skilled workmen and foremen.

The former group was composed of four engineer and ten technician students – the students in the latter group numbered only nine at first. The lectures were mostly given by the (French) heads of the various departments. Education for workers/foremen and technicians was divided into practical instruction (morning) and theory (afternoon).

This shipbuilding school existed for less than two years and was temporarily closed by the order of the Meiji government in 1868, but would be reopened again in 1872 (see below).

Informal Training of the Crew of the Corvette *Fujiyama*

The corvette *Fujiyama*, built in New York, entered Yokohama on 23 January 1866 via London. She was

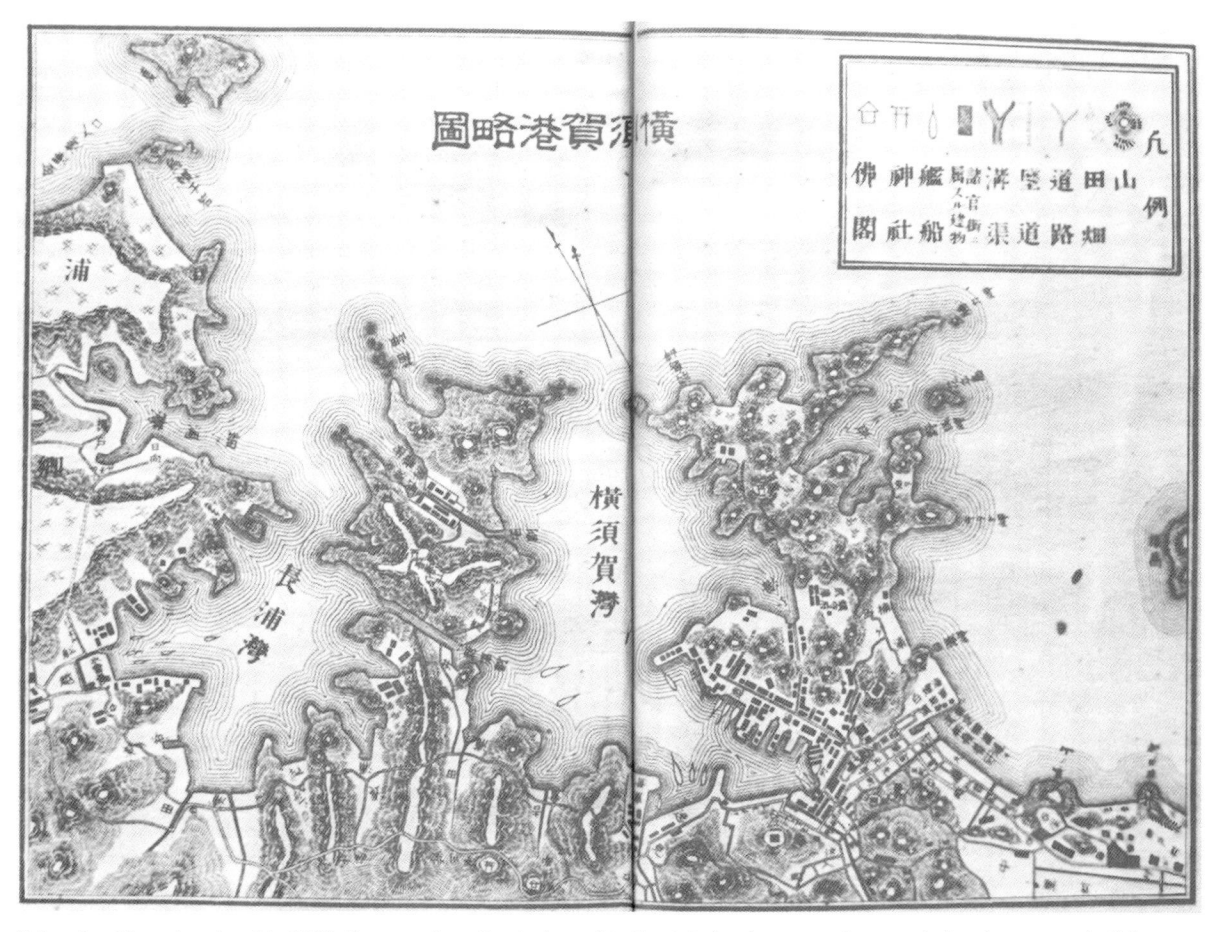

Yokosuka shipyard and port in 1888. The area above the docks and to the right has been greatly expanded and many new buildings can be seen. This map served as a 'guide to the port'. (US Navy)

manned by an American crew. She was handed over to the *Bakufu* on 20 February and anchored in that port. She was the only ship delivered of the three vessels which had been ordered from the USA via the American envoy Pruyn in August/September 1863 and for which the *Bakufu* had prepaid the sum of $637,000. The Shogunate government had announced the second campaign against Chōshū in May 1865 and was impatient for her arrival, and the crew would need to be trained to operate the ship in the shortest possible time. Because all preparations had been made in advance, four men of the crew of the French frigate *La Guerrière* began this duty, probably on the day after delivery. However, training had to be halted in May due to the protest of the British representative.

The instructors were officer 2nd class Bari, a petty officer (named Conquer?), an engineering petty officer (Bège?) and gunnery petty officer Suffren. The instruction of the Japanese officers Nagato Seizō,[32] Kobayashi Bunshirō, Matsumoto Koshirō, Shimazu Bunsaburō, and Matsumoto Kyotarō), petty officers and ratings (total 174 men), was directed by Bari. It appears that no documents relating to this instruction have survived. However, because the crew had to be trained for the operation against Chōshū, data in *History of Japan's Modern Shipbuilding: Meiji Era* indicate that the priorities were gunnery training (handling, clearing for action, direction by the gunnery officer and the gun captains, fighting signals, ammunition handling, the use of broadside guns and end-on firing guns), tactical manoeuvring and daily routine.

Fujiyama was armed with US Parrot guns: one 100pdr, three 30pdr, two 24pdr and two 12pdr. They were the first guns of this type mounted on a Japanese warship, so Suffren had much to do.[33] During the bombardment of Chōshū troops near Kokuraguchi, an accident occurred with one of these guns: a barrel burst, killing the gun crew and wounding others. This accident resulted in the rejection of the Parrot gun due to lack of reliability.

A contemporary article in the *Japan Times* about the *Fujiyama* and the training of the crew by foreigners contained a number of interesting insights. It stated that '… the Japanese government has accepted the *Fujiyama* … but is said to be unhappy with the disproportion between the price and the quality of the ship …' – a common complaint of the period. The writer then cautioned that 'Japan's potential of can be realised only if the preceding instruction by foreigners is overcome, and failure to achieve this will mean that [the *Fujiyama*] cannot be more than an expensive toy ..' However, the time was not yet ripe for the proposed level of independence. Lessons were still being learned from the early breaking off of instruction by the Second Dutch Naval Mission in 1859. The deficiencies in the training of the *Bakufu* navy were now being recognised, and the logical consequence was the perception that the support of foreigners could not yet be dispensed with.

Principal Sources:

Meron Medzini, 'Léon Roches in Japan (1864–1868)' in *Papers on Japan*, Vol 2, Harvard College (1963).

Dr Kajima Morinosuke, *Geschichte der japanischen Außenbeziehungen* (History of Japanese Foreign Relations) Vol 1 *Von der Landesöffnung bis zur Meiji Restauration* (From the Opening Up of the Country until the Meiji Restoration), Ed Horst Hammitzsch, translator Klaus Kracht, Franz Steiner Verlag GmbH (Wiesbaden 1976).

Beasley, WG, *Select Documents on Japanese Foreign Policy 1853–1868*, Oxford University Press (Oxford 1955).

Katsu Awa, *Kaigun Rekishi* (1888).

Honjo Eijiro, 'Léon Roches and the Administrative Reform in the Closing Years of the Tokugawa Regime' in the Kyōto University Economic Review Vol 10, No 1 (July 1935).

Shinohara Hiroshi, *Hired Foreigners of the IJN from the Bakumatsu Era up to the Russo-Japanese War* (*Nippon Kaigun Oyatoi Gaijin Bakumatsu made kara Nichiro-Sensō*) Chuōkōron (1988).

Endnotes:

[1] According to Katsu Awa, *Kaigun Rekishi*, Vol 202, the hiring of Dutch engineers for the building of the iron mill had been considered – the competence of Hendrik Hardes and most of his co-workers had been praised during the building of Nagasaki Iron Mill – but was abandoned because of the condition of the ships built in the Netherlands, which wore out quickly and required costly maintenance repairs.

[2] One exception was the 'punishment' action against Chōshū on 20 July 1863, after the French despatch vessel *Kienchang* was fired on in the Shimonoseki Strait on 8 July.

[3] After the outbreak of the US Civil War American trade was markedly reduced (in 1865 it was only about 2.5 per cent of the value in 1860 at Yokohama), while the volume of British trade increased from 55 percent to 85 percent in the same port during the same period. France's share rose from 0.8 per cent in 1860 to 8.16 per cent in 1865.

[4] During the 'crisis year 1863', Colonel Edward John Neale had represented Britain as *chargé d'affaires*. He had participated in the bombardment of Kagoshima; this was subsequently debated in the House of Commons, where the British Government had to answer some awkward questions.

[5] Medzini, *op cit*, 199.

[6] Kurimoto was appointed Inspector in July 1864 and received by the *Shōgun* the following month. At that audience he was instructed to support foreign commissioner Takamoto Masao in the negotiations about the closing of the ports. Kurimoto had become friends with the French missionary Abbé Mermet Chachon (known in Japan also as Washun), to whom he taught Japanese. Cachon came to Edo in September 1859 together with the first French consul general de Bellecourt and quickly rose to the post of secretary. In his role as Inspector, Kurimoto often met Roches with Cachon acting as mediator.

[7] The *Bakufu* government and various *Daimyate* purchased warships from several countries (mainly via the foreign merchant houses residing in Japan) after the forceful ending of the seclusion policy. However, these ships were mostly old and worn-out, and were sold at exorbitant prices.

[8] The repair was directed by the engineer Jules César Thibaudier, who became Verny's deputy at the Yokohama Iron Mill and the engineering officer of the frigate.

[9] Takahashi states that the following gifts were made: five suits

of clothes for Roches, a sword for Jaurès, a *wakizashi*(?) for the captain of the *La Guerrière* and 100 Koban for the engineering officer.

10. Oguri, born in 1827, had been a member of the diplomatic mission to the USA as Inspector for Foreign Relations (*Gaikoku bugyō*). In December 1860 he was appointed Commissioner for Foreign Relations (*Gaikoku bugyō*), and in July 1862 was appointed Finance Commissioner (*Kanjō bugyō*). He held this post until January 1865.

11. Oguri's view was influenced by the severe defects discovered in *Kanrin Maru* and the behaviour of the American representative following the order of warships and the payment of the larger part of the expenses in advance. The government lost confidence in the Netherlands and the USA, and Britain was rejected as a potential partner due to its close relations with Satsuma and Chōshū. The *Bakufu* therefore turned to France, an acknowledged first-class shipbuilding country with a long tradition and dominant influence. Oguri was of the view that warships had to be built domestically to prevent the expenditure of large sums of money on ships of doubtful military value.

12. Ono Tomogorō's book *Coast Defence of Edo* (*Edo Kaibō Ron*) was submitted to the Cabinet in September 1864 and provided a new impetus. Ono proposed the building of a large shipyard at Edo and identified a suitable location in Yokosuka Bay.

13. During the *Bunkyu* era (1861–63) the political situation had deteriorated and the power of the *Bakufu* had declined. Several of the *Daimyō*s were in open rebellion against the Tokugawa *Bakufu*.

14. The chief engineer of the French Asia Fleet had declined to take on the project, citing his lack of expertise in the building of warships and dockyard infrastructure, and proposed the hiring of a suitable engineer. Jaurès recommended Verny to Roches and suggested his appointment.

15. Soon after beginning the building work, in May 1866, the location was inspected by the British envoy and two officers, among them Lt (Eng) Bond. He highlighted in his report the water depth at the entrance, the favourable location of the bay, which was protected on all sides by mountains, the excellent possibilities for defence of the harbour mouth by the arrangement of coastal batteries on the mountains, and the proximity to Yokohama. Thus the investigations by the French, the Japanese and the British produced identical results.

16. The larger part of the total amount was paid for by the export of copper and other metals, and $1,000,000 were borrowed from France with the export of silk as security, but this back-up was never called upon. Claims by other powers that the agreement violated the principle of free trade, and the reproach 'weapons for silk', were exaggerated and inaccurate.

17. Verny used the term 'arsenal' (Jap *senshō*) but the Japanese translation used 'iron mill' or 'foundry' (*seitetsujō*). It appears that the Japanese central government still considered the production of iron plates, iron pipes and profiles as the primary purpose of the project, with a view to building steam engines and steam boilers, even though Verny's draft listed all shops (from the sail loft and ropery to the foundry) necessary for the construction of modern ships.

18. In the original the terms 'ship repair and shipbuilding yards' were used.

19. In the Japanese documents Verny is referred to as 'steam engineer' (*Yokiyakushi*) – the same term was used for the Dutch Hendrik Hardes at the time of the building of Nagasaki Iron Mill.

20. This enumeration is not complete. There were actually 39 personnel, divided into ten professions (from caulker to material administrator), of which eleven were to operate in leading positions (technicians and foremen) and 26 as skilled workers. At the instigation of Verny one doctor of medicine, one chief of the construction (drawing) section and one foreman were added later. All the hired Frenchmen, total 43 people, had arrived at Yokohama by 9 November 1866.

21. The Japanese source does not explain why 'pounds' instead of the generally-used American currency are stated. Judging from the relative value of US dollars and the British pound sterling and the positions of these two men, 'pounds' may be a printing error.

22. The total responsibility was with Verny; he decided the operation and all other regulations, etc; the Japanese general director was put under his 'command'.

23. This also included the manufacture of steam boilers and engines.

24. This amount covered 3.42 per cent of the total budget, and illustrates how small the Yokohama plant was compared to the Yokosuka enterprise.

25. All buildings were constructed on the same system. On a foundation wall made of square stones frames were erected which supported the roof; the intervals between the frames were then filled with bricks. This system was selected because these materials were readily available.

26. Verny used this wood for the hull of the gunboats built at Ningpo. It rotted comparatively quickly and the service life of these vessels was only 5–6 years. Following this experience use of the wood was restricted; in addition it was costly and the production of parts with large dimensions was difficult.

27. Probably Matsudaira.

28. Due to the lengthening and other measures the calculated costs rose to 19,398 Ryō, and after completion they amounted to 23,607 Ryō

29. 'Note sur l'arsenal Maritime de Yokosuka' in *Revue Maritime et Coloniale*, Vol 73, May 1873, 595–612, states that the building of the dock was facilitated very much by the type of rock (marl), which was resistant but permeable. To use this geological advantage the dock was excavated at a point where a hill had been broken up and the ground had maintained its natural resistance.

30. Before Oguri's offensive the Cabinet had decided to expand the Ishikawajima Shipyard, situated centrally on an island off Edo, instead of the Nagasaki Iron Mill. Hida and his two specialists were sent to the Netherlands in September 1864 to learn about the operation of a modern western-style shipyard and to purchase the necessary materials, machines and tools for the expansion of Nagasaki Iron Mill. However, when Oguri's and Kurimoto's activities resulted in the establishment of the Yokosuka Shipyard, the expansion of the Ishikawajima Shipyard became superfluous, and the rationale for learning how to manage a shipyard from the Dutch ceased.

31. Note that the 'execution regulations' required the learning of the French language even in leisure time. Linguistic proficiency was now recognised as a key qualification for success.

32. Nagato became a teacher at the Naval Academy at Tsukiji in 1877.

33. According to Shinohara the 100pdr and the 30pdr had a gun crew of six (one gun captain and five operators). and in the second campaign against Chōshū the gunnery officer Shimazu Bunsaburō transmitted the orders to the gun captains by whistle signals learned from the French instructors.

ESPLORATORI OF THE REGIA MARINA, 1906–1939

The fast, well-armed 'scout' cruisers built by the Italian Navy prior to and during the Great War were much admired, particularly by the French. **Enrico Cernuschi** looks at the development of these ships, their employment in the confined waters of the Adriatic, and the evolution of the type during the interwar period.

During the time of the so-called *guerres en dentelles* ('Lace Wars') of the 18th century, the captain of a ship-of-the-line could be charged with murder if he attacked a frigate without first giving her a chance to strike her colours. This was because the smaller warship had no chance against her more powerful opponent. For example, on 8 June 1794 the Sardinian frigate *Alceste* (36-gun) seriously damaged the French frigate *La Boudeuse* (32-gun), but was herself seized by the ship-of-the-line *Le Tonnant* (80-gun) after a brief action.[1] It was a matter of structure and punch: a frigate's 12-pounders could not seriously damage a ship-of-the-line's wooden walls, while the 32-pounders of a third rate or larger warship would easily penetrate the hull of a frigate.

For these reasons, from the mid-17th century only ships-of-the-line mattered when it came to measuring the strength of a battle fleet. The smaller and faster sailing vessels, from frigates downwards, served as raiders, trade escorts, despatch vessels or scouts for the squadrons of the main battle fleet. The advent of steam propulsion did not change this state of affairs, as naval guns remained confined to a maximum range of 1,000 yards until the 1880s, when new technologies began steadily to increase battle ranges.

Scouting

A typical case that illustrated the enduring relationship between capital ships (by now ironclads) and scouts occurred during the Battle of Lissa on 20 July 1866, when two Italian *esploratori*, the paddle despatch vessels *Esploratore* and *Messaggero*, accomplished their tasks of scouting the batteries of the island of Lissa and signalled the approach of the Austrian battle fleet. During the following battle they were tasked with repeating the signals of Admiral Persano to the squadrons. Laid down in Britain four years previously, they were among the fastest warships afloat (16–17 knots compared to the 8–13 knots of the ironclads) and armed with two muzzle-loading 4.7in (120mm) guns. They were unarmoured; speed was their only protection. They also had very tall masts for signal flags.

The first cruisers ordered during the 1870s were unprotected iron – later steel – frigates and corvettes mainly intended to be stationed abroad during that great commercial and colonial era. The first generation of armoured cruisers, from the Russian *General Admiral* class laid down in 1870 to the revolutionary French *Dupuy-de-Lôme* ordered in 1887, were the ancestors of the German pocket battleships – ocean raiders able to face the enemy's trade escorts. They were not a true cruising (*ie* scouting) force for the battle line, as they had a mere ⅔-knot advantage at best over their own battleships.

Driven by the increased ranges of the newer main battery guns and the introduction of more powerful explosive shells, tactics evolved towards a doctrine that dictated that actions would be fought by line-of-battle broadsides, favouring the force that opened fire from the better position. By this time warships were faster (18–20 knots). The scouting vessels, which needed to steam at a much greater speed than the battle line, had to deliver, using optical systems, the information the admiral needed to gain the desired position relative to the enemy. Armoured cruisers still had an insufficient speed margin to perform this role effectively, while the protected 2nd and 3rd class cruisers introduced from 1880, which featured a thin armoured deck and guns in open, shielded mountings were essentially 'colonial' warships; they were often slower than battleships and useless for scouting in a fleet action.

It was therefore necessary to invent a new type of scouting platform. Such a vessel had to be fast, but with much better seakeeping qualities than the small torpedo cruisers,[2] torpedo boats and destroyers then entering service. As triple expansion engines (TE) were wasteful in terms of space, weight, water and fuel relative to the power they delivered, the technology of the day limited size, protection and weapons to a scale not too different from the *Esploratore* commissioned in 1863.

The first expression of this new type of warship was the Russian *Novik*, the brainchild of the famous Admiral Stepan Osipovich Makarov. Conceived in Russia and built by Schichau in Germany between 1898 and 1901, she was a 3,038-tonne vessel with a top speed of 25 knots and a main armament of six 120mm/45 guns. Two very tall signal masts gave away the presence of the ship from a distance, but with W/T still in its infancy there was no other way to pass on the information that constituted the very purpose of such a vessel.

In a sudden sortie from Port Arthur on 10 March 1904, *Novik* demonstrated the utility of a fast and (for her size) powerful warship by rescuing a pair of Russian destroyers being engaged by two Japanese cruisers. She was lost in the Far East on 20 August 1904 after being trapped inside a bay of Sakhalin Island by the Japanese protected cruiser *Tsushima*.

The subsequent evolution of the scout cruiser is well known. The next expression of the type was the 1903 British *Adventure* class, the design of which featured turbines instead of reciprocating engines plus a classical 1–2in protective deck able to stop splinters and minimise the effects of 4.7in shell. There followed a series of 'light cruisers' epitomised by the Royal Navy's *Chatham* class, authorised in 1911. Protection now became part of the main structure of this type of warship, which incorporated a short 2in belt. This arguably added little to the ship's survivability or fighting efficiency beyond the ability to protect the machinery from German destroyers and protected cruisers, as 'the placement of the armour, the extent of its coverage and the quality of the plating'[3] were, at best, debatable.[4]

Rapidkreuzer versus *Esploratore*

Warships (including experimental models) are always the fruit of naval policy and thus the expression of a government's choice. In the case of the Italian Navy, its instructions from the cabinet were, from the spring of 1871, quite clear: the interests of Berlin and Rome from 1861 were aligned against Vienna, thereby precluding a Habsburg return to Italy, so France was the most likely enemy of the young Italian kingdom. As France's GDP was twice that of Italy, and the *Regia Marina* needed to protect long and open coastlines, the new kingdom could not possibly match France's *Marine Nationale* in terms of numbers. The answer conceived by the two principal Italian naval thinkers of the day, the future Admiral Simone Pacoret de Saint Bon and General (E) Benedetto Brin, was a revolutionary generation of battleships. The *Duilios* were more powerful, better protected and faster than any ironclad. They were followed by the two *Italias*, a sort of proto-battlecruiser. Speed and hitting power became the trademarks of Italian warships, although at the expense of protection, as the new battle fleet had to trade fighting quality for numbers.

In the longer term this was not a bad bargain. The Italian battleships and armoured cruisers commissioned from the early 1880s and over the next two decades allowed Italian policy at home and abroad to face up to the French threat at a cost approximately half that paid by Paris. Because Italy's battle force (formed by battleships and armoured cruisers) was 2–3 knots faster than the French battle fleet, Italy did not need a faster reconnaissance squadron like other navies of the time.

The situation changed in 1902 when the Austro-

The scout *Quarto* the day of her launch at Venice, 19 August 1911. (Ufficio Storico della Marina Militare – USMM)

Hungarian Navy concluded, after a series of landing exercises, that the time of the coastal battleship was at an end. Habsburg fears of an Italian invasion of Istria and Dalmatia could be countered only by a blue-water battleship force.[5] The first step was the 1906 order of the 'semi-dreadnoughts' of the *Radetzky* class, together with a 3,500-tonne *Turbinen Antrieb Kreuzer* (turbine-powered cruiser), the future *Admiral Spaun*. The battle of Tsushima had recently demonstrated that the range of future naval engagements would be at least 5,000 yards, and the constant evolution of fire control suggested ranges would continue to increase. The British had already begun a new naval race with their *Dreadnought* and both the Austro-Hungarian and Italian constructors were undertaking studies for new all-big-gun battleships. Increased gunnery ranges and the belief that 'crossing the T' of the enemy line's was a tactical requirement – a solution that implied a speed advantage of at least a 5 knots during daylight, and which the greater ranges made almost impossible to achieve – required early reconnaissance by a scouting squadron much faster than the battle force. The British and German response to this new task materialised, after 1905, in the innovative (if soft-skinned) 'battle cruisers', but these very expensive warships were out of the question for the Austro-Hungarian, Italian and even French navies of the time.

The Austro-Hungarians completed *Admiral Spaun* in November 1910. She was the first major turbine-powered warship of the KuK *Kriegsmarine*, but throughout her life an unreliable propulsion system caused problems, and during the Great War the naval command kept her in the Upper Adriatic. She was not, in fact, a true scout but the first AH light cruiser, as she had a short 60mm belt and 50mm box protection for her magazines as well as a 20mm splinter deck.

The Italian response to *Admiral Spaun* was the *Quarto*, ordered in 1907 and commissioned in 1913. That *esploratore* (scout) was the logical complement to the *Regia Marina*'s first dreadnought, *Dante Alighieri*[6] and, unlike *Admiral Spaun*, quickly acquired the reputation of an excellent ship that endured until she was paid off in 1939. Her protection comprised only a 40mm vaulted deck; this was considered sufficient given the long-range nature of the actions against the similar enemy vessels a scout would typically encounter during a reconnaissance mission. *Quarto* had two boilers with mixed firing as her designers envisaged that, despite her relatively limited range, she would be used abroad, mainly in Latin America, to show the flag; habitability and the elegance of the fittings were therefore improved.

During the lengthy spell as Minister of the Navy (11 December 1903 – 12 December 1909) of Admiral Carlo Mirabello, who was followed (until April 1910) by his tutor Admiral Giovanni Bettolo,[7] the *Regia Marina* was a hotbed of new ideas. The search for a faster scout induced staff to conceive and order, before *Quarto* was launched, two ships of a new type of scout, this time designed by Captain (E) Giuseppe Rota: *Nino Bixio* and *Marsala*, launched in 1911 and 1912 respectively; both entered service in 1914.

The *Bixio*s had a cutter bow – which proved to be an unhappy choice. Their protection matched that of *Quarto*, and their five-gun 120mm (4.7in) broadside – the two mountings amidships were disposed *en echelon* – was an improvement over *Quarto*'s four guns. Their machinery, based on Curtis turbines manufactured under licence in Italy, was disappointing, however. Despite modifications made to *Marsala*'s boilers in 1915–1916, these ships were always considered inferior to *Quarto* both for flank speed (26–27 knots *vice* 28) and reliability. The two new Italian scouts were commissioned in 1914, at almost the same time as *Helgoland* and *Saïda*, the first two Habsburg improved successors of *Admiral Spaun*, which were followed in 1915 by *Novara*. These three Austro-Hungarian warships were excellent scouts, without the defects of the prototype, and were able to sustain speeds of 28–29 knots for many hours. Their protection was the same as *Admiral Spaun* except for the magazine box, the thickness of which was increased to 60mm.

After struggling with *Bixio*'s turbines, Italian naval staff undertook the study of a revolutionary propulsion system for the next class of scout: gas turbines. This technology was in use on land at the end of the 19th century, but was it still a novelty at sea. An experimental model of the Armengaud-Lemale type was manufactured at the Spezia yard in 1911 and tested the following year, while Major (E) Filippo Bonfiglietti designed a new class of scout that attempted to combine the best qualities of *Quarto* and the *Bixio*s. This type was unarmoured except for the usual protected deck, but in 1913, following the Austro-Hungarian, British and German examples, the

Admiral Carlo Mirabello. He was still remembered in the 1960s as one of the most forward-looking heads of the Italian Navy, with an ability to consider the military, technological, political and economic aspects of any problem. (USMM)

WARSHIP 2022

The scout *Quarto* in 1925. (USMM)

The scout *Nino Bixio*. On paper she was an improvement over *Quarto*, but only two units were built; they were commissioned in 1914 and paid off in 1927 (*Marsala*) and 1929 (*Bixio*) respectively. (USMM)

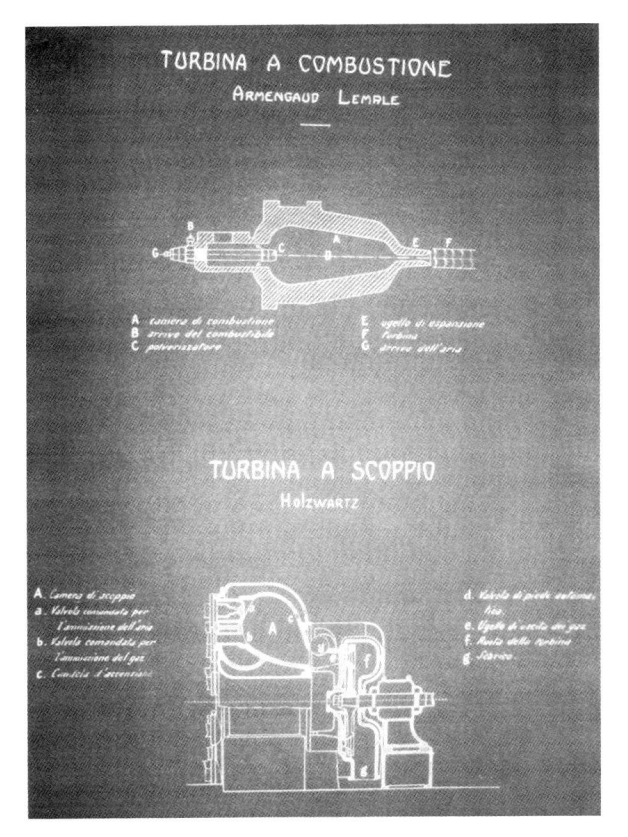

ESPLORATORI OF THE *REGIA MARINA*, 1906–1939

Plans of the gas turbine tested at La Spezia in 1912 for the class of scouts that would follow the *Bixio*s. The technology was too advanced for its time, and the *Regia Marina* did not again attempt development until 1938. The Italian Air Force, which was always independent of the Navy, preferred during those same years a Swedish solution for its own future gas turbines. (USMM)

upgraded armour and a heavier main armament based on a newly-designed 152mm/50 gun with high muzzle velocity that the Ansaldo company was promising to deliver in 1918.

In 1914 three new *esploratori* of the *Mirabello* class were ordered. The press announced they would be 5,000-tonne warships armed with eight 152mm/50 in single mountings (see accompanying artwork), and with general characteristics as follows: length 145.80m wl, 137.40 pp; beam 14.40m and draught 4.68m; 4-shaft Parsons turbines with steam supplied by four oil-fired boilers and two with mixed firing; designed speed was 29 knots with 42,000shp.

The KuK *Kriegsmarine* followed that same year with a similar solution for its three planned *Rapidkreuzer* (fast cruisers) of the *Ersatz Zenta* class. This design adopted a 120mm gun and had upgraded protection with a 20mm belt and a 38mm deck. The *Rapidkreuzer* were to support the four projected 21-knot battleships of the *Ersatz Monarch* class, which were to be armed with ten 35cm (13.8in) guns. On the other side of the Italian peninsula the French, finally abandoning their faith in armoured cruisers as a scouting force, were planning to build ten scout cruisers of the *Lamotte-Picquet* class, which featured a 28mm belt that covered the vitals only. This was part of a major 1912 programme that also included nine 22/23-knot battleships of the *Normandie* and *Lyon* classes – all in an effort to recoup the lost years between 1906 and 1911 spent constructing the six 'semi-dreadnoughts' of the *Danton* class.[8]

War in Europe

When the Great war began the three 'Admirals', *Carlo Mirabello*, *Carlo Alberto Racchia* and *Augusto Riboty*, had yet to be laid down. The immediate British and French naval blockade stopped the deliveries of high-tensile steel for their hulls, leaving enough material for only a single ship. In the summer of 1914 Admiral Paolo

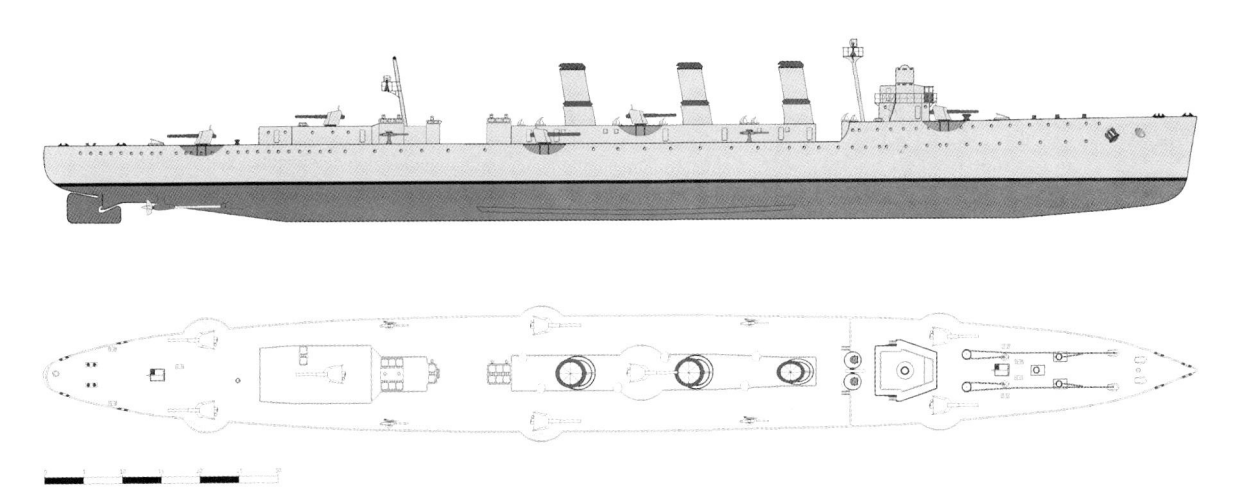

Italians decided to incorporate a thin waterline belt into their new designs. However, the weight and space gains promised by the gas turbines did not materialise, and Bonfliglietti undertook a new, larger project with

The 5,000-tonne *Mirabello* as originally designed in 1913. She would be superseded by a ship of the same name but displacing less than 2,000 tonnes. (Drawn by Fabrizio Santi Amantini)

151

Thaon di Revel, the *Regia Marina*'s Chief of Staff, decided to start from scratch a new type of *esploratore leggero* inspired by the previous *esploratori leggeri* (light scouts) of the *Poerio* class – *Alessandro Poerio*, *Cesare Rossarol* and *Guglielmo Pepe* had originally been ordered in 1913 as long-range destroyers to serve as escorts for the German Mediterranean Squadron in operations against French traffic with North Africa.[9] The new ships would undertake scouting tasks for the battle fleet originally planned for the 5,000-tonne *Mirabello*s. In a *tour de force* Captain (E) Nabor Soliani proposed a sort of super-destroyer, fast, with a powerful armament and, of course, no protection whatsoever. Three new warships, bearing the same names as the original *Mirabello*s, were ordered and laid down between November 1914 and February 1915. Commissioned in 1916 and 1917 they were an excellent solution, performing well during the Great War in a variety of different roles.[10]

This innovative project was then repeated, in an upgraded version, with the five ships of *Leone* class ordered in January 1917. *Leone*, *Tigre* and *Pantera* were completed in 1924, again delayed for lack of steel, as the German submarine war curtailed Italian imports during the final two years of the Great War, while *Leopardo* and *Lince*, on which little work had been done, were scrapped on the slipway during the early 1920s.

Carlo Mirabello, an imaginative 1,800-tonne replacement for a 5,000-tonne scout. (USMM)

The four *esploratori leggeri* of the *Aquila* class had different origins. Conceived as a private venture by engineer Luigi Scaglia, they were ordered by Romania in 1913 from the Pattison Yard of Naples and requisitioned in June 1915 by the *Regia Marina*. Conceived for hit-and-run missions in the Black Sea against the Russians (Bucharest was considered, until 1914, a probable ally of the Triple Alliance), they were completed between 1917 and 1920.

The scout *Poerio* during the Great War. She was actually little more than a destroyer. *Racchia* was sunk by a mine in the Black Sea on 21 July 1920, and the two surviving warships of this class, *Poerio* and *Pepe*, by now worn out, were sold to the Spanish Nationalist Navy in 1938 becoming, respectively, *Huesca* and *Turel*. (USMM)

The KuK *Kriegsmarine*, badly affected by the Allied blockade, could only dream of obtaining similar vessels. As a paper exercise, a series of 2,000–2,440-tonne warships were projected in 1918. They were to be armed with 15cm guns and to be capable of 34 knots, but the Empire could not hope to build such a vessel during the war.[11]

The French, whose destroyer flotilla in the Adriatic was plagued by an excessive number of machinery defects between 1915 and 1917 and was retired by the spring of 1918, observed the Italian *esploratori* and *esploratori leggeri* and the Habsburg scouts with the greatest attention, and found inspiration for the fast and unprotected light cruisers of the *Duguay-Trouin* class and the *contre-torpilleurs* of the *Jaguar* class authorised under the 1922 Programme

Allies, Not Friends

The Italians, having only three *esploratori* facing four Austro-Hungarian *Rapidkreuzer*, asked for British help in early May 1915. The Royal Navy agreed to deploy a pair of light cruisers to Brindisi, but refused Rome's requests to buy two of its ships. At first the Italian Navy had thought to purchase the two former Greek cruisers of the *Katsonis* class, requisitioned by the British on the stocks and completed as *Chester* and *Birkenhead*. Then the Italians asked for two cruisers of the *Arethusa* class, and later that same year declared that they would be satisfied with the even older scouts of the *Active* class, but the British refused to contemplate any such transfers. A key factor was that such a transaction would have cost London all influence in the Adriatic theatre of war. The naval partnership of the two allies was far from ideal: the Anglo-Italian alliance in the Lower Adriatic was characterised by formal correctness and numerous disagree-

The newly-commissioned light scout *Aquila* at Brindisi in February 1917. Ordered by Romania in 1913 as *Vifor*, *Vijelie*, *Vartez* and *Viscol*, the four ships of this class were renamed *Aquila*, *Sparviero*, *Nibbio* and *Falco*. The last two were sold to Romania again in 1920 and renamed *Mărăşti* and *Mărăşeşti*, while in October 1937 the first pair became the Spanish nationalist destroyers *Melilla* and *Ceuta*. Prior to their transfer a fourth (dummy) funnel was added so that they could be mistaken for the smaller Nationalist destroyer *Velasco*. (USMM, Aldo Fraccaroli collection)

ments, foreshadowing the unhappy partnership between the *Regia Marina* and the German *Kriegsmarine* in the Second World War.

On 29 December 1915, following a lengthy pursuit up and down the Adriatic, the Austro-Hungarian light cruiser *Helgoland* escaped an Allied trap formed by the *Dartmouth* and *Quarto* on one side and *Bixio* and *Weymouth* on the other with only light damage. According to the Italians, *Weymouth* made a wrong turn, thereby squandering the only chance the slower Allied cruisers had to cut off *Helgoland*. The British pointed the finger at the Italians, blaming slow and unreliable communications between the Allies.

Worse was to follow on 15 May 1917, when the crippled *Saïda* and *Novara* were saved from a stronger force

Augusto Riboty, one of *Mirabello*'s two sisters. The third ship of the class, *Cesare Rossarol*, was sunk by a mine on 16 November 1918. *Mirabello* and *Riboty* served in both world wars. (USMM)

Admiral Paolo Thaon di Revel, the head of the Italian Navy during the Great War. He declined the role of Navy Minister in 1925 in protest against Mussolini's decision not to proceed with the carrier he had promised Thaon di Revel in 1922 to induce him to join his newly-appointed government. Believing in the decisive value of reconnaissance, the Italian admiral had from the outset been an advocate for naval aviation and the *esploratori*. (USMM)

by a lucky hit that struck the British cruiser *Dartmouth*. The Italian commander at sea, Admiral Alfredo Acton, who had embarked in *Dartmouth* because she was the first cruiser ready to sail, reported that, following a direct hit, the ship's CO had ordered a turn to port and the flooding of the forward magazine, and claimed that this effectively prevented the destruction of the Austro-Hungarian cruisers before the arrival from Cattaro of armoured cruisers. The British, on their side, stated that the order to abandon the chase had been given by the Italian admiral. For the *Regia Marina* this was direct confirmation that the 'Nelson Touch' was a myth, and that there was an opportunity to reshuffle the respective cards in the Mediterranean after the war.

In 1917 the Italian Navy was studying a new 6,500-tonne *esploratore* armed with twelve 152mm/50 guns and able to make 36 knots; protection was, as usual, negligible. However, there was no real opportunity to build even a single warship of this type – a pair was proposed – during the war. The Navy then explored the possibility of purchasing from the United States two of the future light cruisers of the *Omaha* class that would been laid down in 1918, but no serious effort was pursued before the end of the war.

Since the main purpose of this article is to describe the naval policy behind the evolution of the Italian scouts, the technical data for these warships is limited to the table on page 156.

The Strategic Picture in the Lower Adriatic, 1915–1918

The three best Austro-Hungarian *Rapidkreuzer* were based at Cattaro as a raiding force from 1915. They had the advantage of the initiative, and could sail by night to attack enemy sea traffic and coasts, and retreat at dawn at flank speed to avoid interception by the enemy's cruisers before arriving under the protection of the

The scouts *Quarto*, *Marsala* and *Bixio* during the Great War. Between 24 May 1915 and 11 November 1918 the Italian scouts conducted a total of 710 offensive sorties and 164 escort missions for a total of 27,139 hours at sea. (USMM)

Austro-Hungarian shore batteries and of the 24cm guns of the two armoured cruisers based at Cattaro. It was a safe, well-established tactic that nevertheless paid few dividends. In twenty-eight KuK *Kriegsmarine* raids in the Otranto Channel between 24 May 1915 and 23 April 1918, Austrian surface vessels sank only two transports (*Japigia* on 21 April 1917 and *Carroccio* on 15 May 1917). During that same period Austro-Hungarian surface warships also sank an additional 17 coasters, tugs and small sailing vessels along the Albanian coast. The total losses recorded out of the 1,076 sailings between Italy, Albania and Greece during the Great War were 13 merchant vessels lost. The Allied trawlers and drifters patrolling the Otranto barrage between 1915 and 1918 suffered a little more, with losses from all causes of 4 per cent (mostly British).

The damage caused by the 10cm guns of the Austro-Hungarian scouts and destroyers during the eight shore bombardments they conducted against small towns in the Lower Adriatic, all of which were chosen as targets because they were undefended by gun emplacements, produced virtually no tangible results.[12] The bombardments were discontinued after December 1917, as the efficiency of the raiding force had by then been compromised by wear and tear.

On the other side of the hill the Allied surface vessels, including the *esploratori* and *esploratori leggeri* of the *Regia Marina*, conducted during that same period seventeen heavier (305mm/12in, 254mm/10in, 190mm/7.5in and 152mm/6in included) and longer-range shore bombardments in the Lower Adriatic, most of them against the Austro-Hungarian front in Albania. There were also fifteen raids against the Habsburg maritime traffic with Albania, which sank three merchant vessels and four sailing ships. Ten Italian incursions by MAS boats (MTBs) into enemy harbours in the Lower Adriatic caused the sinking of a further two merchant vessels. For a valid comparison it would be necessary to consider bombardments by Allied scouts and destroyers, but the hard truth was that Austria-Hungary could deploy only a fast raiding force in the Lower Adriatic, while the

The Austro-Hungarian scouts *Helgoland* and *Novara* at Cattaro during the Great War. The most important strategic task accomplished by the KuK *Kriegsmarine* raiding force formed by two armoured cruisers, three scouts and up to a dozen destroyers based in the poorly-equipped base of Cattaro, was to pin down double the number of comparable Allied warships in the Lower Adriatic bases of Brindisi and Vlora. (USMM)

Entente forces in that area included battleships and armoured cruisers. This was Sea Denial versus Sea Power, and in the long run the result was inevitable.

During the related eighteen actions fought periodically by Allied and Austro-Hungarian surface forces in the Lower Adriatic the final record was: two Italian and two Austro-Hungarian destroyers lost; seven KuK *Kriegsmarine* scouts, thirteen destroyers and two torpedo boats damaged, as against two British light cruisers, one Italian *esploratore* (*Bixio*) and one *esploratore leggero* (*Aquila*), two British, two French and one Italian destroyers. For the damaged warships the score in terms of direct hits was 39:15 in favour of the Allies.

The *Helgoland*s proved to be more robust than envisaged: they sustained 26 direct hits but all nevertheless survived the war, even though in poor condition. Leaks in the bottom plating caused *Novara* to founder in 1920 while she was under tow from Brindisi to Toulon.

The Postwar Renaissance

In 1919 Thaon di Revel requested two *esploratori oceanici* (oceanic scouts) for the new Italian fleet. The project was for a 4,100-tonne cruiser, with six single 152mm/50, a 50mm belt and a 25mm armoured deck, and a speed of 29 knots. These warships were to be able to serve abroad or, better, as ambassadors for the Italian shipyards and industries, as Italy was currently fully engaged in exporting to Latin America in an attempt to reduce its balance of payments deficit. Regarding armour, the theory was that protection on this scale would be sufficient at the long ranges (16–19,000 metres) envisaged for future naval actions. Budgetary constraints caused the programme to be deferred, as the allocation to Italy in 1920 of three German light cruisers and of two former Austro-Hungarian *Helgoland*s (not to mention the *esploratori leggeri Premuda* and *Cesare Rossarol II* and the two large former German destroyers *V 116* and *B 97*) meant there was no urgency. By May 1920 the two future *esploratori* had become a pair of 5,000-tonne warships that evolved, six years later, into the *grandi esploratori* (large scouts) of the 'Condottieri' class, in

The Austro-Hungarian scout *Novara* was the prototype for a series of fast light cruisers which proved a thorn in the side of the Allies during the Great War. She is seen here at Cattaro in 1915–16. (NHHC, NH-87448)

Esploratori: Table

Class	No of ships	Displacement	Horse-power	Speed (knots)	Main Guns (as built)	Protection Deck	Belt
Quarto	1	3,271t	29,215	28	6 x 120mm, 6 x 76mm	38mm	Nil
Bixio	2	3,575t	23,000	26	6 x 120mm, 6 x 76mm	38mm	Nil
Poerio	3	1,028t	20,000	31	6 x 102mm	Nil	Nil
Mirabello	3	1,784t	44,000	33	8 x 102mm[1]	Nil	Nil
Leone	5	2,195t	42,000	33	8 x 120mm	Nil	Nil
Aquila[2]	4	1,594t	40,000	36	3 x 152mm, 4 x 76mm	Nil	Nil
Di Giussano	4	5,110t	95,000	36.5	8 x 152mm	20mm	24 + 18mm
Cadorna	2	5,232t	95,000	36.5[3]	8 x 152mm	20mm	24 + 18mm
'Navigatori'	12	1,628t	50,000	38	6 x 120mm	Nil	Nil
Tashkent[4]	1	2,893t	110,000	39	6 x 130mm	Nil	Nil
'Capitani romani'	12[5]	3,686t	110,000	40	8 x 135mm	Nil	Nil

Notes:
1. The armament was modified in 1917 to 1 x 152mm, 7 x 102mm. After the Great War the ships reverted to 8 x 102mm.
2. Ordered by Romania; two were later again sold to Romania and two to Spain. The armament is the one initially planned in Italy; it was later modified to 4 x 120mm.
3. In 1939, following modernisation, ten of the ships were able to steam at a maximum 32 knots; the two umodified ships, *Usodimare* and *Da Recco*, were capable of 33 knots.
4. Built for the USSR.
5. Only three were completed during the Second World War; a fourth was completed to a revised design after the war.

which punch (now eight more ambitious 152mm/53 in twin turrets) and even higher speed was purchased at the expense of protection, which was designed to resist light shells and splinters only (belt 24mm and deck 20mm).[13]

The basic strategic concept, supported from 1923 by Admiral Costanzo Ciano, a Navy hero and the only member of that service to become, following retirement, an important member of the fascist regime, was to resurrect in the Western Mediterranean the old Adriatic Austro-Hungarian hit-and-run tactics, this time in big letters with twelve 5,000-tonne *grandi esploratori* of the 'Condottieri' class and 24 *esploratori leggeri* of the 'Navigatori' type. The new *grandi esporatori*, which were informally referred to as *contro-cacciatorpediniere* (destroyers of destroyers) were a response to the French *contre torpilleurs* of the *Jaguar* and *Guépard* classes. The two new types of warship, designed by Generals (E) Giuseppe Rota and Giuseppe Vian, were to constitute an expendable 38-knot force, with numbers offsetting their soft skins. Mussolini was enthusiastic, but the budget was halved following the world financial crisis of 1929 and only six 'Condottieri' and twelve 'Navigatori' were ordered. Both classes were, by the late 1930s, worn out and dated despite their impressive modern silhouette. By 1937 it was planned to convert the first four 'Condottieri' into antiaircraft ships or, as late as the winter of 1939–40, to sell them abroad (*Di Giussano* to Sweden). On 30 April 1940 the Italian Navy Command concluded that the first six 'Condottieri' were not fit to serve in the battle squadrons, but were expendable vessels to be used for special missions. In 1938 the 'Navigatori' were reclassified as *cacciatorpediniere* (destroyers), being by now

The light scout *Leone* during trials in 1924. Fast and powerful (eight 120mm guns in twin mountings) but with a modest range, these ships were well suited to the Red Sea theatre, where they served from 1935 until their loss in April 1941. They were reclassified as destroyers on 5 September 1938. By 1940–41 their age limited top speed to 28–30 knots. (USMM)

The scout *Quarto* at Massawa in 1936. Between the two world wars this ship spent much of her time stationed in China, East Africa and Spain. (USMM)

ESPLORATORI OF THE *REGIA MARINA*, 1906–1939

The light cruiser *Bartolomeo Colleoni* at Venice in 1932. The first four '*Condottieri*' were, from the mid-1930s, much criticised by the new generation of officers, as they were considered unseaworthy, poor gunnery platforms in rough seas and above all too vulnerable, while their range was by now little more than that of a destroyer as their Belluzzo turbines consumed too much distilled water and fuel. During the action of Cape Spada on 19 July 1940 they were unable, during the 24 minutes of the first phase of that action, to catch a faster flotilla of British destroyers of the recent 'H' and' I' classes , which they sighted and pursued at a range that increased from 17,500 metres to 24,000 metres. (USMM)

The '*Navigatori*'-class light scout (later destroyer) *Alvise Da Mosto*. The twelve '*Navigatori*' were built to accompany the first and second series of the '*Condottieri*', forming homogeneous fast and heavily-armed squadrons supported by land based torpedo-bomber floatplanes. (USMM)

similar to the more recent, larger foreign destroyers such as the British 'Tribals'. Trials of the first group of '*Condottieri*' in 1930 had been undertaken at displacements that ranged from 5,565 tonnes (*Colleoni*) to 6,028 tonnes (*Bande Nere*) with an average speed of 38–39 knots. However, by 1940 displacement had increased considerably: *Colleoni* now displaced 7.670 tonnes and *Bande Nere* 8,040 tonnes. Flank speed declined accordingly to 32 knots at best, while as gunnery platforms the ships lacked stability.[14]

The thinking that underpinned the *esploratori* was still considered sound in Italy, but the expression of this thinking became confused. The four destroyers of the *Maestrale* class commissioned in 1934 were ordered and rated as *cacciatorpediniere*, but formed an *esploratore* flotilla and had their funnel decorated with the bands typical of the scouts before being incorporated in September 1938 into a destroyer flotilla. It seemed the new Chief of Staff, Admiral Domenico Cavagnari, had an initial tendency to label all the new destroyers as scouts, limiting the *cacciatorpediniere* category to the earlier, and slightly smaller units of the classes between the *Sella* and the *Freccia* types. In 1934 Cavagnari requested a project for a *Maestrale potenziato*, a 2,100-tonne warship armed with six 120mm guns and with extended range, but Mussolini rejected this proposal for budgetary reasons. The following year the ever-persistent Cavagnari ordered a study for a new, fast scout incorpo-

The light cruiser *Armando Diaz* in1935. She was a slightly improved version of the first series of '*Condottieri*'. The following ships of the *Muzio Attendolo*, *Eugenio di Savoia* and *Duca degli Abruzzi* classes, even though they were grouped by *Jane's* under the same '*Condottieri*' category, were totally different warships and true light cruisers. (USMM)

The destroyer *Scirocco* at Riccione, Summer 1935. The funnel bears the band typical of the *esploratori leggeri*. This was the first photo taken by the then 14-years-old and future illustrious naval photographer and collector Aldo Fraccaroli. (USMM, Collezione Aldo Fraccaroli)

rating the innovative underwater hull-form devised by general (E) Umberto Pugliese. The design of an *esploratore veloce* was ready by late Spring 1935, but Pugliese and Cavagnari were both dissatisfied; they concluded it would have been possible to get much more with just a little extra effort. The first ship of this type was sold to the Soviet Union on 9 September 1935, becoming the *Tashkent*, built in Italy and delivered to the Russians on 6 May 1939.[15]

By early 1936 the project by Pugliese for what was now an *esploratore oceanico* – reclassified in 1938 as a 'cruiser' after the new deputy Chief of Staff of the *Regia Marina*, Admiral Inigo Campioni, revised the classification of warships, abolishing the term *esploratore* – was ready. In the same year the twelve ships of the future '*Capitani romani*' class were authorised. According to the original requirements they were to be powerful, fast and unprotected warships for the Mediterranean and Red Sea, and sterling examples of Italian design and economy, with interiors inspired by the most recent and beautiful aeronautic style.[16] In the original project they were to embark a floatplane (though without a catapult) to suit them for an independent role like the French, Portuguese and Dutch sloops.

However, budgetary constraints again intervened. Italy had to reduce expenditure after the Ethiopian conflict and from 1937, because the United States was the only foreign country still granting credit in hard currency to Rome, had to accept quantitative quotas for imports. The high-tensile steel for the hulls of the new battleships *Roma* and *Impero* ordered in December 1937 could be purchased in America, but as Italian industry was heavily reliant on the recycling of scrap iron it had little capacity for steel

Palma de Mallorca, 10 January 1938. The old *Quarto* was still active during the Spanish Civil War. On 1 August 1938, at Pollensa Bay, the ship suffered a boiler explosion which killed seven seamen. She was able to sail to La Spezia on 18 August, but a technical commission considered her not worth repairing and she was paid off on 5 January 1939. (USMM)

ESPLORATORI OF THE *REGIA MARINA*, 1906–1939

requiring high heat treatment. The high-tensile steel necessary for the first unit of the 'Capitani romani' class (*Giulio Germanico*, laid down on 3 April 1939) was secured in 1938 only by scrapping the elderly scouts *Venezia* (ex-*Saïda*) and *Brindisi* (ex-*Helgoland*), thereby completing a cycle begun thirty years previously. This measure, together the scrapping of the old armoured cruiser *Pisa*, the light cruiser *Ancona* (ex-German *Graudenz*) and thirteen old destroyers, was a necessary transitional solution before the modernisation of the existing steel works and the completion of the huge new works of Cornigliano, near Genoa, which would be accomplished by late 1941. However, this was not a good omen for Italy in view of the future European crisis and war.

The era of the *esploratori* was, in any event, over. The few apparently similar foreign ships, such as the IJN *Yubari* launched in 1923, the Dutch *Tromp*s and the Japanese *Agano*s of the late 1930s, were designed as destroyer flotilla leaders, not scouts.

The first three 'Capitani romani' were commissioned in 1943. On the night of 16/17 July 1943, in the Strait of Messina, the *Scipione* encountered the British MTBs *260, 313, 315* and *316*. Steaming at nearly 40 knots the former *esploratore* ran down the enemy boats, sinking *MTB 316* and damaging *313*.[17] It was the swan song of

La Spezia, 18 June 1941. Three photos of the paid-off *Quarto*, now used as a target for new Italian shells and underwater weapons. The first trial involving an MTM explosive boat was undertaken in November 1940. (USMM)

159

The brand-new small cruiser *Attilio Regolo* during the summer of 1942. She represented the ultimate evolution of the scout design. (USMM)

the scout, a type of warship that aircraft had made obsolete some years previously.

Sources:

Vincent P O'Hara, W David Dickson and Richard Worth (Eds), *To Crown the Waves*, Naval Institute Press (Annapolis, 2013).

Enrico Cernuschi, *Sea Power the Italian Way*, Ufficio Storico della Marina Militare (Rome, 2017).

Vincent P O'Hara and Leonard R Heinz, *Clash of Fleets: Naval Battles of the Great War, 1914–18*, Naval Institute Press (Annapolis, 2017).

Giorgio Giorgerini, Augusto Nani, *Gli incrociatori italiani*, Ufficio Storico della Marina Militare (Rome, 2017).

Franco Bargoni, *Esploratori italiani*, Ufficio Storico della Marina Militare (Rome, 1996).

Erminio Bagnasco and Enrico Cernuschi, *Le navi da guerra italiane*, Albertelli (Parma, 2009).

Enrico Cernuschi, *Battaglie sconosciute*, In Edibus (Vicenza, 2014).

Luigi H Slaghek Fabbri, *Con gli inglesi in Adriatico*, Ardita (Rome, 1934).

Enrico Cernuschi, 'Gli occhi della flotta', *Storia Militare*, Dec 1999 and Jan 2000.

Enrico Cernuschi, 'Nomen numen', *Storia Militare*, Dec 2001.

Enrico Cernuschi, 'Acciaio e cantieri', *Rivista Marittima*, Jun & Jul 2011.

Enrico Cernuschi, 'Capitani romani', *Storia Militare*, Feb 2015.

Enrico Cernuschi and Fabrizio Santi Amantini, 'I super Bixio', *Rivista Marittima*, Mar 2018.

Endnotes:

1. Pierangelo Manuele, *Il Piemonte sul mare: La Marina sabauda dal medioevo all'unità d'Italia*, L'Arciere (Cuneo, 1997), 114–116.
2. A type of torpedo-boat flotilla leader with limited endurance whose forerunner was the Italian *Pietro Micca*, laid down in Venice in 1875.
3. Richard Worth, *In the Shadow of the Battleship: Considering the Cruisers of World War II*, Nimble Books (2008), 2.
4. David K Brown, 'Second World War Cruisers: Was Armour Really necessary?', *Warship 1992*.
5. In 1907 the Italian Navy concluded, following similar exercises, that the new rapid-fire weapons meant that major landings against a modern well-equipped army were impossible, and that only regiment-sized raids were feasible.
6. A twin of *Dante*, to be named *Conte di Cavour*, was planned and laid down in June 1909, but a better-armed project had just been adopted, and a week after the first frames of *Cavour* had been laid on the slipway she become the first of a new class of three battleships.
7. Bettolo had previously twice held the post of Navy Minister between 1899 and 1903.
8. The Italian Navy minister Admiral Leonardi Cattolica had on his desk, since October 1910, courtesy of the Italian Navy's intelligence service, a copy of that programme, which was debated by the parliament in Paris behind closed doors, and his choices about the *Caracciolo*s were made accordingly.
9. The *Poerio*s were originally to have eight 450mm torpedo tubes and four 102/35 guns. They were an enlarged version of the destroyer *Liz*, a private-venture warship built by Ansaldo and sold to Portugal but finally purchased by the Royal Navy in March 1915 as HMS *Arno*.
10. In 1904 Admiral Fisher had followed a similar line of thinking, ordering what would finally become the *Swift*, a 2,180-ton flotilla leader planned as an oversize destroyer; the design has been generally regarded as a failure.
11. René Greger, 'The Last Austro-Hungarian Destroyer Projects', *Warship* No 21.
12. Achille Rastelli 'I bombardamenti sulle città', in *La Grande Guerra 1915–1918*, Gino Rossato (Valdagno, 1994), 183–249. Austro-Hungarian naval bombardments in the Lower Adriatic were as follows: 24 May 1915 against Manfredonia, Vieste and Barletta; 24 May 1915 against Tremiti Island and Torre Mileto; 13 July 1915 Pelagosa Island; 17 August 1915 Pelagosa Island; 23 June 1916 Giulianova; 2 August 1916 Molfetta; 5 November 1916 Sant'Elfisio a Mare. The average was about 50 rounds on each occasion; no damage to military targets or railways was recorded.
13. Until the mid-1920s the Italian classification of the first two Washington cruisers *Trieste* and *Trento* was *incrociatori leggeri* (light cruisers); the *incrociatori maggiori* classification was used only for battlecruisers, a type of warship the Italian Navy dreamed of building until 1934, when the first two fast battleships of the *Littorio* class were ordered. In 1928 the 10.000-tonne cruisers were reclassified as *incrociatori pesanti* (heavy cruisers) and the '*Condottieri*' become *incrociatori leggeri*.
14. Giuliano Colliva, 'La Marina fa la spesa', *Rivista Marittima*, Nov 2002, 104.
15. In March 1939 Cavagnari tried to obstruct the delivery of *Tashkent*, claiming that the European situation was so serious that it was not in the interests of the Italian Navy to deprive itself of such a powerful vessel, but Mussolini ordered that the partnership should proceed as the partnership with the USSR was a key factor in his greater plan for a general rebalancing of the political and economic situation in the Balkans and the Middle East; a further factor was that since 1937 Stalin's Russia had become a major supplier of oil to Italy.
16. *Navires et Histoire*, Hors-Série No 5, 110.
17. The British appear to have been reluctant to admit any damage from Italian warships' fire; the loss of this MTB was officially recorded as 'caused by torpedo from Italian cruiser off Reggio, S. Italy', Ships of the Royal Navy: Statement of Losses during the Second World War, HMSO, 1947. No torpedo had been launched by *Scipione*.

MODERN EUROPEAN FRIGATES

Conrad Waters assesses how recent design trends in large surface combatants have been reflected in the latest generation of European frigates.

The thirty or so years that have elapsed since the end of the Cold War have seen a generational change in warship design. Some of the many and varied factors – geopolitical and financial, as well as technological – that have marked this period were outlined in the author's 'Design Trends in Modern Surface Combatants', published in *Warship 2020*. This sequel describes some of the warship classes that have been completed in recent years, with a primary focus on European frigates.

Although naval power – and, consequently, naval design and construction – has shifted steadily eastwards to the emerging Asia-Pacific economies, it is European shipyards that have remained at the forefront of warship design throughout much of the post-Cold War period.[1] European leadership during this time has been strengthened both by the hiatus in Russian warship development in the aftermath of the collapse of the Soviet Union, and seemingly poor choices by the United States. Here, both the *Zumwalt* (DDG-1000) class and the littoral combat ship variants have proved to be design 'dead ends'. As a consequence, the US Navy's surface combatant force remains dominated by the missile destroyers of the *Arleigh Burke* (DDG-51) class, originally conceived in the later stages of the Cold War.

Although the US Navy may have fallen behind the curve in warship design, American technological prowess in the naval sphere has remained a dominant influence. This has, arguably, been most clearly evidenced by the impact of its Aegis combat system and AN/SPY-1 series of radars, which first entered service on board the cruiser *Ticonderoga* (CG-47) as long ago as January 1983.[2] Designed to counter a new generation of Soviet missiles

The immediate post-Cold War era saw many leading European navies commission large and sophisticated air defence escorts, such as the British Type 45 destroyer *Diamond* pictured here. Ways were subsequently found to incorporate many of the advanced capabilities they brought into a much wider range of European surface combatants. (Crown Copyright 2014)

that could evade or saturate existing defences, the *Ticonderoga*s and their *Burke*-class successors were to be the primary drivers of a series of European air defence escorts that trace their origins to the late 1980s but were ordered well after the Cold War had ended. Combining advanced combat systems with multi-function, phased array radars, these large and sophisticated ships currently provide core 'high end' air defence capabilities in six of Europe's leading fleets.[3]

Fridtjof Nansen Class (Norway)

A major drawback of this new generation of ships was the size and complexity (and hence cost) associated with their multi-function radars and associated combat systems and weapons. This meant that, at least outside the US Navy, they were restricted to a relatively small number of large warships, typically operated by the major fleets. The Royal Norwegian Navy's order for five new frigates equipped with the Aegis system in June 2000 therefore represented a major step forward. It demonstrated that the new generation of advanced air defence equipment could be installed in a moderately-sized vessel operated by one of NATO's smaller navies.

The Royal Norwegian Navy's requirement for new frigates to replace the 1960s-era *Oslo* class was established during the early 1990s. In similar fashion to their predecessors, the new ships were to be primarily configured for the anti-submarine warfare (ASW) mission but be capable of general-purpose operation. Although they were conceived after the break-up of the Soviet Union, Norway's proximity to Russia and the still-potent submarine flotillas of its Northern Fleet appear to have been a key driver of the Navy's requirements. These factors mandated an effective air defence capacity, both to counter land-based aircraft equipped with stand-off weapons and also submarine-launched 'pop-up' missiles. The selection of a Spanish Navantia design equipped with a variant of Aegis over a number of rival bids when a preferred contractor was announced in May 1999 was reportedly influenced by the system's potential to meet this need. All the new ships were to be built in Spain at Navantia's Ferrol yard.

The successful Navantia proposal for the Norwegian frigate was derived from its construction of the Aegis-equipped F-100 *Álvaro de Bazán* class for the Spanish Navy. These large, 6,300-tonne ships were part of the series of European air defence escorts referred to above and were the smallest Aegis-equipped vessels at that time. The Spanish shipbuilding group was able to adapt this design to the Norwegian requirement for an even smaller, 5,300-tonne ASW-optimised frigate by adopting a lighter and more compact variant of the Lockheed Martin-built AN/SPY-1 series of radars.

Like its sister arrays, the AN/SPY-IF is a passive electronically-scanned or phased array that uses electronic

Norway's *Fridtjof Nansen* class introduced the Aegis combat system and SPY-1F radars into a relatively small hull. This is *Otto Sverdrup*, the third unit of the class, off the Norwegian coast in 2016. (Jakob Østheim via Forsvaret)

MODERN EUROPEAN FRIGATES

Fridtjof Nansen

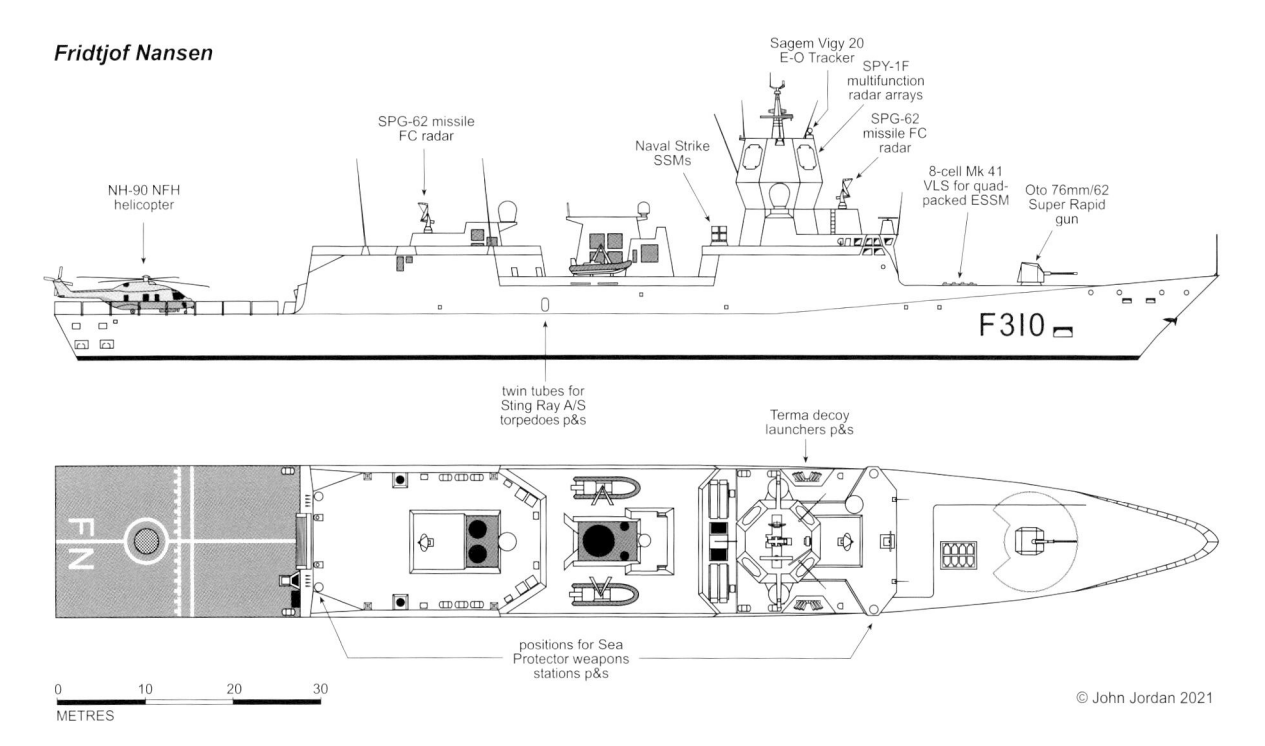

Diagram labels:
- NH-90 NFH helicopter
- SPG-62 missile FC radar
- Naval Strike SSMs
- Sagem Vigy 20 E-O Tracker
- SPY-1F multifunction radar arrays
- SPG-62 missile FC radar
- 8-cell Mk 41 VLS for quad-packed ESSM
- Oto 76mm/62 Super Rapid gun
- F310
- twin tubes for Sting Ray A/S torpedoes p&s
- Terma decoy launchers p&s
- positions for Sea Protector weapons stations p&s

0 10 20 30
METRES

© John Jordan 2021

software rather than mechanics to direct its radar beams. Each of its four fixed antennae is 8ft (2.4m) in diameter compared with the 12ft (3.7m) of other arrays in the series; the number of radar elements is reduced from 4,350 to just over 1,850 in consequence, giving the SPY-1F less power and range than earlier models. The fact that its lighter weight allows it to be located higher in the ship provides partial compensation. The Norwegian frigates are equipped with the quad-packed Evolved Sea Sparrow Missile (ESSM) for close-range and local air defence (housed in an eight-cell Mk 41 vertical launch system) rather than the longer-range Standard Missiles used for area defence that typically equip larger Aegis combatants. The missiles are directed by two fire control radars.[4] The Aegis command and control functions that integrate the SPY-IF arrays with other sensors and weapons are broadly similar to those found in other Aegis ships of the period. However, the 'customisation' inherent in the system is demonstrated by the integration of ASW and anti-surface warfare components provided by Norway's Kongsberg group.

The resulting *Fridtjof Nansen* class frigates are otherwise notable for their anti-submarine capabilities. These include the combination of a Spherion MRS 2000 bow-mounted sonar with a CAPTAS-2 towed array sonar. A single NFH variant of the NH Industries NH90 helicopter is the main means of prosecuting underwater targets; there are also two twin fixed torpedo launchers equipped with the BAE Systems Stingray lightweight torpedo for close-range defence. Other armament includes the ubiquitous 76mm Oto gun and Kongsberg NSM (Naval Strike Missile) surface-to-surface missiles. Weapons stations for Kongsberg's Sea Protector are also provided for self-defence against asymmetric threats (see drawing).

Structurally, the Norwegian ships evidence the emphasis on radar stealth typical of most recent surface combatant designs, including the careful shaping of external surfaces and the enclosure of much equipment. The hull incorporates a degree of ice-strengthening to take account of deployment in Arctic waters. Other elements of 'Arcticisation' include the provision of enclosed bridge wings and the extensive use of de-icing systems. One interesting design choice is the combined diesel and gas turbine (CODAG) propulsion arrangement. This forgoes the use of the more acoustically stealthy diesel-electric propulsion elements that are now a common feature of modern ASW combatants.

While noteworthy for introducing advanced radar and combat system capacity into a comparatively small hull, the *Fridtjof Nansen* class largely predates the increased emphasis on modularity and adaptability that have become typical in modern frigates. This has been one of a number of trends that have tended to drive ship size upwards and could explain why SPY-1F has not been more widely adopted. However, the frigates have given good service since joining the fleet from 2006 onwards, increasing the Navy's ability to undertake distant operations. One of the class, *Helge Ingstad*, was lost following a collision in 2018. An independent investigation effectively excluded the design from blame for this outcome, concluding that the damage sustained was over and above that which the ship was designed to withstand.

Absalon/*Iver Huitfeldt* Classes (Denmark)

Dating from roughly the same time as the *Fridtjof Nansen* class are the Royal Danish Navy's frigates of the *Absalon* and *Iver Huitfeldt* classes. Their construction

The Danish *Absalon* class frigate *Esbern Snare* pictured during a NATO exercise in 2018. The *Absalon*'s combination of modular weapons stations with a flexible mission deck has had an influence on many subsequent European frigate designs. (Christian Valverde via Forsvaret)

was driven by the re-orientation of Denmark's defence forces from territorial defence towards international stabilisation after the end of the Cold War and the consequent need for ships that could deploy to more distant waters. This resulted in a decision to build two types of new warship – a flexible support ship combining logistics, amphibious and frigate-type functions and a frigate like patrol ship – to a largely common design. These were to become, respectively, the *Absalon* and *Iver Huitfeldt* classes. A project office for the ships was stood up around the turn of the millennium and orders for two *Absalon* class ships were placed in 2001; contracts for three frigates of the *Iver Huitfeldt* class followed in 2006.

All five ships ordered under the project were assembled by the now-closed Odense Steel Shipyard.[5] However, to help maintain spending within a strict budget, outfitting of sensors and weaponry was carried out by the Danish Defence Acquisition and Logistics Agency (DALO) and the Navy. This resulted in a delay between initial delivery in 2004–05 and entry into full operational service. Further economy was achieved by large-scale use of commercial building standards, albeit overlaid where appropriate with naval specifications to ensure a high level of survivability. In common with other warships of the era, stealth was another important design consideration.

The two initial *Absalon*-class vessels featured two characteristics that are becoming common in modern surface combatant design. One was a modular approach that permitted the use of containerised weapons and equipment based on the Standard Flex (or StanFlex) system introduced with the earlier patrol boats of the *Flyvefisken* class. Similar to the *Mehrzweck Kombination* (MEKO) concept pioneered by the German Blohm & Voss shipyard, StanFlex uses pre-prepared positions that interface with the ship's combat and other systems and into which the appropriate containers can be slotted. *Absalon* is fitted with positions for five StanFlex modules, which are typically used for containers housing ESSM and Harpoon surface-to-surface missiles. The modules were taken from existing stocks, generating additional savings. The other innovation was the specification of a large 900m^2 'flex deck' running forward from beneath the flight deck and hangar. The deck is accessible by a RO-RO ramp and can house vehicles, troops, personnel landing craft or additional containerised equipment; this can even extend to a modular hospital.

Other aspects of the *Absalon* design reflect more traditional general-purpose frigate requirements. Armament includes a 127mm Mk 45 Mod 4 main gun, Rheinmetall Millennium close-in weapons stations and fixed ASW torpedo tubes. There is an Atlas ASO series bow-mounted sonar but no towed array. The hangar has sufficient capacity to house two Merlin-sized helicopters. The importance of fuel economy is reflected in the twin-shaft diesel propulsion system, which provides a modest

MODERN EUROPEAN FRIGATES

The Danish *Iver Huitfeldt* class – the lead ship is seen here in 2012 – demonstrated how the flexibility inherent in the basic *Absalon* design allowed its adaptation to form the basis for a powerful yet cost-effective air defence frigate. The ships were fitted out incrementally, and *Iver Huitfeldt* did not have her full outfit of weapons and systems at the time this photo was taken. (Torbjørn Kjosvold via Forsvaret)

maximum speed of 24 knots but good endurance. The use of existing StanFlex modules and the other economy measures referenced above meant that the ships were relatively affordable despite being quite large vessels (full load displacement is in the region of 6,600 tonnes).

The subsequent *Iver Huitfeldt* class – delivered from 2011 onwards – demonstrates the inherent flexibility in the Danish concept in that it has adapted the same hull as the basis for a powerful air defence frigate. This has essentially involved forgoing the large flex-deck found in

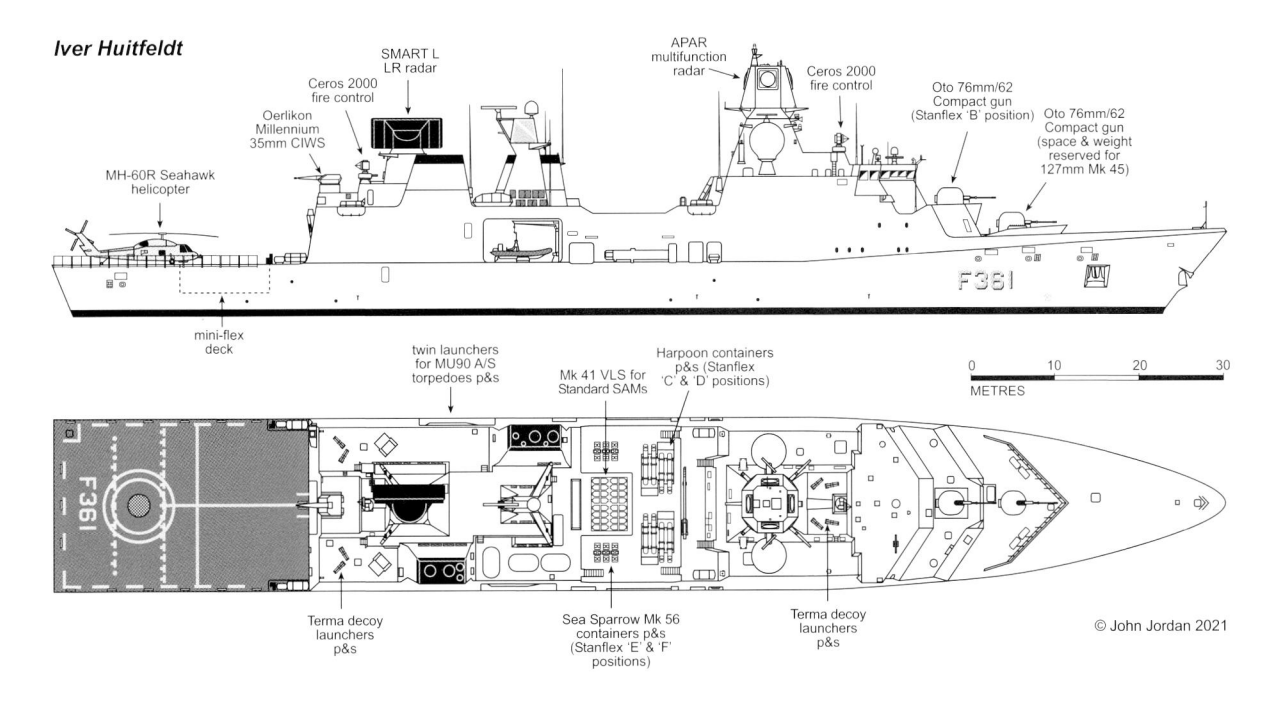

Iver Huitfeldt

© John Jordan 2021

165

the early ships to allow the installation of an area air defence capability. This is based on the Thales Nederland APAR/SMART-L radar combination and a Mk 41 vertical launch system (VLS) for Standard Missiles. SMART-L is a long-range surveillance radar providing an overall 'picture' of the surrounding airspace, whist APAR is used to track and engage any threats that are identified.[6] One of the new generation of active phased array radars – with each of its elements individually energised – APAR is able to use interrupted continuous wave illumination (ICWI) to guide a ship's missiles on to target, thereby obviating the need for separate missile fire control radars. The size and volume provided by the large hull used by the Danish ships allowed full-sized versions of these systems to be incorporated in the design, avoiding some of the compromises inherent in Norway's choice of the pared-back SPY-1F.

Another significant change in the revised design is a doubling of the diesel engine arrangement specified for the *Absalon* class. This permits a boost in maximum speed to 28+ knots. Overall displacement is similar to the preceding ships. The frigates continue to make full use of the StanFlex system for additional weaponry, which is broadly similar to that found on board the *Absalon*s once the area air defence system is discounted. The ability to house containerised equipment and deploy a variety of boats and unmanned systems from the large boat bays means that much of the flexibility of the base design is maintained. Extensive automation permits operation by a crew of *c*100, about the same level needed by *Absalon* but fewer than those found aboard Norway's *Fridtjof Nansen* class. This was an important consideration in an era of defence economies, given the high proportion of a ship's operating cost accounted for by personnel.

Further evidence of the platform's versatility was provided in 2019: in September of that year it was announced that Babcock's 'Arrowhead 140' design – based on the *Iver Huitfeldt* hull – had been selected to meet the British Royal Navy's Type 31 patrol frigate requirement. 'Arrowhead 140' uses the inherent adaptability of the design to provide the basis for a customised

The Babcock International 'Arrowhead 140', which forms the basis for the British Type 31 frigate, is yet another illustration of the adaptability in the Danish *Absalon/Iver Huitfeldt* hull. (Babcock International)

offering tailored to a specific navy's needs. While hull and propulsion remain fixed, other equipment – armament, sensors and the combat management system – can be adapted according to individual preference. For example, the five warships ordered for the Royal Navy are lightly armed for presence/constabulary operations. Equally, more heavily equipped 'Arrowhead 140' variants are being considered by overseas navies to meet specifications for more high intensity duties. Meanwhile, in yet another example of the design's flexibility, the pair of existing *Absalon* class vessels are currently being reconfigured as anti-submarine frigates.

FREMM Series (France/Italy)

By far the most numerous of the post-Cold War European surface combatants are the Franco-Italian FREMM multi-mission frigates. As of mid-2021 a total of 19 vessels are in service with the navies of France, Italy, Egypt, and Morocco; additional ships remain under construction or on order for the two original partner nations as well as for Indonesia. The FREMM design also forms the basis for the US Navy's new frigates of the *Constellation* (FFG-62) class (see below).

The FREMM programme – the acronym stands for *frégate européenne multi-missions* and *fregata europea multi-missione* respectively – arose as a result of the impending block obsolescence of many French and Italian surface warships that had first been delivered in the 1970s and 1980s. The two nations had previously worked together building a pair of 'Horizon'-type air defence destroyers for each of their navies, establishing the foundations for further collaboration. A cooperation agreement for the new project was signed towards the end of 2002. It was initially envisaged that the programme would encompass 27 ships: 17 for France and ten for Italy. However, the French order was progressively cut back to just eight units as a result of reductions in defence spending.

The early years of the collaborative programme were marked by considerable difficulties reconciling the conflicting requirements of the partner navies against a backdrop of financial constraints. The solution ultimately adopted was to take advantage of the trend towards design modularity by agreeing the utilisation of a common hull and broadly similar propulsion system architecture that could otherwise be configured to meet specific national priorities. This approach also allowed the respective navies to optimise their own ships to perform particular missions. This has resulted in specific air defence, anti-submarine, land-attack and general-purpose FREMM variants all being proposed over the course of the programme. A design and development contract for the new ships was placed via the joint European procurement agency OCCAR in November 2005. An order for eight French ships – to be built by what is now Naval Group at Lorient – was placed on the same date. Agreement for an initial batch of two Italian ships was subsequently confirmed in May 2006. These –

MODERN EUROPEAN FRIGATES

and the remaining Italian frigates – have been assembled at Fincantieri's Riva Trigoso shipyard prior to outfitting at Muggiano near La Spezia.

French FREMMs: The first FREMM to be completed was the French *Aquitaine*. Fabrication commenced in March 2007 and she was delivered in 2012. Displacing around 6,000 tonnes in full load condition, her overall design is evidently heavily influenced by the stealth frigates of the *La Fayette* class laid down during the 1990s and successor designs built for export. Considerable attention has been paid to minimising radar cross section (RCS) through the shaping of hull and superstructures, the enclosure of working decks and the integration of weapons and sensors into the superstructures. As a result, the design's overall RCS is claimed to be substantially reduced compared with these previous ships. Another interesting design feature – shared with a number of other recent surface warships – is the placement of the ship's diesel exhausts close to the waterline. Here they can be cooled by seawater, minimising infra-red signature.

Aquitaine was ordered as a French anti-submarine FREMM variant, which inevitably placed an emphasis on reducing acoustic signature. This requirement influenced the propulsion system used across all the FREMMs: it combines a gas turbine for a swift sprint to the location of a potential submarine with diesel generator-powered electric motors for silent, low-speed target prosecution. Pioneered in the British Type 23 frigates, this arrangement has increasingly become the gold standard for warships focused on ASW operations. In the French ships, a combined diesel-electric or gas turbine (CODLOG) system uses power drawn from the ship's

The French FREMM *Languedoc* in dry dock. The acoustically stealthy hull design of the FREMMs owed much to Naval Group's expertise in both surface combatant and submarine design. (Raphaël Demaret via Naval Group)

France's *Alsace* is visually similar to the French Navy's other FREMM type frigates but incorporates a number of modifications to equip her for the air defence role. (Naval Group)

Aquitaine

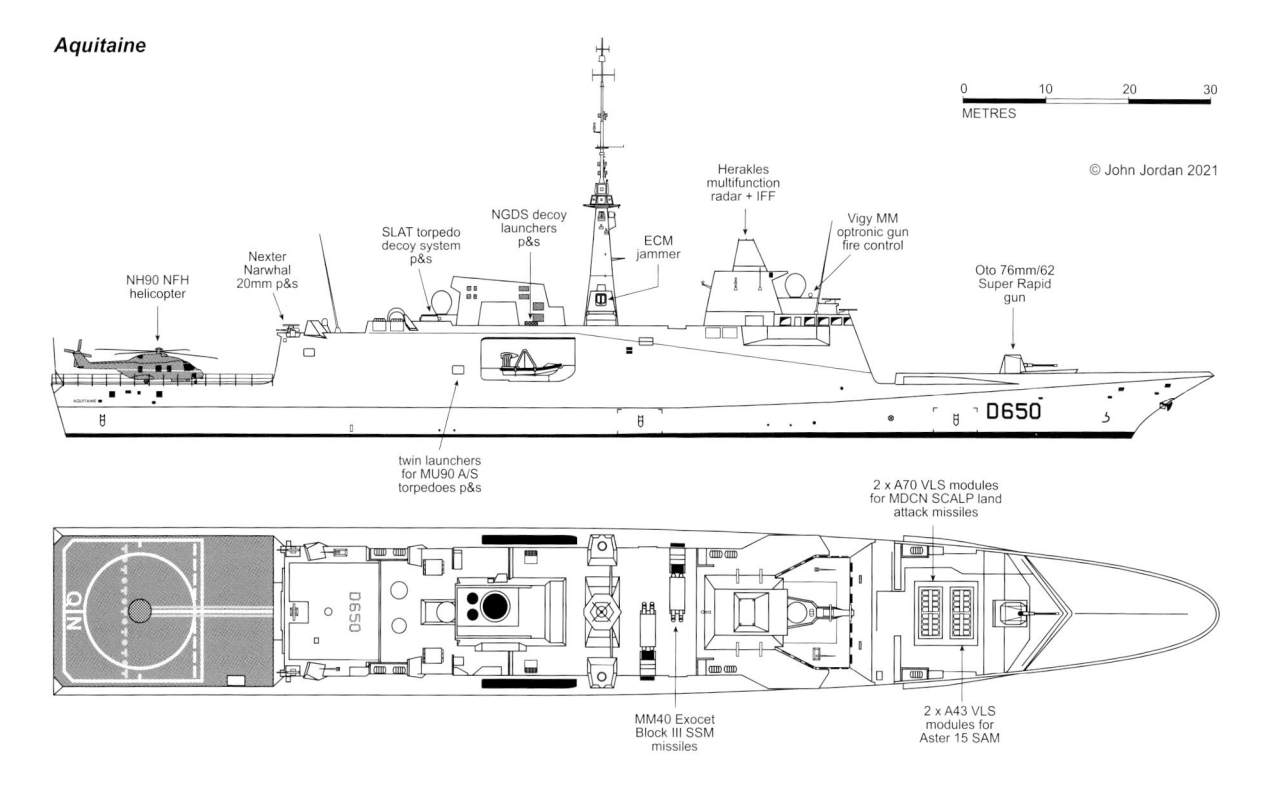

electrical distribution system to supply electric motors wrapped around the propulsion shafts for low-speed running. A single GE LM2500+ turbine provides the power for higher speeds of up to 27 knots. Other aspects of acoustic stealth include the use of noise dampening techniques and the elastic mounting of some equipment. France's considerable experience in both surface warship and submarine design was also drawn upon to optimise hull and propeller hydrodynamics.

Aquitaine's ability to detect and track submarines is provided by a Thales sonar suite comprising a UMS 4110 bow-mounted sonar and a UMS 4249 (CAPTAS-4) towed array sonar.[7] A single NFH helicopter similar to that embarked in the *Fridtjof Nansen* class is the primary means of attacking any targets detected. Two twin fixed torpedo launchers and the SLAT torpedo defence system are provided as a close-range back up. Although optimised for the ASW role, other elements of the equipment fit provide a wide range of general-purpose capabilities that reflect the need for versatility during an age of reduced overall hull numbers. Air defence is in the hands of a Herakles multi-function passive phased array, which performs surveillance, tracking and fire control functions. It is paired with short/medium-range Aster 15 surface-to-air missiles housed in two eight-cell Sylver A43 VLS modules. Two additional, strike length eight-cell Sylver A70 modules house long-range MdCN cruise missiles, thereby providing some of the land attack functions that were to be incorporated in a planned land-attack FREMM that was never built. The cruise missiles are supplemented by launchers for Exocet MM40 surface-to-surface missiles and a 76mm gun. Naval Group's fully distributed SETIS performs combat management system functions.

In line with other contemporary European frigates, economy in operation and maintenance were key design drivers. *Aquitaine* and her sisters can be operated by a core crew of 108, although sufficient accommodation is provided for 145 personnel and this can be further expanded due to the modular nature of the cabins. In addition to the extensive use of automated ship systems, commercial operating practices have been adopted from the merchant marine in areas such as bridge layout and watch-keeping arrangements. The ship's structure, particularly internal passageways, has been carefully sized to facilitate the removal and replacement of equipment for maintenance or upgrade.

Although all eight French ships have been built to the same basic design, later units of the class have been subject to progressive upgrades. For example, it has been reported that the fifth and sixth French FREMMS – *Bretagne* and *Normandie* – have two Sylver A50 modules and two Sylver A70 modules to allow the longer-range Aster 30 missile, which features a larger booster, to be deployed. The seventh and eighth frigates – *Alsace and Lorraine* – have been more extensively modified to act as specialised FREMM-DA (*frégate européenne multi-missions de defence aérienne*) variants. Modifications include a uniform armament of four eight-cell Sylver A50 VLS modules, enhancements to the SETIS combat management system, and a more powerful version of the Herakles multi-function radar. Another variation, seen in the single French FREMMs acquired by Egypt and Morocco, has been the deletion of the Sylver A70 VLS cells in ships for export due to restrictions on the overseas sale of sensitive technology.[8]

MODERN EUROPEAN FRIGATES

Carlo Bergamini

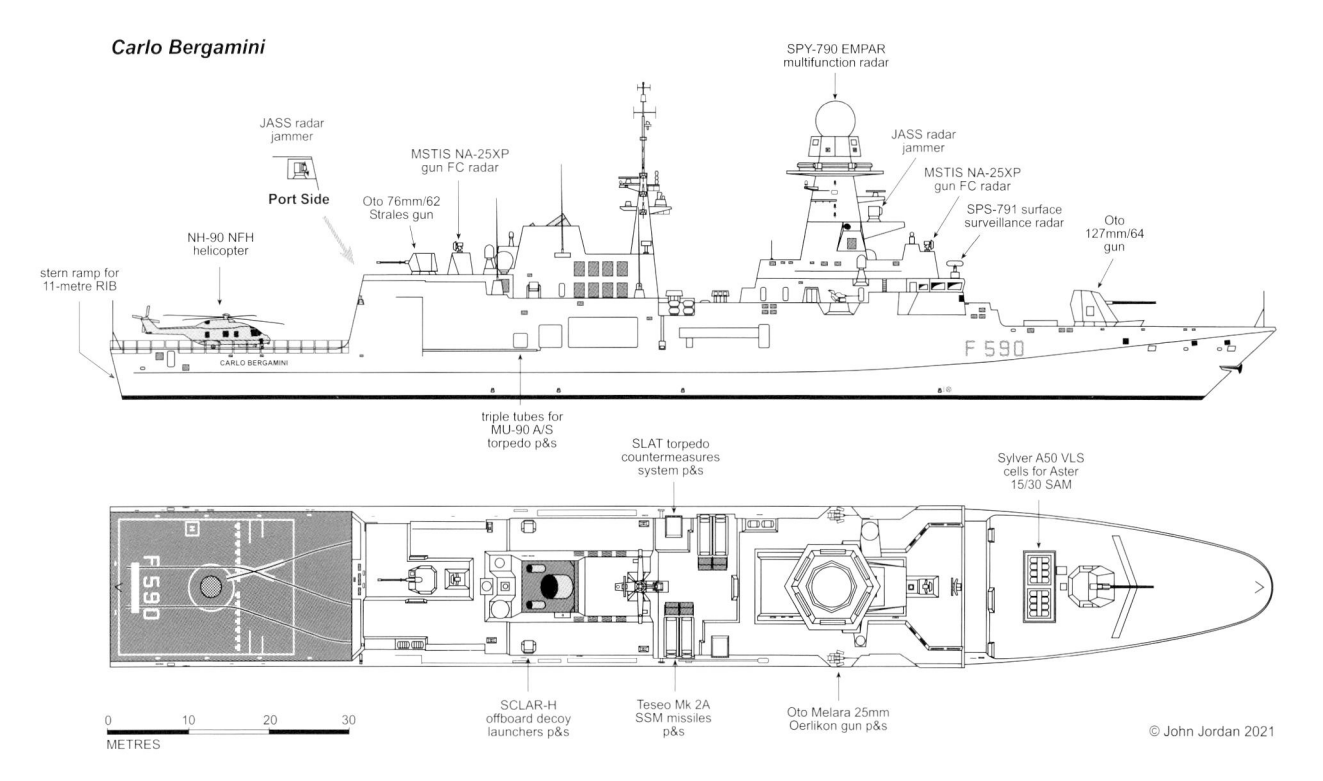

[Diagram labels: stern ramp for 11-metre RIB; NH-90 NFH helicopter; JASS radar jammer Port Side; MSTIS NA-25XP gun FC radar; Oto 76mm/62 Strales gun; SPY-790 EMPAR multifunction radar; JASS radar jammer; MSTIS NA-25XP gun FC radar; SPS-791 surface surveillance radar; Oto 127mm/64 gun; triple tubes for MU-90 A/S torpedo p&s; SLAT torpedo countermeasures system p&s; Sylver A50 VLS cells for Aster 15/30 SAM; SCLAR-H offboard decoy launchers p&s; Teseo Mk 2A SSM missiles p&s; Oto Melara 25mm Oerlikon gun p&s; © John Jordan 2021]

Italian FREMMs: The Italian FREMMs have a noticeably different appearance from their French counterparts, despite sharing the same basic design. This is in part driven by the Italian Navy's selection of the indigenous EMPAR (European Multi-function Phased Array Radar), which is housed in a prominent radome atop a tower foremast similar in configuration to that of the RN's Type 45 destroyer. The radar serves the same purpose as the Herakles on board the French FREMMs but is regarded as a more capable system, particularly as modified with individual active elements. Although largely driven by industrial considerations, its selection was also influenced by Italy's requirement for improved air defence capabilities in its own frigates.

Conversely, the Italian FREMMs have only two eight-cell Sylver A50 modules for Aster 15/30 missiles, as additional launch cells for the deployment of land attack cruise missiles was not part of their own design specification. Despite this, space has been reserved for two further A50 modules, and the forecastle of the Italian ships is a deck higher as a result to accommodate their additional length. These differences help to account for the Italian FREMMs being heavier than the French frigates; full load displacement is in the region of 6,700 tonnes. They are also slightly longer due to the incorporation of a 3.6 metre hull plug in the vicinity of the flight deck. Reports suggest this was required to counter excess forward trim caused by the additional weight in the forward part of the ship.

Although sharing a broadly similar propulsion system with their French counterparts, Italy's FREMMs use a combined diesel electric and gas (CODLAG) arrangement under which their electrical motors supplement the gas turbine during high-speed operation. The Italian-designed diesel generators use traditional exhausts in the forward mast and funnel rather than the waterline system adopted for France's FREMMs. Interestingly, the Italian Navy has also found that it needs more sailors to crew its ships than the *Marine Nationale*, the core complement being in the region of 130 plus helicopter detachments.

The Italian Navy elected to build their FREMMs in general-purpose and anti-submarine variants, again reflecting the adaptability provided by modern, modular designs. The lead ships of both types were delivered in 2013. In practice, the two iterations are broadly similar, differing only in specific equipment fit. The *Carlo Bergamini* GP variant is optimised for anti-surface and land attack missions, incorporating a fully automated Oto 127mm gun in 'A' position capable of firing extended-range ammunition. Other armament includes an Oto 76mm Strales mount in 'X' position for second-line anti-air and anti-missile defence, four twin Teseo Mk2A surface-to-surface missile launchers and two triple A/S torpedo tubes. A large hangar can house up to two helicopters, while the Thales UMS 4110 bow-mounted sonar is common with the French ships. A Leonardo ATHENA combat management system integrates weapons and sensors. Six of Italy's ten FREMMs are to be built to this configuration, including two replacement units for a pair of ships sold to Egypt in 2020. It also appears likely that an additional six FREMMS ordered by Indonesia in mid-2021 will be completed to the general-purpose configuration.

The *Virginio Fasan* ASW FREMM trades some of the GP variant's strike capabilities for an enhanced outfit of anti-submarine sensors and weaponry. This includes the

The Italian FREMM frigates share a common hull and a broadly similar propulsion system to their French counterparts but have a significantly different appearance due to national design preferences. This is the ASW variant *Virginio Fasan*. (Fincantieri)

CAPTAS-4 towed array sonar and SLAT torpedo defence system fitted in the French *Aquitaine* class, as well as two twin launchers for MILAS anti-submarine missiles. The armament of Teseo anti-shipping missiles is halved in consequence. The Italian ASW FREMMs also have to make do without the large 127mm gun, being equipped with a second 76mm Strales mount in compensation. Four ships have been completed to this configuration.

US Navy *Constellation* (FFG-62) Class: Another FREMM iteration has resulted from the US Navy's 2020 decision to select a proposal from Fincantieri to use the type as the 'parent design' for the new FFG(X) frigate. Essentially a replacement for the troubled littoral combat ship programme, the United States envisages building as many as 20 of these frigates as it seeks to expand force numbers as part of its Distributed Maritime Operations (DMO) concept.[9] Fincantieri currently has contracts and options to build the first ten of what will become the *Constellation* (FFG-62) class frigates at its Marinette Marine facility in Wisconsin. This yard was previously responsible for building the *Freedom* (LCS-1) variant of the littoral combat ship. Deliveries are currently scheduled for FY2026 onwards.

The *Constellation* (FFG-62) class arguably demonstrates both the flexibility and limits of modern modular design. The American intention is effectively to take the Italian FREMM platform and overlay it with a suite of US Navy weapons systems and electronics. The combat management system will be a derivative of Aegis, while the main multi-function radar will be a scaled-down Enterprise Air Surveillance Radar (EASR) variant of the AN/SPY-6 system first developed for the latest Flight III *Burke* class destroyers.[10] Weaponry will include a 32-cell Mk 41 VLS for ESSM and Standard Missiles, a Mk 110 57mm gun, a RAM close-in-weapons system and canister-launched Naval Strike Missiles. All this capability is likely to require a further increase in crew, with berthing for up to 200 provided.[11] The changes have also

A computer-generated graphic of the lead US Navy FFG(X) programme frigate *Constellation* (FFG-62). She is essentially an Italian FREMM overlaid with US Navy sensors and equipment, although it has proved necessary to expand the hull to accommodate the desired modifications. (US Navy)

MODERN EUROPEAN FRIGATES

The German F125 class frigate *Nordrhein-Westfalen*. The design represents an unusual attempt to optimise a frigate-sized warship for global constabulary operations. (ThyssenKrupp Marine Systems)

influenced a decision to stretch hull length and beam beyond that of the parent FREMM, with full load displacement increasing to 7,300 tonnes in consequence. The US Navy has stated that this will have only a modest impact on the revised ship, as internal layout will remain largely the same. However, it is hard to avoid the conclusion that this will increase technical risk and the associated danger of cost overruns.

F125 *Baden-Württemberg* Class (Germany)

A different approach to modern warship design is represented by the German Navy's F125 *Baden-Württemberg* class. The class echoes many of the trends previously discussed, including the modularity derived from its MEKO heritage and the resulting increase in size over traditional frigates to support this flexibility. However, the F125 has a number of unique characteristics. These include an emphasis on the duplication of key equipment to ensure an ability to maintain operations after action damage or breakdown, as well as optimisation for enduring but lower-intensity operations.

The F125 class traces its origins to attempts to reconfigure the German Navy from Cold War combat missions in the European region towards a greater capacity to perform expeditionary stabilisation duties across the globe. This required a vessel that had the size, endurance and reliability to undertake prolonged deployments. More specifically, the new frigates were to be able to operate for up to two years away from their home base and sustain double the operating hours at sea achieved by their predecessors. A dual-crewing system helped provide justification for just four ships replacing eight existing vessels. The ships would also be better equipped to combat asymmetric threats such as piracy and terrorism that had replaced the Cold War dangers of fleet-on-fleet combat and be able to support a wide range of humanitarian missions. The class was ordered from the ARGE-125 consortium of ThyssenKrupp Marine Systems and Lürssen in 2007, However, fabrication of the first ship only commenced in 2011 and a number of problems during build meant that it was not until 2019 that she was accepted into service.

Displacing some 7,300 tonnes in full load condition, *Baden-Württemberg* and her sisters are among the largest ships considered by this article. In addition to the volume requirements of modular designs, size was driven by the duplication of key systems and sensors. This is most visibly evidenced by the adoption of a twin 'island' superstructure to distribute this equipment forward and aft but also extends, for example, to the wide dispersion of the diesel generators used in the CODLAG propulsion system. A key consideration behind this approach was the lower likelihood of fatal damage occurring during lower-intensity operations. However, it is worth noting that high levels of damage resistance have been a recurring feature of all recent German frigate designs,

Baden-Württemberg

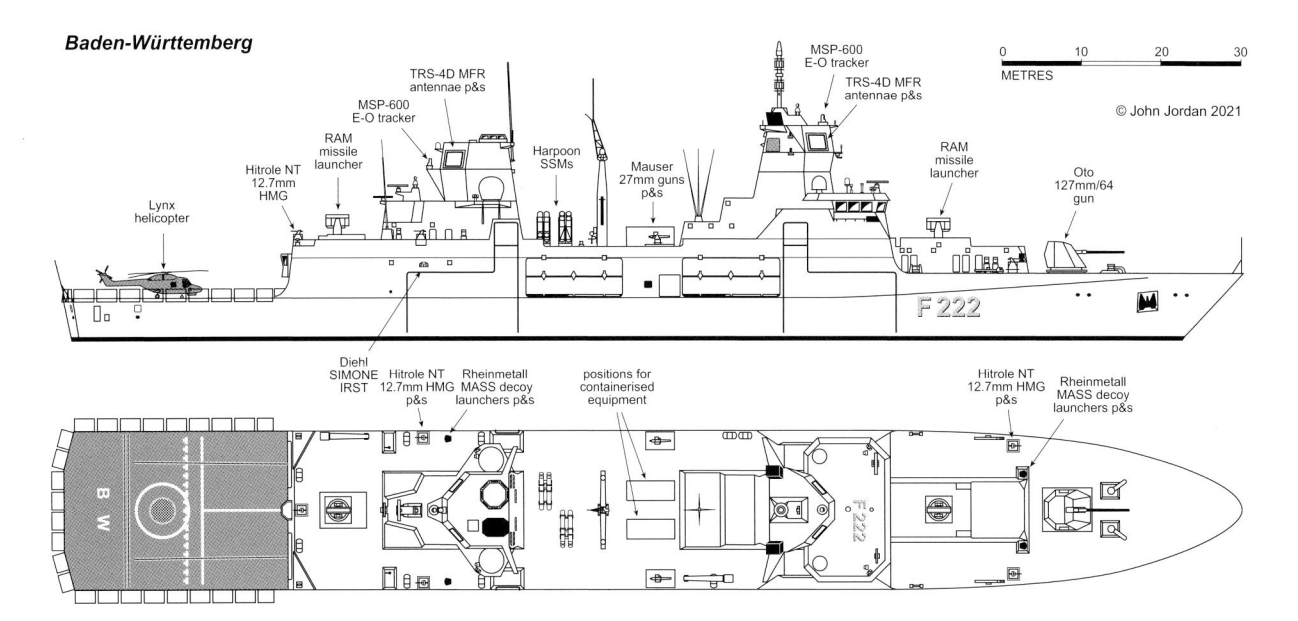

extending to the use of armoured box girders at strength deck level, blast-resistant bulkheads and stringent shock standards. The customary emphasis on stealth includes a reflection-minimising 'X'-shaped cross section.

The F125 class otherwise combines sophisticated command and control facilities with a more limited suite of conventional weaponry. The main radar is the Hensoldt TRS-4D multi-function active phased array. In common with similar systems already described, it is able to perform a range of surveillance, tracking and fire control functions. It is claimed to have particularly good performance in littoral conditions and incorporates a high-resolution surface search function optimised for detecting small surface targets. However, air defences are limited to a pair of RAM launchers for short-range protection and there is no organic ASW capability beyond that provided by the ability to deploy up to two helicopters.

Instead, the principal armament is provided by a range of gun systems. These encompass a single Oto 127mm mounting, two Rheinmetall 27mm MLG cannon and five positions for Leonardo Hitrole weapons stations equipped with 12.7mm machine guns. Together, they can assist a range of stabilisation requirements from the provision of naval gun fire support for landing operations through to countering swarm attacks by fast boats or drones. Space is also provided for up to eight Harpoon surface-to-surface missiles and a range of non-lethal systems such as water cannon. Command and control is provided by an ANCS Atlas Naval Combat System. The increasing use of containerised systems is reflected in provision for embarking two standard TEU (twenty-foot equivalent unit) containers between the two islands.

Although reflecting many recent design trends, *Baden-Württemberg* represents an attempt to design a sophisticated frigate-sized warship for a set of missions usually performed by much cheaper constabulary patrol vessels.[12] This would appear to be a luxury given the increase in international tensions in the period since the class was ordered, and the experiment seems unlikely to be repeated. Indeed, the follow-on F126 class frigates that have recently been ordered are notable for their emphasis on warfighting potential.

Future Developments

In general terms, the latest European frigates currently in build reflect many of the trends evident in the ships that have been described above. These include a continued emphasis on stealth of all kinds, the flexibility to accept a broad range of equipment and systems through the use of a modular design approach, and a tendency towards increased size and displacement. A particularly common theme is the ability to deploy and house the new generation of unmanned and autonomous systems that seem likely to play an ever-growing part in future naval warfare.

Type 26 Global Combat Ship: The British Type 26 frigate, also known as the Global Combat Ship, arguably represents the standard for recent surface combatant design trends. Tracing its origins to concept work on replacements for the Royal Navy's Type 22 and Type 23 frigates that commenced during the mid-1990s, the Type 26 is a 'high end' ASW warship. Following a long and convoluted development history, an order for an initial batch of three Type 26 frigates was placed in July 2017. A total class of eight ships is ultimately envisaged, all of which will be built at BAE Systems' shipyards on the River Clyde. The lead Type 26, *Glasgow*, is expected to enter service in the middle of the current decade.

Displacing some 7,000 tonnes and with an overall length of 150 metres, the Type 26 clearly demonstrates the size growth seen in recent warship designs.[13] Her conventional mono-hull is optimised for acoustic stealth whilst hunting submarines, a priority also reflected in her

MODERN EUROPEAN FRIGATES

A graphic of the British Type 26 Global Combat Ship. With a highly flexible configuration and a large mission deck for containerised and other equipment, the Type 26 represents the epitome of modern surface combatant design. (BAE Systems)

A BAE Systems graphic illustrating the different Type 26 variants developed for the navies of (front to back) the United Kingdom, Australia and Canada. (BAE Systems)

MODERN EUROPEAN FRIGATES: PRINCIPAL CHARACTERISTICS

	Fridtjof Nansen	*Absalon*	*Iver Huitfeldt*	*Aquitaine*
Country	Norway[1]	Denmark	Denmark	France
Number	5	2	3[2]	8[3]
Commissioned[5]	5 April 2006	19 October 2004	21 January 2011	23 November 2012
Full Load Displacement	5,300 tonnes	6,600 tonnes	6,600 tonnes	6,000 tonnes
Principal Dimensions	133.2m x 16.8m x 4.9m	137.0m x 19.5m x 5.3m	138.7m x 19.8m x 5.3m	142.2m x 19.8m x 5.4m
Propulsion	CODAG	CODAD	CODAD	CODLOG
	26 knots max	23 knots max	28 knots max	27 knots max
	4,500nm range	11,500nm range	9,000nm range	6,500nm range
Main Armament	1 x 76mm Oto gun	1 x 127mm Mk 45 gun	2 x 76mm Oto guns	1 x 76mm Oto gun
	Mk 41 VLS (8 cells)	3 x Mk 56 VLS (36 cells)	MK 41 VLS (32 cells)	Sylver A43 VLS (16 cells)
			2 x Mk 56 VLS (24 cells)	Sylver A70 VLS (16 cells)
	2 x quad NSM SSM	2 x quad Harpoon SSM	2 x quad Harpoon SSM	2 x quad Exocet SSM
	2 x twin A/S TT	2 x twin A/S TT	2 x twin A/S TT	2 x twin A/S TT
Modular Systems	–	Large Flex Deck	Mini Flex Deck	–
Aircraft[6]	1 x NH90 NFH	1/2 x MH-60R Seahawk	1 x MH-60R Seahawk	1 x NH90 NFH
Principal Radar	AN/SPY-1F passive MFR	SMART-S 3D	APAR active MFR	Herakles passive MFR
			SMART-L 3D	
Principal Sonar	Spherion MRS 2000 hull	Atlas ASO-94 hull	Atlas ASO-94 hull	UMS 4110 hull
	CAPTAS-2 TAS	TAS to be fitted		CAPTAS-4 TAS
Crew	120 core (145 berths)	100 core (170 berths)	100 core (165 berths)	110 core (145 berths)

Notes:

1. Designed and constructed in Spain.
2. A further five ships are being built to a modified 'Arrowhead 140' design for the British Type 31 programme. Indonesia has also acquired rights to build two 'Arrowhead 140' frigates under licence.
3. One additional ship with slightly modified equipment supplied to each of Egypt and Morocco.
4. Two additional ships have been supplied to Egypt and four ASW variants to Italy. Six further ships have been ordered by Indonesia, and the design also forms the basis of the US Navy's *Constellation* class.
5. Due to differing national practices, date sometimes refers to the date of delivery.
6. Typical helicopter detachment.

CODLOG propulsion arrangement and extensive silencing of onboard equipment. A sonar suite comprising an Ultra Electronics Sonar 2150 bow-mounted array and Thales Sonar 2087 (CAPTAS-4) is similar to that found in many of the combatants already discussed. Other equipment includes the combination of a Type 997 Artisan surveillance and tracking radar with Sea Ceptor SAMs for air defence. A 24-cell Mk 41 VLS – to be equipped with the Anglo-French Future Cruise/Anti-Ship Weapon (FCASW) being developed by MBDA – and a 127mm/62 Mk 45 Mod 4 gun, both of US Navy origin, provide meaningful anti-surface warfare capabilities.

The recent trend towards adopting modular systems is reflected in the incorporation of an adaptable mission bay immediately forward of (and linked with) the ship's single helicopter hangar. This will house and deploy a wide range of equipment depending on the particular mission in hand. This could include a second helicopter, remotely operated vehicles and/or containerised equipment. There is space for up to ten TEU containers. Less apparent but of equal importance is the hosting of combat and ship systems on BAE Systems' shared infrastructure. This common computing architecture replaces the separate hardware and associated networks used to support specific functions in previous generation ships. This eases the integration of a wide range of new equipment and software into the network, making ongoing upgrade more straightforward than in the past.

The Global Combat Ship design has been selected by Australia and Canada to meet their own navies' future frigate needs. In both cases, the basic Type 26 hull and propulsion system will be overlaid with equipment that reflects national priorities, both industrial and operational. For example, the nine Australian *Hunter*-class frigates will utilise the indigenous CEAFAR 2 phased array in conjunction with the Aegis combat management system and a Saab Australia tactical interface. Meanwhile, Canada's fifteen planned Canadian Surface Combatants will see the Lockheed Martin Canada CMS 330 combat management system interface with an Aegis module and the company's new AN/SPY-7 active phased array to control three layers of protective missile systems. These revisions are likely to push displacement significantly above that of the baseline British ship, thus demonstrating the importance of ensuring adequate design margins if the flexibility provided by modular equipment is to be utilised effectively.

F-110 Class Frigate: Spain's new F-110 class is another modern European frigate design optimised for ASW but

MODERN EUROPEAN FRIGATES

Carlo Bergamini	*Baden-Württemberg*
Italy	Germany
6[4]	4
29 May 2013	17 June 2019
6,700 tonnes	7,300 tonnes
144.0m x 19.7m x 5.5m	149.6m x 18.8m x 5.4m
CODLAG	CODLAG
27 knots max	26 knots max
6,500nm range	4,000nm range
1 x 127mm Oto gun	1 x 127mm Oto gun
1 x 76mm Oto gun	
Sylver A50 VLS (16 cells)	2 x RAM Mk49 CIWS
4 x twin Teseo SSM	2 x quad Harpoon SSM
2 x triple A/S TT	
–	space for 2 x containers
2 x NH90 NFH	2 x Sea Lynx
EMPAR active MFR	TRS-4D active MFR
UMS 4110 hull	–
130 core (200 berths)	120 core (190 berths)

Navantia's F-110 frigate design is similar to the British Type 26 in that it incorporates a wide range of modern design features in a warship optimised for ASW. (Navantia)

with considerable multi-mission potential, and is intended to replace the existing *Santa María* (FFG-7) class. A contract for five of the new frigates was signed in April 2019. All the ships will be built by Navantia at Ferrol, where lead ship *Bonifaz* is expected to commence construction in 2022.

The *Bonifaz* class is a little smaller than the Type 26 frigates, with a displacement of 6,100 tonnes and a length overall of 145 metres. The design appears to owe much to the previous F-100 class air defence frigates and the hull may therefore not enjoy the same degree of underwater stealth as the British Type 26. The ASW capabilities of the type appear otherwise similar to the other European frigates previously described: these include use of an acoustically silent CODLAG propulsion system and a detection suite based around Thales' bow-mounted UMS 4110 and CAPTAS-4 towed array sonars.

The importance of being able to embark specialised systems is reflected in the incorporation of a mission bay similar to that found in the Type 26. This is located to starboard, parallel with the helicopter hangar. It can be used to house containerised systems, small boats or unmanned vehicles. The use of an Aegis air defence module and SPY-7 phased array with data coordinated by an indigenous SCOMBA combat management system has parallels with the approach adopted in the Canadian Surface Combatant. Spain has high hopes of following previous success by exporting variants of the class, and the design has already been shortlisted in the competition for Poland's new surface combatant.

Pattugliatore Polivalente d'Altura (PPA): A further example of current design approaches is the family of PPA oceanic 'patrol vessels' developed by Fincantieri for the Italian Navy. With an overall length of 143 metres and a displacement in the region of 6,000 tonnes, these are essentially frigate-like vessels that are intended to be outfitted for missions of varying intensity. More specifically, the Italian Navy envisages them being completed in 'Light', 'Light Plus' and 'Full' iterations, each with steadily increasing combat capability. The PPA can therefore be outfitted to perform stabilisation duties similar to those envisaged for the German F125 class but can be readily adapted to undertake more intensive operations if the need arises. A total of seven of the class have been ordered from Fincantieri, encompassing two Light, three Light Plus and two Full variants, under Italy's 'Naval Law' of 2014/15.[14] Like the Italian FREMMs, these are assembled at Riva Trigoso and outfitted at Muggiano. The first, *Paolo Thaon di Revel*, is expected to commission before the end of 2021.

All the PPAs will be equipped with a 127mm main gun, 76mm secondary gun, lighter calibre close-in weapons and a flight deck and hangar for one heavy or two smaller helicopters. The more fully-specified versions will benefit from Sylver VLS modules and upgraded multi-function radars for use with the Aster surface-to-air missile, while the capabilities of the Full variants will be further broadened by equipment such as towed array sonar and torpedo defence systems. A complex propulsion system supplements a primary CODAG arrangement with the use of electrical motors drawing power from the ship's distribution system for low-speed operation up to ten knots.

The class incorporates a number of innovations that may influence the future direction of warship design. One of these is an integrated bridge that incorporates a cockpit-like navigation position inspired by the aerospace sector to reduce overall crewing requirements. There is also much greater integration of combat, platform management and navigation systems than previously seen. An innovative wave-piercing bow (see photo) improves hydrodynamic performance, adding around a knot to maximum speed.

There are two separate zones for modular systems. One, amidships, is equipped with davits and a crane to handle small boats and containerised equipment. It can accommodate up to eight containers or two RHIBs. The other, at the stern, can handle additional boats and has sufficient space to be configured as a base for Special Forces, as a hospital, as accommodation for migrants, or as a command post for autonomous systems.

Frégate de défense et d'intervention (FDI): The French Navy's *Amiral Ronarc'h* class represents a somewhat different approach to modern European frigate design. Originally known as the *frégate de taille intermédiaire* ('medium-sized frigate'), the design was specifically intended to be smaller than the French FREMM in order to attract export orders. The underlying thinking appears to have been based around the difficulties faced by smaller and developing navies in operating and sustaining the larger combatants in service with the major fleets. Launched in 2015, the programme will initially see five ships built by Naval Group at Lorient. Fabrication of the lead ship began in 2019; she will be operational around the end of the decade.

Displacing some 4,500 tonnes and with a length overall of 122 metres, the FDI is the smallest ship considered in this article. Despite this, the design is still around 700 tonnes heavier than the *La Fayette* class it is destined to replace, reflecting the ongoing upward pressures on warship size. Needless to say, the ships benefit from the attention to RCS stealth that has been a hallmark of recent French warship design, and they will be well-equipped for their size. Key sensors include the new Sea Fire multi-function array and a compact variant of the Thales sonar suite installed in the FREMMs. Armament comprises an Oto 76mm gun, Sylver VLS modules for Aster surface-to-air missiles, and canisters for Exocet anti-ship weapons. There are also anti-submarine torpedo tubes and a hangar for a medium helicopter and/or drones. A wave-piercing, ram-type bow is indicative of the attention being given to hull form in many recent ships. However, some compromises have been inevitable. These include relatively noisy diesel propulsion and the lack of a dedicated mission deck for modular systems. Possibly for industrial reasons, there seems to be little flexibility in the equipment options available to customers for the type's 'Belh@rra' export variant. This may hinder overseas sales given the availability of more modular ships, although the design has recently been selected to meet the Hellenic Navy's requirement for new frigates.

Looking farther into the future, it is interesting to speculate on what further innovations lie ahead. The focus on

The lead Italian PPA-type patrol vessel *Paolo Thaon di Revel*, a frigate in all but name. The design can be fitted out to perform operations of various levels of intensity and incorporates two separate zones for accommodating and deploying modular systems. (Fincantieri)

MODERN EUROPEAN FRIGATES

A graphic of the 'Belh@rra' export variant of France's FDI. The well-equipped design is more compact than many recent European frigates, but appears to lack flexibility for ready adaptation. (Naval Group)

unmanned or minimally-crewed vessels – likely operating in concert with manned warships – is one key trend. The US Navy, in particular, is well advanced in developing large and medium unmanned surface vessels that will act as platforms for modular payloads of weapons and sensors as part of its distributed approach. It seems possible that the already considerable evolution in warship design being seen in the first quarter of this century is little more than a precursor to a period of much more rapid change ahead.

Sources:

This chapter has largely been researched from contemporary industry and government literature, as well as press releases and news reports. The following sources provide more extensive reading material:

Friedman N, *The Naval Institute Guide to World Naval Weapons Systems*: Fifth Edition, US Naval Institute Press (Annapolis, 2006).

O'Rourke R, *Navy Constellation (FFG-62) Class Frigate Program: Background and Issues for Congress*, Congressional Research Service (Washington DC, 2021).

Toremans G, 'Iver Huitfeldt Class Frigates' in Waters C (ed), *Seaforth World Naval Review 2014*, Seaforth Publishing (Barnsley, 2013), 104–19.

Various specific ship reviews in editions of the editor's *Seaforth World Naval Review* and information in the annually updated *Janes Fighting Ships*, Janes (Coulsdon) provide additional information.

Endnotes:

[1] This is a simplification. Some major Asian nations, notably Japan, have a considerable track record of designing their own warships, although there has often been considerable reliance on overseas weapon systems and sensors. More recently, an increasing number of navies have completed the transition from constructing vessels with overseas design assistance through producing local designs incorporating foreign equipment to producing ships that are largely designed and outfitted on the basis of indigenous expertise.

[2] The Aegis combat system terminology initially appears to have been used to refer to the combination of computerised command and control, sensors and weapon systems that provided Aegis's integrated capability. However, the use of Aegis's command and control capabilities with an ever-broadening range of missiles and sensors means that the term is seemingly increasingly used to refer to the core combat management system functionality.

[3] Spain (F-100 class), The Netherlands (*De Zeven Provinciën* class), Germany (F124 *Sachsen* class), France and Italy ('Horizon' class) and the United Kingdom (Type 45 *Daring* class).

[4] The fire control radars are required to direct the missile in the later stages of an engagement as the initial ESSM Block 1 was not fitted with an active homing capability. The subsequent ESSM Block 2, which is currently entering service, includes an active radar homing capability that does not depend on such target illumination.

[5] The three *Iver Huitfeldt*-class frigates were to be the last vessels built by the shipyard.

[6] The SMART-L surveillance radar and its derivatives have been a popular choice for European air defence vessels, equipping the majority of the ships of this type that are currently in service. APAR is also used in the latest Dutch and German air defence escorts.

[7] CAPTAS-4 was first introduced as Sonar 2087 in the British Type 23 frigates.

[8] The two French-built FREMMs that were exported were essentially diverted from the French Navy's production run, the Egyptian *Tahya Misr* (ex-*Normandie*) being transferred only very shortly before her scheduled delivery to the *Marine Nationale*.

[9] Distributed Maritime Operations (DMO) is a concept under which technology is used to integrate the total warfighting potential of diverse, geographically dispersed ships, submarines and aircraft into a war-winning weapon. It is tending to drive US Navy force structure towards larger numbers of small ships, including unmanned vessels.

[10] The Raytheon SPY-6 series of active phased array radars adopts a system of 'building blocks' known as radar modular assemblies (RMAs) that can be stacked together to provide increased power and capacity. Each of the *Constellation* class's three fixed SPY-6(V3) arrays comprises nine RMAs. By comparison, the four larger SPY-6(V4) arrays in the Flight III *Arleigh Burke* class destroyers have as many as 37 RMAs.

[11] The Italian FREMMS also have accommodation for around 200 personnel, but this has only been achieved by converting the space reserved for additional Sylver VLS launchers for this purpose.

[12] The closest equivalent would appear to be the Royal Netherlands Navy's *Holland* class oceanic patrol vessels. However, these are far less expensive ships.

[13] By way of comparison, the preceding Type 23 frigates currently displace around 4,800 tonnes.

[14] The Naval Law was enacted to allow a 'one-off' boost to the Italian fleet's modernisation in the face of the block obsolescence of many ships built in the Cold War era.

THE AUSTRALIAN *BATHURST*-CLASS MINESWEEPER CORVETTE

The minesweeper corvettes of the *Bathurst* class were the largest group of warships ever built in Australia. This study by **Mark Briggs** looks at the role played by the class in the revival and expansion of Australian shipbuilding.

The *Bathurst* class formed the backbone of the Royal Australian Navy (RAN) during the Second World War. Although originally classified as minesweepers they were used, like many small ships in wartime, in a wide variety of roles for which they had not been intended. Apart from minesweeping and anti-submarine work they supported special operations among the islands to the north of Australia and performed a key role in surveying landing sites, often under the noses of the enemy. Less appreciated has been the central part they played in rejuvenating Australian shipbuilding. Several shipyards that had become virtually derelict during the economic depression of the 1930s were restored to life during the Second World War in order to build them.

The Origins of the Design

In the years following the Great War of 1914–18, with Germany's High Seas Fleet at the bottom of Scapa Flow and the Treaty of Versailles limiting the German Navy to little more than a coastal defence force, the greatest naval threat to Britain and her empire seemed to come from Japan. This was the assessment presented by the Admiralty at the 1923 Imperial Conference in London.[1] The Japanese possessed a powerful fleet of battleships and cruisers, but they also had a sizeable force of submarines. From 1924 to 1940 Japan built thirty-five large ocean-going submarines, some with hangars to house a light reconnaissance aircraft, others fitted to lay mines.

These submarines were primarily intended to work alongside the main fleet in its much anticipated showdown with the US Navy, but they clearly could also be used to attack Britain's maritime trade. Under a League of Nations mandate Japan controlled a number of former German Pacific territories such as the Caroline and Marshall Islands. From here large Japanese submarines could reach Australian waters. With road and rail communications still underdeveloped, Australia relied heavily on coastal shipping to move people and goods between its major cities, all of which were on the coast.

The Australian Commonwealth Naval Board (ACNB), roughly equivalent to the British Board of Admiralty, was aware of the threat. A 1930 assessment estimated that Japan had twenty-nine submarines that could reach Australia's east coast from bases in the mandated territories, with Palau and Truk the most likely.[2] When Britain had faced German submarine attacks on its coastal shipping in the First World War the Admiralty had mobilised a fleet of fishing boats, pleasure craft, ferries and other small commercial vessels to mount minesweeping and anti-submarine patrols around the entrances to Britain's ports and along the coastal shipping lanes. By 1918 this Auxiliary Patrol, as it was known, had grown to 3,174 vessels of all descriptions.[3]

Australia, however, did not have anywhere near as many fishing boats and other small ships that could be requisitioned for these tasks as Britain. Aware of the difficulty of obtaining suitable craft in Australia, Rear Admiral Sir Percival Hall-Thompson, a Royal Navy officer on secondment as head of the ACNB, wrote to London in September 1925 requesting up to fifty trawlers be sent to Australia should there be a war in the Far East.[4] As the Admiralty came to appreciate the lack of small craft suitable for minesweeping and anti-submarine work outside of home waters, not just in Australia but across British ports from Aden to Hong Kong, they developed plans to despatch up to 200 trawlers from Britain within the first few months of war. Twenty-five were allocated to Australia.[5]

The problem with this strategy was that it hinged on there being no simultaneous threat in European waters. By 1935 this was no longer the case. Under Adolf Hitler the German Navy, including its U-boat arm, was being rebuilt. War with Germany, or even the possibility of war, meant Britain would not be sending hundreds of trawlers to the Far East should hostilities break out with Japan.

Belatedly the ACNB came to recognise its predicament. In 1938 a need was identified for a training vessel for the Navy's newly-established anti-submarine warfare school at HMAS *Rushcutter* in Sydney. About £100,000 also became available after it was decided that the last of three boom defence vessels already approved was no longer required. Proverbially, two birds could be killed with one stone. The design of a small minesweeper/anti-submarine vessel could be developed and a prototype built using the money for the no-longer-needed boom defence vessel. This prototype could then serve as the training ship at HMAS *Rushcutter*, allowing any design problems to be ironed out in time for large-scale wartime construction.

With war clouds on the horizon both in Europe and the Far East, the ACNB drew up a requirement for a 'simple and easily constructed general purpose local defence vessel' that could be built in Australia outside of the Cockatoo Island shipyard in Sydney Harbour. Cockatoo Island was the only facility in Australia with experience in naval shipbuilding, and it was assumed that it would be fully committed to larger naval work.

The new vessel was intended for the so-called 'outer patrol' (later renamed the 'anti-submarine patrol') – that is on the open sea outside the entrances to Australia's major ports and along the coastal shipping routes. The ACNB envisaged a vessel of around 500 tons armed with a single 4in gun, two depth charge throwers and two depth charge chutes with stowage for 25 depth charges. Top speed should not be less than 10 knots, preferably higher, with a capacity for rapid acceleration. A minimum endurance of 2,000 miles was required. To facilitate building in Australia the vessel should be powered by reciprocating engines with a single Yarrow-type boiler. While the vessel was to be primarily for anti-submarine work it also had to be able to operate as a minesweeper, the depth charges being replaced by the sweep gear, though the design was not to be 'closely restricted' to a minesweeper's shallow draught.[6]

Having settled on the sort of vessel it wanted, the ACNB wrote to the Admiralty in London to see if there were any existing British designs that might fit the bill. The Admiralty obliged by sending Australia the drawings for two ships: the anti-submarine trawler *Basset* and the slightly larger fishery protection and minesweeping trawler *Mastiff*. Both ships were prototypes based on trawler designs that had been developed for large scale production in wartime. *Basset* was 160ft 6in overall with a maximum draught of 11ft 7in. Displacing 596 tons deep, *Basset* had a single-shaft reciprocating engine and a top speed of 12 knots. *Mastiff* was 163ft 6in in length with a draught of 12ft 3½in. She had a full load displacement of 690 tons and a slightly more powerful engine than *Basset*, giving her an extra half knot of speed. Both ships were armed with a single 4in gun and two machine guns and could be fitted to carry either depth charges or minesweeping gear.[7]

On the surface the two designs looked promising; they broadly met the specifications set down by the ACNB. The designs were handed over to the RAN's senior engineer, Rear Admiral PE McNeil, for assessment. After working as an engineer in the merchant navy and at the John Brown shipyard in Scotland, McNeil had joined the RAN, serving aboard a number of ships including the battlecruiser HMAS *Australia* until appointed Director of Engineering at the Navy Office in Melbourne in 1931. McNeil dismissed both designs: '*Mastiff* and *Basset* appear to be too slow and unhandy for A/S work,' McNeil explained, 'and of too deep a draught for minesweeping.'[8]

As McNeil highlighted, the requirements for minesweeping and anti-submarine operations were contradictory. Minesweepers needed a shallow draught to lessen the chance of striking mines while carrying out their dangerous work, but an anti-submarine vessel needed a deeper draught for its all-important Asdic to operate effectively. Too shallow a draught and there was a danger of 'quenching', where the Asdic apparatus on the keel suffered from coming too close to the surface as the ship rolled. In the Royal Navy this had led to the construction of separate minesweepers and anti-submarine sloops, but Australia's limited resources meant it was essential that a single ship was capable of carrying out both roles.

Designing the *Bathurst* Class

With no suitable vessel available from Britain it fell to Rear Admiral McNeil and his staff to develop a design that met the ACNB's requirements. It was a major ask. In July 1938 the Engineering and Construction Branch at the Navy Office in Melbourne had a permanent staff of just seven along with three temporary draughtsmen.[9] Inevitably, as McNeil and his team started working on the first draft of a new design, changes to the original ACNB requirements began to arrive. Admiral Sir Ragnar Colvin, Chief of the Naval Staff and First Naval Member of the Naval Board, now wanted two 4in guns and as much speed as could be obtained 'without raising size and cost unduly.'[10]

The first sketches were for a vessel 180ft on the waterline, 186ft overall, with a 31ft beam and a maximum draught of 10ft. In accordance with Admiral Colvin's request there were two 4in guns. Asdic was included and depth charges were interchangeable with minesweeping gear. In accordance with the ACNB's initial specifications there was a single Yarrow-type boiler with two reciprocating engines and 1,550ihp. There were two diesel generators to provide electrical power but no steam set, and a small evaporator to supply boiler water only. Top speed was estimated at 15½ knots with an operational radius of 2,850 miles at 12 knots. The ship would be built to commercial standards to keep down costs and to facilitate construction in dockyards unused to naval work, though some concessions were made to their fighting roles: bullet-proof plating, for example, was included for the wheelhouse and bridge.

Displacement was 680 tons standard, 850 tons full load. This was rather more than the ACNB had initially indicated, and McNeil felt obliged to explain the increase in size. The design, he argued, should be regarded as a small sloop and represented 'the smallest type in which reasonable sea going qualities and speed for the purpose in view can be obtained.' Draught could not be reduced any further without greatly affecting seaworthiness. He estimated the cost at £110,000, ten percent more than had been found for the project. On a positive note McNeil believed that, except for guns, gunnery instruments, electric cables and boiler tubes, the ship could be built in Australia from local resources in wartime.[11]

Admiral Colvin was delighted with the design, famously describing it as a 'very useful little ship.' Construction of a prototype, he urged, should begin immediately without waiting for the usual nod from London. His one concern was how quickly the vessel could accelerate from slow to full speed, rapid acceleration being 'most desirable' in an anti-submarine vessel.[12] 'With a speed of 15½ knots,' McNeil responded, 'the craft, considering her size, would be generally of fairly fast type.' The Yarrow oil-fired boiler with which she would be fitted was the best available design to quickly meet sudden demands for steam, and the vessel should be capable of accelerating to 10 knots within a minute of the order for full speed being received in the engine room. 'In general manoeuvrability,' McNeil went on to claim, 'she would class about midway between the average small merchant vessel available to be taken up for A/S purposes and the destroyers.'[13]

Inevitably, as the design was fleshed out further changes were made. McNeil pointed out that with only a single boiler the vessel would be out of service for forty-eight hours every four weeks for boiler cleaning. With two boilers the vessel would require two boiler cleaning periods in the month but would still be available for service, albeit with reduced effectiveness.[14] Although two boilers would increase the cost and complexity of the vessel, the proposal was adopted and the final design included two Yarrow boilers side by side in a single boiler room. A third diesel generator was also included along with a larger evaporator and distiller.

General Characteristics

In keeping with the original ACNB requirements, the *Bathurst* class was a very simple design. There was a raised forecastle that extended half the length of the ship ending just abaft the funnel with the deckhouse extending to the quarter deck where the minesweeping gear was fitted. Accommodation for ratings was in hammocks forward with the galley below the bridge. Officers' accommodation was aft with the ward room and officers' bathroom in the deckhouse and cabins on the upper deck below them. Total complement was between eighty-five and ninety including six commissioned and twelve non-commissioned officers.

The shallow draught required for minesweeping (mean draught was 8ft 3ins) and cramped accommodation made the *Bathurst* class uncomfortable ships to serve in. 'Living in a corvette was like living in a concrete mixer mounted on a roller coaster,' recalled Frank Walker, who served aboard HMAS *Latrobe*. 'Even in dry dock they would roll if a heavy dew fell. At sea they were vixens: petulant, cantankerous, moody and capricious.'[15]

The two Yarrow watertube boilers were located side by side amidships with the two triple expansion engines,

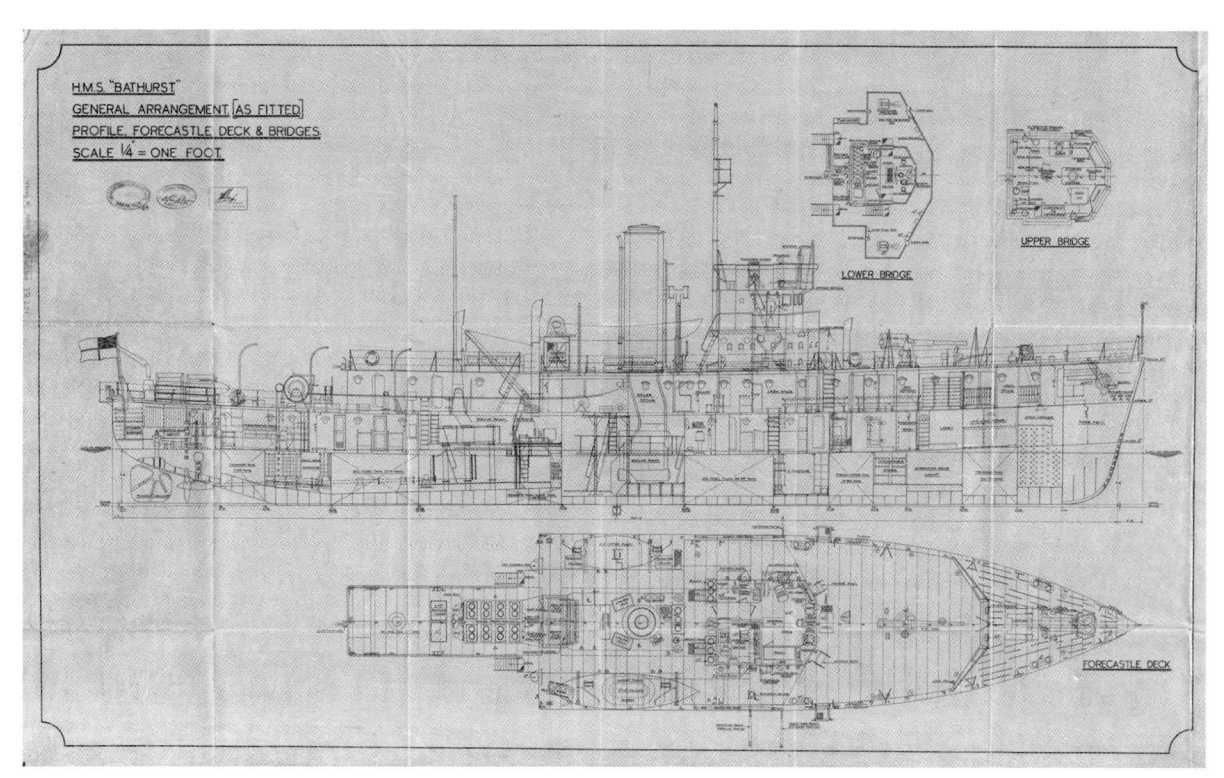

HMAS *Bathurst*: General Arrangement Plans (As Fitted)

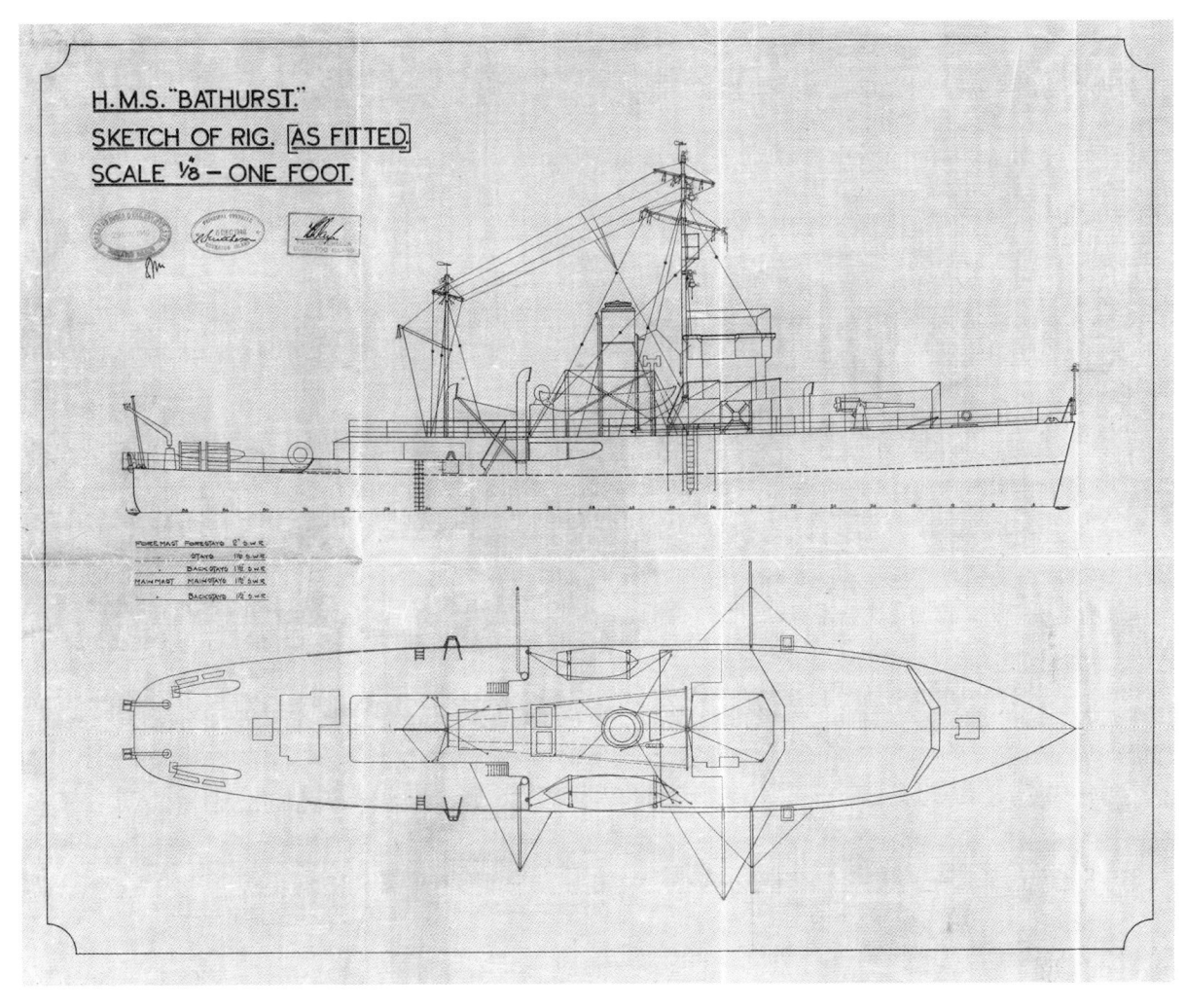

ns
THE AUSTRALIAN *BATHURST*-CLASS MINESWEEPER CORVETTE

HMAS *Bathurst*: Sketch of Rig (As Fitted)

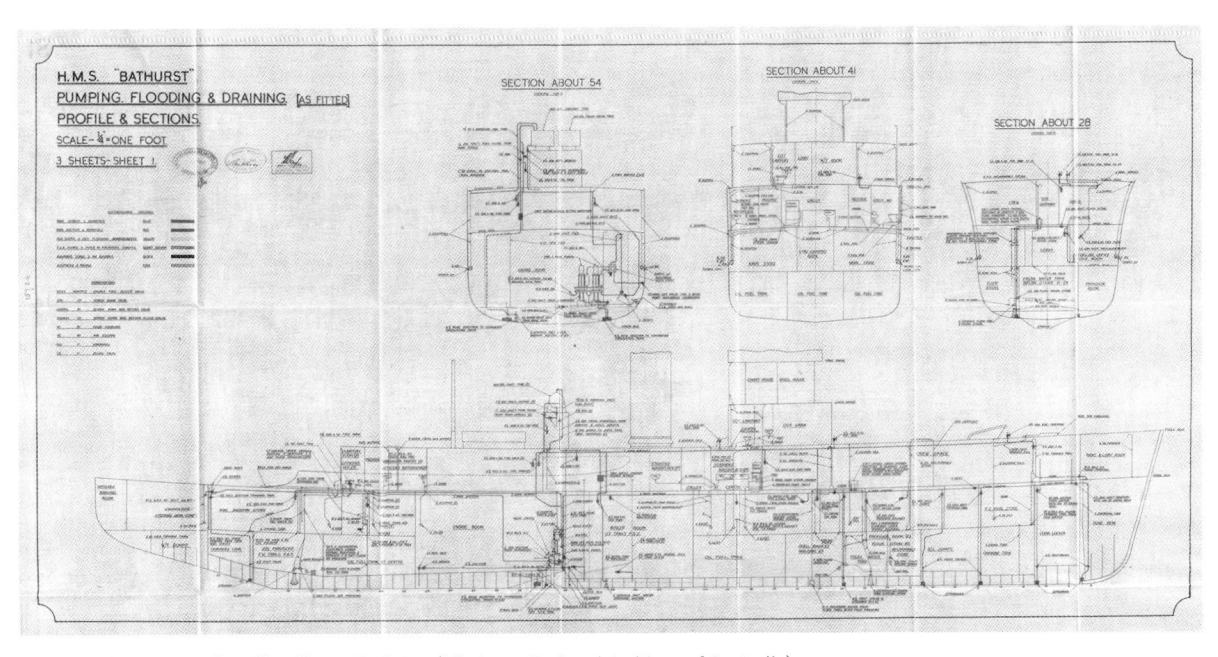

HMAS *Bathurst*: Pumping, Flooding & Draining (All plans: National Archives of Australia)

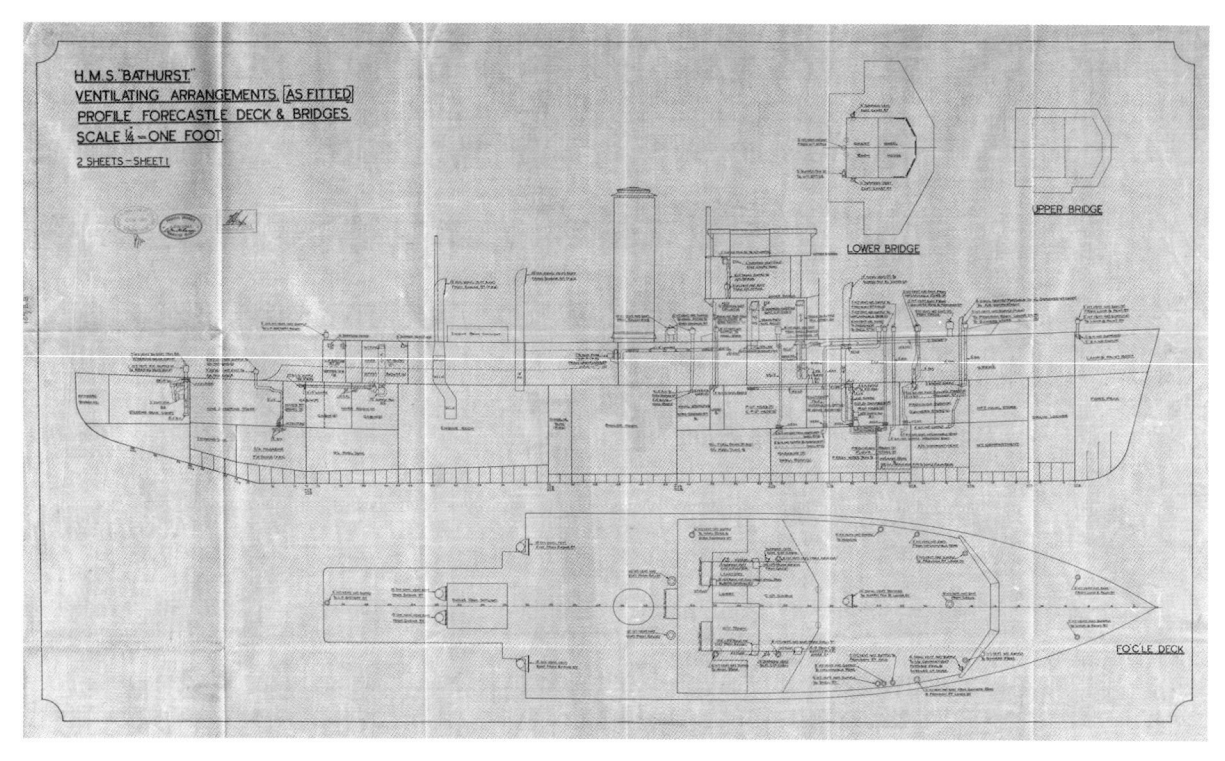

HMAS *Bathurst*: Ventilating Arrangements

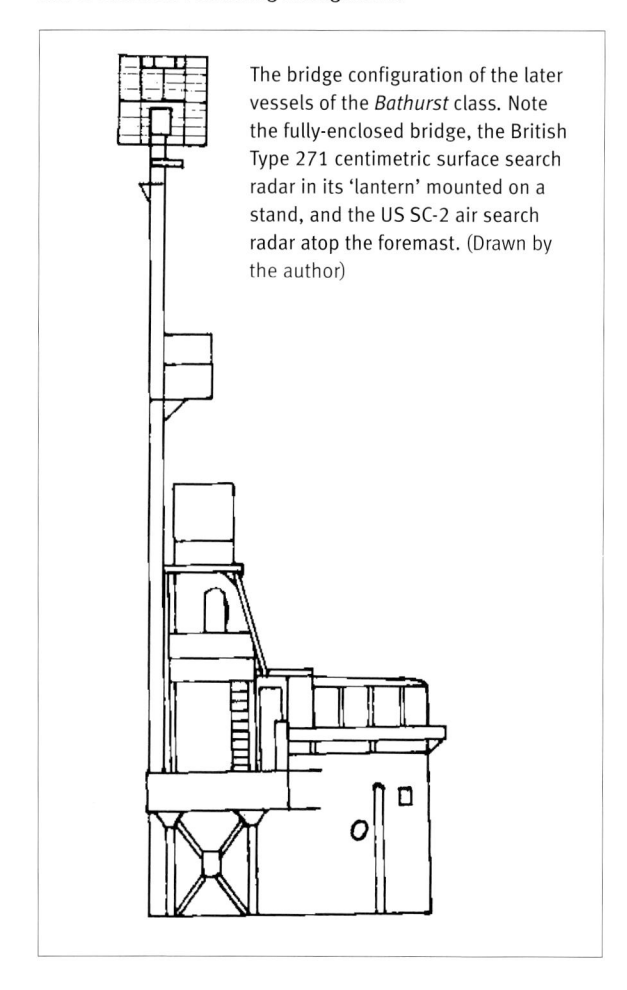

The bridge configuration of the later vessels of the *Bathurst* class. Note the fully-enclosed bridge, the British Type 271 centimetric surface search radar in its 'lantern' mounted on a stand, and the US SC-2 air search radar atop the foremast. (Drawn by the author)

each driving a separate shaft, behind them. The first group of twenty-four vessels had engines rated at 1,750ihp; in the remainder of the class this was raised to 2,000ihp, with larger boilers. This gave the later group a top speed of about 16 knots, around a knot more than the first group. 134 tons of oil was carried, providing a radius of just over 3,500 miles at 9 knots.

Although originally intended to carry two 4in guns none of the *Bathurst* class ever did so. From *Bathurst* on the after gun was replaced by a single 20mm Oerlikon to provide better anti-aircraft defence. This in turn was replaced during the war with a single 40mm Bofors, and 20mm Oerlikons replaced the light machine guns in the bridge wings. Due to difficulties in obtaining 4in guns three of the early vessels – *Bathurst*, *Katoomba* and *Launceston* – were armed with a single 3in 12pdr gun. The early ships had their main guns in open mountings but these were all later fitted with shields.

The original *Bathurst* class design included a three-level bridge structure. At forecastle deck level was the commanding officer's cabin and bathroom. Above this was an enclosed wheelhouse with a chart room and radio room to the rear. This in turn was topped by a British destroyer-style open bridge with a small enclosed hut for the Asdic operator. By the fourth ship, HMAS *Ballarat*, a new arrangement was introduced, probably to reduce topweight, something that became increasingly important as radar and additional anti-aircraft weapons were fitted. There was now a two-level bridge structure with the chartroom and later a radar room alongside the commanding officer's cabin at forecastle deck level. Above this was the bridge, which in some early vessels

THE AUSTRALIAN *BATHURST*-CLASS MINESWEEPER CORVETTE

was left open with glass windshields and an awning for shade in the tropics. Later ships were completed with a fully-enclosed bridge and earlier vessels had their open bridges enclosed with a permanent roof. A platform for a 20in searchlight was installed at the back of the new bridge in front of the foremast.

As the war progressed the amount of electronic equipment aboard the ships increased. From the outset the *Bathurst* class carried Type 128 Asdic. Initially no radar was carried but ships were gradually fitted with British Type 271 surface search radar with its characteristic lantern covering, mounted on a stand above the searchlight platform at the back of the bridge. Following the outbreak of the war against Japan Australia gained access to American equipment and ships began to sport a US Navy SC-2 air search radar atop the foremast.

Building the *Bathurst*s

The ACNB's plan had been to build a single prototype of the *Bathurst* class that would serve as the training ship for the anti-submarine warfare school at HMAS *Rushcutter*. This prototype would be built at Cockatoo Dockyard in Sydney, thereby allowing any problems to be ironed out in time for large-scale construction in war.

The launch of the first of class, HMAS *Bathurst*, at Cockatoo island Dockyard, Sydney, 1 August 1940. (RAN)

Despite Admiral Colvin's enthusiasm for beginning construction immediately, an order for the first unit was not placed until December 1939. The delay was partly due to design changes, such as including a second boiler, but also because questions had been raised about the funding. The ACNB had intended to use the money already allocated to a third boom defence vessel to build

HMAS *Bathurst*, presumably running trials prior to commissioning as an Australian flag flies from the mainmast but there is no White Ensign. The photograph shows the original British destroyer-style open bridge and the single 4in gun forward without a shield. (RAN)

HMAS *Lithgow*, showing the new lower bridge with searchlight platform behind it. The bridge is still open but with a glass windshield. The 4in gun forward has a shield and the ship has been fitted with the British Type 271 surface search radar with its distinctive lantern cover on a stand above the searchlight platform. The antenna of a US Navy SC-2 air search radar can be seen at the head of the foremast. (Allan Green collection, State Library of Victoria)

the ship. As the new vessel was clearly not a boom defence vessel, the Director of Naval Accounts argued it would be inappropriate to use this money to fund it.[16]

War intervened before the funding issue could be resolved. A third boom defence vessel was ordered using the money already appropriated for this purpose. New wartime funding provided for the construction of the minesweeper corvettes. Cockatoo Dockyard submitted a tender for the lead ship in October 1939 and an order was placed in December. HMAS *Bathurst* was finally laid down at Cockatoo on 10 February 1940. The opportunity for a prototype to test the design, however, had long passed. The government approved an initial batch of seven vessels for the RAN, ordering four, with the second vessel, HMAS *Lismore*, laid down at Mort's Dockyard in Sydney just sixteen days after HMAS *Bathurst*.

Mort's Dock was the obvious next choice to build the minesweepers after Cockatoo Island. Established in the 1850s it had struggled through the interwar period to emerge with a virtual monopoly on commercial shipbuilding and repair work in Sydney. The Williamstown dockyard in Victoria was another obvious candidate. The dockyard had originally been owned by the state government, but protracted industrial unrest during the First World War had led to the Commonwealth taking control. After the war it had been sold to the Melbourne Harbour Trust, which concentrated on building and repairing its own small vessels: tugs, dredgers, barges and the like. Williamstown had the best shipbuilding facilities outside of Sydney but it was already engaged in converting merchant vessels into troop ships for the Commonwealth. With a total workforce of just 178 it simply did not have enough skilled people to begin new construction until this work was completed. The first *Bathurst* was not laid down until 16 October 1940.[17]

Williamstown's problem was universal. 'In every district in which vessels are being built,' the War Cabinet was told, 'output is being limited by the amount of labour available, which is in no case equal to that which could be employed on the work.'[18] One option was to bring other enterprises with established workforces into the programme. Another Sydney shipyard, Poole & Steele, was awarded a contract. It had once operated a substantial shipbuilding enterprise in South Australia but a lack of orders during the depression years had seen this close. Thereafter the firm had focused on building mining dredgers for export at its small Balmain shipyard. It laid down its first *Bathurst* on 9 September 1940.

Finding other prospective shipbuilders was a bigger challenge. Like Poole & Steele's South Australian works, many had either closed down or had abandoned shipbuilding after the war. Walkers Shipyard at Maryborough in Queensland was a case in point. Walkers had a long, if intermittent, history of shipbuilding dating back to 1877. It had built three merchant ships for the Commonwealth during the First World War but shipbuilding operations had ceased in 1928. Much of the equipment had been sold off and the slipways had become 'little more than an open paddock running down to a tidal river.'[19] Walkers, however, continued to build steam locomotives and rolling stock. With the outbreak of war the ACNB asked Walkers if it could again build ships for the Commonwealth. The slipway was cleared of silt and undergrowth and an electric hammerhead crane, bought from the Walsh Island shipyard at Newcastle, was installed. The first *Bathurst* class corvette, HMAS *Maryborough*, was laid down at Walkers on 16 April 1940 and launched just six months later, a remarkable achievement given that the yard had not built ships for over a decade.[20]

Walkers, Cockatoo Island and Morts Dock all received follow-on orders. Build time for the first group across the three shipyards averaged around seven months, with a similar amount of time required for fitting out. This was on a par with construction times for the similar *Bangor*-class minesweepers in Britain. The need for the new ships, however, was acute. Soon after the outbreak of the war Australia had offered to build *Bathurst*-class vessels for the Royal Navy once the RAN's initial requirements had been met. The Admiralty jumped at the chance, requesting ten as soon as possible. Minesweeping had

A minesweeping paravane being launched from the quarterdeck of HMAS *Bathurst*. (RAN)

become an urgent requirement for the Royal Navy, with German mines taking a heavy toll of Allied shipping in the first few months of the war. Twenty-seven ships had been sunk by mines in November 1939 and thirty-two in December totalling 203,000 tons.[21] The need was so urgent that the Admiralty asked if the four vessels currently building and three others ordered for the RAN could be released for service with the Royal Navy as soon as they were completed, the Australian requirement being made up from later construction.[22]

With seventeen vessels now on order there was a desperate need to find new builders. The country was

The launch of HMAS *Maryborough* at the Walkers shipyard in Queensland, 17 October 1940. The slipway had been an overgrown paddock just a few months earlier. In the foreground is the hammerhead crane that had been purchased from the NSW State Dockyard in Newcastle to enable the building of corvettes, much to the anger of Newcastle businesses and residents. (RAN)

scoured for derelict shipyards or general engineering works that might be able to build ships. Robinson Brothers in Melbourne had the workforce and equipment but could not get access to a site on the Yarra River for a slipway. Evans Deakin in Brisbane was better placed. Like Robinson Brothers, Evans Deakin was a general engineering firm rather than a specialist shipbuilder, but it had taken over the Brisbane City Council's slipway at Kangaroo Point on the Brisbane River in 1939. Its first vessel was the 1,200-ton oil fuel lighter *Rocklea*, built for the RAN and completed on 23 November 1940. Its first corvette, HMAS *Launceston*, was laid down one month later.

Newly opened facilities by the miner and steel maker Broken Hill Pty Ltd next to its steel works at Whyalla in South Australia also added much-needed capacity. BHP's General Manager, Essington Lewis, had visited Japan before the war. Shocked by the level of militarism and the rapid expansion of Japan's steel and armaments industries, he was convinced Australia needed to do more to strengthen its defences. Perhaps aided by the prospect of lucrative government orders, Lewis convinced the BHP board to develop a shipyard at Whyalla. The company was already spending £1,500,000 to build a blast furnace at the site, and following the outbreak of war in Europe decided to throw in a further £500,000 to add a rolling mill to produce steel plates, expand and deepen the harbour and develop a shipyard. Progress was rapid and the slipways were completed by the middle of 1940. The first corvette, named HMAS *Whyalla* after the steel town, was laid down on 24 July 1940, the first of four to be built there.

One shipyard that had not received any contracts was the New South Wales State Dockyard at Walsh Island in Newcastle, north of Sydney. Although considered early on for possible minesweeper construction the site was badly run down. It had ceased shipbuilding operations in 1933 and much of its equipment, including its hammerhead crane, had been sold. Described as presenting 'a most unattractive and depressing spectacle' the slipways 'were covered with a shroud of scrub extending to wharves so dilapidated as to form a fitting frame to a dreary picture of industrial desolation.'[23]

With the outbreak of war, the Newcastle Chamber of Commerce began agitating for the shipyard to be reopened. The New South Wales government toyed with the idea of selling or leasing the operation, but with uncertainty about Commonwealth contracts there were no takers. A Newcastle Citizens Shipbuilding Committee was formed to press all levels of government for action, and the Premier of New South Wales promised to revive the island's facilities if Navy orders were forthcoming. It became a thorny political issue for the Commonwealth government, the war cabinet sourly admitting that the failure to place orders with the State Dockyard had seriously undermined the government's claim that Australia's shipbuilding facilities were fully committed to war work.[24]

HMAS *Ballarat* in 1941 with the fully enclosed bridge and searchlight platform behind it that was to become standard on most corvettes during the war. As yet no radars have been fitted. (RAN)

HMAS *Tamworth* (B250) is tied up alongside several of her sisters in reserve after the war. They display the final iteration of the *Bathurst* class with a shielded 4in gun forward, a fully enclosed bridge with 20mm Oerlikons in the bridge wings, and a 40mm Bofors abaft the mainmast, just visible in this image. Both the British Type 271 and the US SC-2 radar are fitted. (Allan Green collection, State Library of Victoria)

Driven by the public agitation, the government agreed to place an order with the Walsh Island shipyard, but only when the site was 'a going shipbuilding concern.'[25] Following an assessment of the facilities, the state government decided to move whatever plant and equipment could be salvaged to a new location at Carrington on the mainland. The Commonwealth agreed to loan the state government £257,000 to rebuild the shipyard at the new site.

Construction at Carrington began in February 1942 and an order for the first *Bathurst* class corvette was placed with the NSW government in June 1942. Much ingenuity was shown in establishing the yard. With no cranes remaining the vertical supports of the old engineering shop were set up on either side of the new slipway with crane girders placed where the roof trusses had been. These were spaced to suit the span of the travelling crane from the former ship construction shop, which was then used to move heavy elements into place.[26]

Despite these innovations progress was slow. The future HMAS *Strahan* was laid down in October 1942 but as late as December 1943 only a few tons of steel were in place. The slipway itself remained unfinished, Engineer Captain GID Hutcheson reported to the ACNB, and a lack of labour meant there was barely enough steel coming out of the shops to keep the construction workforce employed.[27] *Strahan* was not completed until March 1944 and she was the only one of the *Bathurst* class to be built there.

With the Walsh Island order the last of the shipyards to build the *Bathurst* class corvettes was brought into the programme. Demand for the little ships continued to grow. The Royal Navy wanted additional vessels and the Commonwealth agreed to build a further ten along with four ships for the Indian Navy. Reassessing its needs, the RAN also increased its requirement to twenty-four vessels, raising the total orders to forty-eight. This was stretching Australia's shipbuilding capacity to its limits, and a request from the government of India for an additional six ships was rejected. 'By the fullest use of Australian shipbuilding resources and good organisation,' the War Cabinet was advised, it might just be possible to complete the forty-eight vessels before the end of 1941.[28] This proved overly optimistic; only fifteen vessels had been completed by that time.

As production ramped up, however, the number of ships coming off the slips increased. By the beginning of 1942 a corvette was being launched somewhere in Australia roughly every two weeks, a remarkable achieve-

Table: Average Building Times for Individual Shipyards

Builder	No of units built	Ave time LD>L (months)	Ave time LD>C (months)
Cockatoo Island	10	7.5	11
Mort's Dock	14	5.5	10
Williamstown	6	9	16
Walkers Ltd	8	6.8	13
Poole & Steele	6	9.5	14.6
BHP Whyalla	4	9.7	18.5
Evans Deakin	11	6.8	14.3
NSW State Dockyard (Walsh Island)	1	10	18
Total	60	8.1	14.4

LD laying down
L launch
C completion

Starboard-side depth charge projector aboard HMAS *Castlemaine*. A mine can be seen in the background (Author).

ment given the run-down state of the Australian shipbuilding industry at the start of the war. Ultimately sixty ships were completed, making them the biggest class of warships ever built in Australia. The largest number, fourteen, were built by Morts Dock, with Evans Deakin accounting for eleven and Cockatoo Island contributing ten. The accompanying table shows the production figures for the eight shipyards involved in the programme and the average construction times for each yard.

The greatest bottleneck, apart from labour shortages, was the difficulty in obtaining specialised equipment such as guns, navigation and gunnery control instruments, plus the all-important Asdic apparatus. At the beginning of the programme these items all had to be imported from Britain. As the war progressed, however, more and more items were manufactured locally. In 1942 the Garden Island Naval Depot in Sydney Harbour had begun constructing the single 4in Mk XVII gun used aboard most of the *Bathurst* class corvettes. By August 1943 the Maribyrnong Ordnance Depot in Victoria was manufacturing 40mm Bofors anti-aircraft guns at the rate of twenty-five a month. Asdic sets too were now being produced in Australia by Airzone Ltd of Camperdown in Sydney.

The first twenty-four corvettes cost on average £187,500. The final thirty-six were more expensive, with an average price of £250,000. The increase in cost was due to the larger engines and boilers as well as improved

The bridge of HMAS *Castlemaine*. Beautifully restored, HMAS *Castlemaine* is preserved as a museum ship at Williamstown in Victoria close to the shipyard where she was built (Author).

A view looking forward along the port side of HMAS *Castlemaine* showing the steps leading up to the bridge. The galley chimney can be seen next to the funnel. A 20mm Oerlikon can just be made out in the bridge wing while atop the bridge can be seen the lantern cover for the antenna of the British Type 271 radar (Author).

anti-aircraft defences. Later vessels were fitted from the outset with a single 40mm Bofors and two 20mm Oerlikon anti-aircraft guns whereas the first group originally carried only a single 20mm Oerlikon.

Winding Down Construction

As early as April 1941 the Admiralty was urging Australia to shift its shipbuilding efforts toward a larger, faster and more capable escort. Intended for operations in Australia's coastal waters, the *Bathurst* class were quite small. Their limited anti-aircraft armament was a particular concern. The Royal Navy's operations in Norway and off Dunkirk had highlighted the serious risk aircraft posed to even small vessels, and the tightness of the *Bathurst* design meant there was little scope to increase its anti-aircraft firepower. The Admiralty suggested Australia build the new 'River' class frigate, a 1,370 ton design with a top speed of 20 knots armed with two 4in guns and up to ten 20mm cannon.

The ACNB understood the need for larger and better-armed escorts. In the middle of 1941, however, many Australian shipyards were just getting into their stride building the *Bathurst* class. The 'River' class would require shipyards that, in some cases, had only just emerged from grass paddocks to adjust to a new design that was both bigger and more complex. 'After overcoming considerable initial difficulties in connection with the construction of the current AMS [*Bathurst* class] vessels in Australia on the present comparatively large scale, production is now proceeding steadily,' the War Cabinet was told. 'To switch over to a new design at this stage may be expected to cause a slowing down of the rate of deliveries until production of the new type could be got well under way.'[29] In July Cabinet approved construction of a further twelve *Bathurst*-class ships as a priority while placing orders for an initial batch of six larger frigates when slipways at suitable yards became available.

Events confirmed the wisdom of the ACNB's decision. By the end of the year Australia was at war with Japan and soon under direct attack. On 20 January 1942 the *Bathurst*-class corvette HMAS *Deloraine* sank the Japanese submarine *I.124* off Darwin.[30] By the middle of 1942 Japanese submarines were operating off Australia's east coast and *Bathurst*-class ships finally found themselves escorting convoys between Australia's major ports.

Convoy escort and minesweeping roles apart, the *Bathurst* class were versatile ships used for many other purposes. They were regularly used to support army and special forces units in operations among the islands to Australia's north; a corvette could carry 100 troops for four days and up to 300 troops in an emergency. They also undertook surveying and charting missions around the northern coast of New Guinea and among the islands of the southwest Pacific preparing the way for amphibious landings. These often involved operating close to Japanese positions, taking shelter from enemy air attacks under steep foreshores or hiding among mangroves, the ship festooned with local vegetation for camouflage.[31] HMAS *Benalla* was part of the survey group that prepared the way for the US landing at Leyte Gulf in the Philippines in November 1944.

Britain allowed Australia to retain the twenty ships it had ordered but many served as part of the Royal Navy. *Gawler*, *Ipswich*, *Lismore* and *Maryborough* formed the 21st Minesweeping Flotilla in the Mediterranean; *Cairns*, *Wollongong*, *Cessnock* and *Geraldton* all participated in the invasion of Sicily in 1942. With the twelve additional vessels ordered in 1941, construction of the *Bathurst* class continued in Australia into 1944. The last vessel, HMAS *Parkes*, was launched by Evans Deakin in Brisbane on 16 March 1943 and completed in May the following year. Only one ship was sunk by direct enemy action: HMAS *Armidale* was lost to Japanese aircraft in December 1942 while attempting to insert Dutch and Australian forces into Timor. Two others were lost to collisions during the war and HMAS *Warrnambool* was sunk while clearing mines in 1947.

Construction of the *Bathurst* class had led to a resurgence of shipbuilding in Australia. Many of the participating shipyards continued to operate after the war, but only one now remains. The Commonwealth continued to support its dockyards at Cockatoo Island and Williamstown until the 1980s. Cockatoo Island was closed in 1991. Williamstown was sold to Tenix Defence in 1987 and earned praise for its building on time and within budget ten *Anzac*-class frigates based on the German Meko 200 design during the 1990s. Tenix won the contract to complete two large amphibious ships of the *Canberra* class based on a design by Spanish shipbuilder Navantia in 2007. Their hulls were built at Ferrol in Spain with Tenix fabricating the superstructure and completing the fitting out. Tenix was taken over by BAE Systems in 2008, who continue to operate the Williamstown facilities.

Poole & Steele stopped building ships in 1945 and Mort's Dock closed in 1958. The State Dockyard in Newcastle closed in 1987. After building seven *Attack*-class patrol boats for the RAN in the 1960s and eight *Balikpapan*-class heavy landing craft in the early 1970s, Walkers ceased operations in 1974. Evans Deakin, which also built *Attack*-class patrol boats, continued until 1976. BHP continued to build ships, mostly for its own use, at Whyalla until 1978.

Endnotes:

[1] Admiralty Memorandum on Naval Policy, 11 June 1923, National Archives of Australia (NAA), A981 IMP 111.
[2] 'Japanese Submarine and Trade Operations', 1930, NAA MP1185/8 1846/4/363.
[3] Dwight R Messimer, *Find and Destroy: Antisubmarine Warfare in World War I*, USNIP (Annapolis MD, 2001), 22.
[4] David Stevens, *A Critical Vulnerability: The impact of the submarine threat on Australia's maritime defence 1915–1954*, Sea Power Centre (Canberra, 2005), 62.
[5] Ibid, 84–85.
[6] Australian Commonwealth Naval Board (ACNB) to Admiralty, 15 July 1938, NAA MP1049/5 2026/2/191.
[7] Admiralty to Navy Office, Melbourne 31 December 1938, NAA MP1049/5 2026/2/191.
[8] Minute by Rear Admiral PE McNeil, 14 March 1939, NAA MP1049/5 2026/2/191.
[9] Stevens, *op cit*, 112.
[10] Marginal note by Admiral Sir Ragnar Colvin on ACNB minute 'General Purpose Local Defence Vessel', 3 August 1938, NAA MP1049/5 2026/2/191.
[11] McNeil minute, 11 January 1939, NAA MP1049/5 2026/2/191.
[12] Colvin minute, 15 February 1939, NAA MP1049/5 2026/2/191.
[13] McNeil minute, 17 February 1939, NAA MP1049/5 2026/2/191.
[14] Ibid.
[15] Frank Walker, *Corvettes: Little Ships for Big Men*, Kingfisher Press (Budgewoi, NSW, 1995), 25.
[16] NAA MP151/1 603/271/4.
[17] John Sullivan, 'Williamstown Naval Dockyard – Part 1', *Naval Historical Review*, September 1986, https://www.navyhistory.org.au/williamstown-naval-dockyard-part-1/3/.
[18] 'Production in Australia of Minesweepers for the Admiralty', 10 June 1940, NAA A5954 514/1.
[19] DP Mellor, *Australia in the War of 1939–1945: The Role of Science and Industry*, Australian War Memorial (Canberra, 1958), 455.
[20] JA Concannon (ed), *Shipbuilding at Walkers Limited Maryborough, Qld, Australia, 1877–1974*, Maryborough District Family History Society (Maryborough, 2009), 46.
[21] John Terraine, *Business in Great Waters*, Wordsworth Military Library (London, 1989), 699.
[22] Admiralty signal, 29 March 1940, NAA A5954 514/1.
[23] Government of NSW, *The State Dockyard Newcastle, NSW: Its Wartime Establishment and Production, January 1942 – December 1945* (Sydney, 1946), 14.
[24] War Cabinet minute, 29 August 1940, NAA A5954 514/1.
[25] Ibid.
[26] Government of NSW, *The State Dockyard Newcastle, NSW*, 18.
[27] Report by Engineer Captain Hutcheson to ACNB, 11 December 1943, NAA MP138/1 603/201/1128.
[28] War Cabinet minute 17 August 1940, NAA A5954 514/1.
[29] War Cabinet minute, 3 June 1941, NAA A5954 514/1.
[30] Two other *Bathurst*-class corvettes, *Lithgow* and *Katoomba*, and the US destroyer *Edsall* were also involved in the action but it seems probable that *I.124* had been sunk by the time they arrived on the scene.
[31] Lieutenant-Commander GCI, RAN, 'Charting New Guinea', *HMAS Mk III* (Canberra, 1944), 27.

C 65 *ACONIT*: FRANCE'S PROTOTYPE OCEAN ESCORT

Inspired by the US Navy's 'ocean escorts' of the *Bronstein* and *Garcia* classes, the French *Aconit* was to have been the first of a series of five antisubmarine corvettes, but ended up being the only unit of her class. **John Jordan** looks at the origins of the type and focuses on some of the key systems incorporated in the design.

During the early 1950s, with the support of US Mutual Defense Assistance Program (MDAP) funding, the *Marine Nationale* had built a series of *escorteurs rapides* of the E 50 and E 52 types, derived from the US *Dealey* (DE 1006) class and designed primarily for convoy escort in the Atlantic. These were due to decommission from about 1970, and discussions on their replacements began in 1962. The final ship of the T 47/53 series of 'fleet escorts' (*escorteurs d'escadre*), *La Galissonnière* (T 56 type), had recently been completed to trial a new generation of antisubmarine sensors and weaponry, and these would form the basis of the armament of the new ships.

The project went through many iterations, with various options being considered, including drone antisubmarine helicopters and gas-turbine or diesel propulsion machinery. In the end the French opted for a ship with similar capabilities to the new US 'ocean escorts' of the *Bronstein* (DE 1037) and *Garcia* (DE 1040) classes, which had single-shaft turbine machinery capable of delivering 27 knots – adequate to counter a 20-knot nuclear submarine – and were fitted with the large low-frequency SQS-26 bow sonar and armed with a combina-

Aconit moored inboard of the *escorteur rapide Le Normand* (E 50 type) on South Railway Jetty, Portsmouth, on 19 February 1977. The disparity in size of the two ships is readily apparent. Note the non-standard FNFL flag at the jack staff. One of nine 'Flower'-class corvettes lent by the Royal Navy, the previous *Aconit* was one of the most famous ships of the Free French Navy. During the Second World War she escorted 116 convoys, spending 728 days at sea. She was awarded the Croix de la Libération and the Croix de Guerre 1939–1945, and was cited by the British Admiralty. (All photographs by John Jordan)

Profile and Plan of *Aconit* as completed

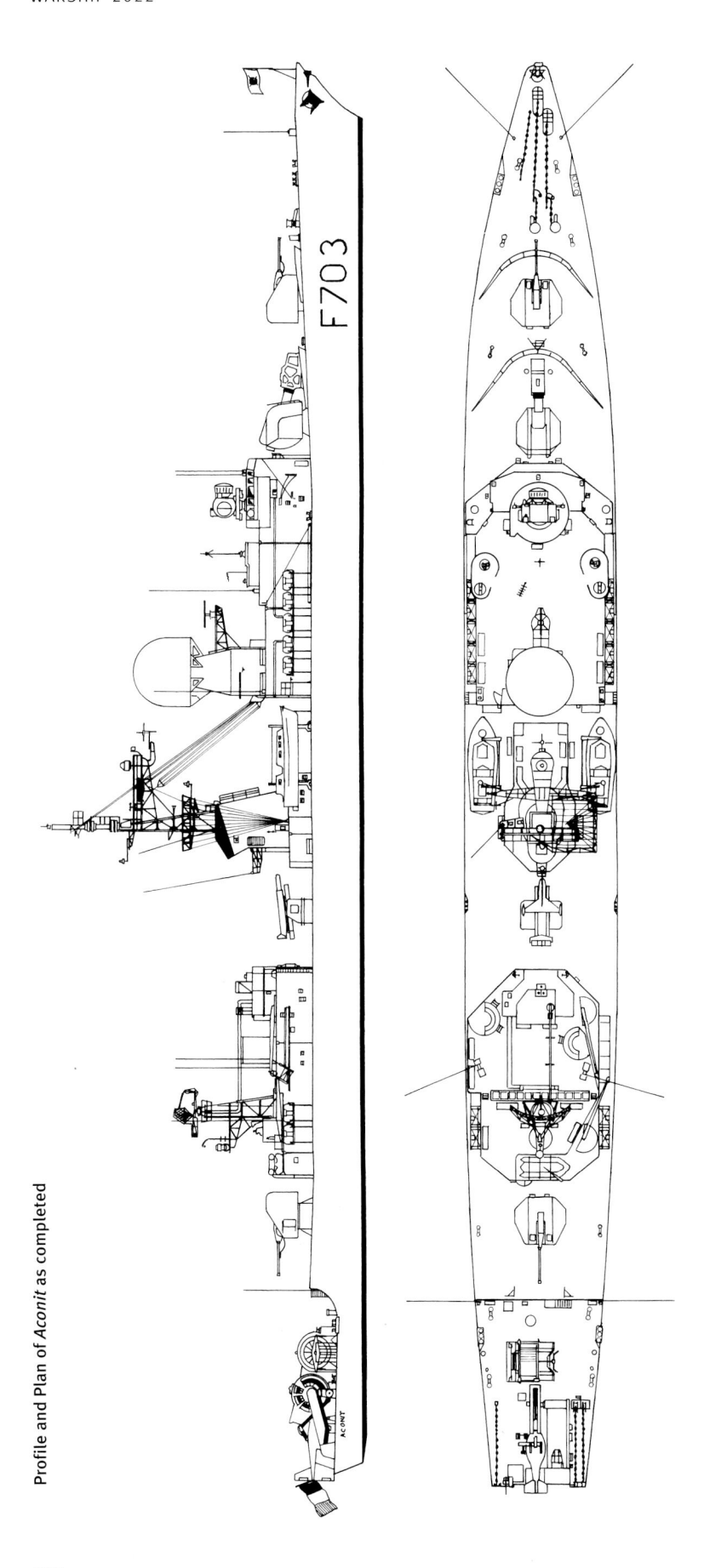

Characteristics (as completed; dimensions from plans)

Displacement	3,510t trials; 3,840 full load
Dimensions	118m pp, 127m oa x 13.4m x 5.4m max
Machinery	Single-shaft steam turbines: two boilers, 45km/cm², 450°C; Rateau geared turbines; 28,650shp = 27 knots
Oil fuel	620 tonnes
Endurance	4,878nm at 18 knots
Armament	2 x 100mm Mle 68 DP
	1 x Malafon A/S missile system (13 missiles)
	1 x 305mm A/S mortar (57 rounds)
	2 x catapults for eight L5 A/S torpedoes
Electronics	SENIT 3 data system
	DRBV 13 air/surface surveillance & target designation radar
	DRBV 22 air surveillance radar
	DRBN 32 navigation radar
	DRBC 32B fire control radar
	DUBV 23 bow sonar
	DUBV 43 variable depth sonar
	ARBR 15 ESM
Complement	22 officers, 230 men

Note: All the drawings of *Aconit* have been adapted from the official DCAN plans held by the Archives de l'Armement, Châtellerault.

© John Jordan 2008

tion of ASROC, DASH and acoustic homing torpedoes. Anti-air capability was limited to dual-purpose guns and 'passive' electronic warfare (EW) defences.

A programme for five 'corvettes'[1] was agreed in December 1964, and staff requirements for a ship to be designated C 65 were sent to the Head of the DCAN on 10 January 1965. The first unit was included in the 1965 Estimates and was subsequently named *Aconit*. On 20 January 1966 construction was assigned to Lorient Naval Dockyard and the machinery was ordered from ECAN Indret.

Work on the plating and frames began on 15 November 1966, and *Aconit* was laid down in the ship hall at Lanester on 22 March 1968. The launch on 7 March 1970 was a low-key affair due to the recent loss of the submarine *Eurydice*. *Aconit* began her sea trials on 3 February 1971, some fifteen months later than anticipated, and was commissioned on 1 September. Trials were prolonged due to technical problems with the pressure-fired boilers and with the SENIT 3 action information system.

Hull & Propulsion Machinery

The shell was of 60 SS steel, with 50 SS steel employed for the remaining structures and AG 4 for the internal bulkheads. There was a pronounced clipper bow with negative sheer which served to take the stem anchor clear of the large sonar dome, and a low quarterdeck from which the large 'fish' for the variable depth sonar was deployed. Below the main deck there were fourteen watertight transverse bulkheads that divided the ship into fifteen compartments. The working and accommodation spaces on the main deck were linked by a central passageway running from fore to aft. Two electro-hydraulic stabilisers powered by 140kW motors ensured a steady platform for the ship's weapons, and there was a single balanced rudder with a surface area of 9.57m² which was specially designed for going astern.

As in the US Navy's ocean escorts of the contemporary *Garcia* class, there was a single-shaft propulsion system, with steam for the turbines supplied by two small-tube boilers with superheating and automated pressure firing rated at 45kg/cm² with a maximum steam temperature of 450°C. The boilers were located side by side at the after end of a single boiler room and enclosed in air-tight boxes. The ACB Rateau machinery, comprising one high-pressure (HP) and one low-pressure (LP) impulse turbines, drove the single shaft with its 3.9m-diameter, four-bladed fixed-pitch propeller via double reduction gearing. Designed horsepower was 28,650shp for 27 knots. There was an auxiliary boiler (10kg/cm²) for ship's services, and a 450V/60Hz electrical circuit was supplied by two groups of turbo-alternators each rated at 1,000kW (boiler room) and two groups of diesel alternators each rated at 480kW (auxiliary machinery room abaft the engine room), for a total of 2,960kW.

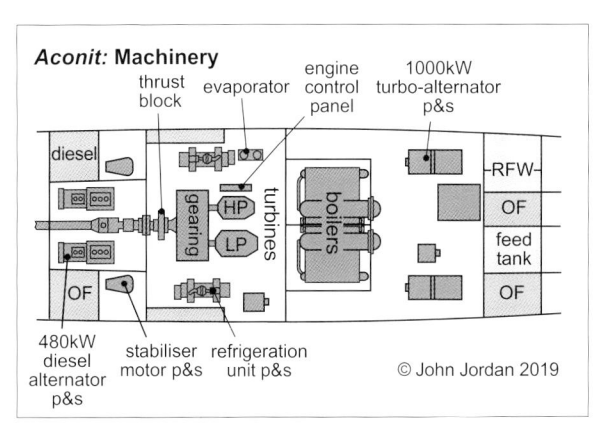

Armament

Aconit was equipped with two of the new 100/55 Mle 1968 dual-purpose guns (one forward, one aft), which were credited with a theoretical rate of fire of 60rpm. The gun could fire semi-armour piercing (OPF), high explosive (OEA) or OXL practice rounds, and had a maximum theoretical range of 17,000m against surface targets and 6,000m against aircraft. A ready-use carousel held 10 fixed rounds, and the hoist had a capacity of 20 rounds per minute. Elevation was from -5 to +80 degrees.

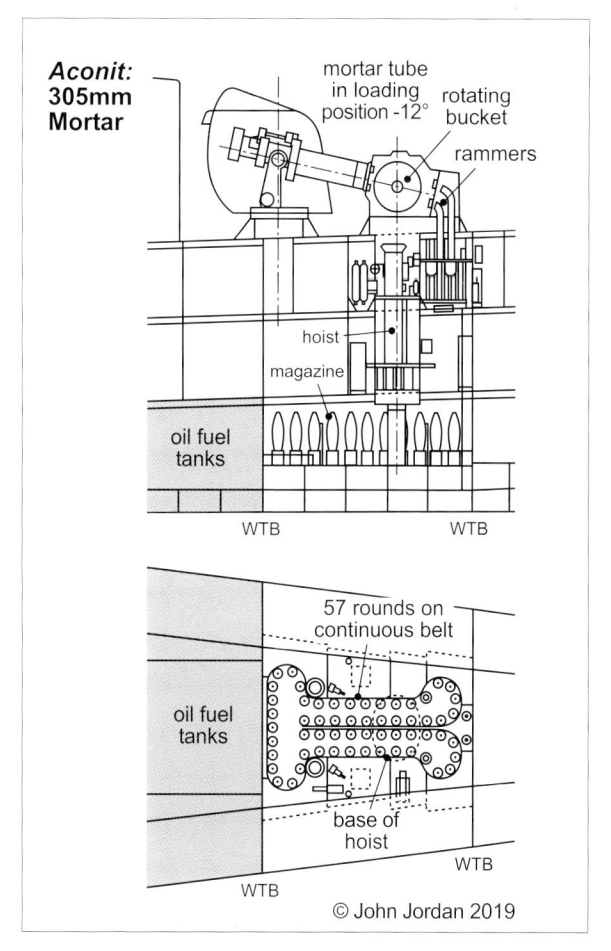

The principal weapons mounted in *Aconit* were for her primary antisubmarine role, and had been trialled in *La Galissonnière*. The 305mm mortar forward of the bridge had been a feature of the last three *escorteurs rapides* (E 52B type) and the nine *avisos-escorteurs* of the *Commandant Rivière* class. Designed in-house, the mortar was manufactured at Ruelle but had an ammunition supply system by CAF Loire. There were four tubes in a single cradle, which were reloaded with the tubes depressed to 12 degrees from a glacis forward of the mounting. The vertical hoist from the magazine, which was located in the hold, could lift four rounds simultaneously, the rounds being rammed into the muzzles of the tubes via a tilting bucket at the top of the hoist. The projectiles weighed 228kg and had a time delay, the fuze being set within the launch tube; 57 were stowed on a continuous belt in *Aconit* (see drawing). The mounting had remote power control for training and elevation, with manual back-up. It required only four personnel: the mounting commander (right), a firing operative (left), plus two men in the magazine. A four-round salvo could be fired every 25 seconds out to a range of 2,700m; minimum range was 400m.

The Malafon antisubmarine missile had been tested on the missile trials ship *Ile d'Oléron* and on *La Galissonnière*; development was completed 1966. Like the Australian Ikara, Malafon was a cruise missile which carried a short-range homing torpedo; twin detachable booster rockets were slung beneath the airframe. Built by the aircraft manufacturer Latécoère, the Mk I Mle 13 had an overall length of 6.15m and the missile body a diameter of 0.56m; it weighed 1,500kg and flew at a maximum speed of 830km/h; range was between 3,000m and 12,000m. Malafon was remotely controlled (for direction only) from the operations centre, using shipborne sensors; in *Aconit* it was tracked by the DRBV 13 radar. The boosters were ditched 4 seconds after launch; the missile then cruised at an altitude of 200m, releasing its torpedo 800m from the target (see graphic). The torpedo descended on a parachute, then followed a helical search pattern, using its acoustic homing head to locate the submarine. The L4 torpedo had a length of 3.13m and a diameter of 533mm, and weighed 540kg including a 150kg warhead. It had a maximum range of 5,000m at 30 knots and was effective down to a depth of 300m.

A missile could be fired from the launch ramp every 90 seconds. Thirteen missiles were embarked in *Aconit*, and

Aconit: ASW Capabilities

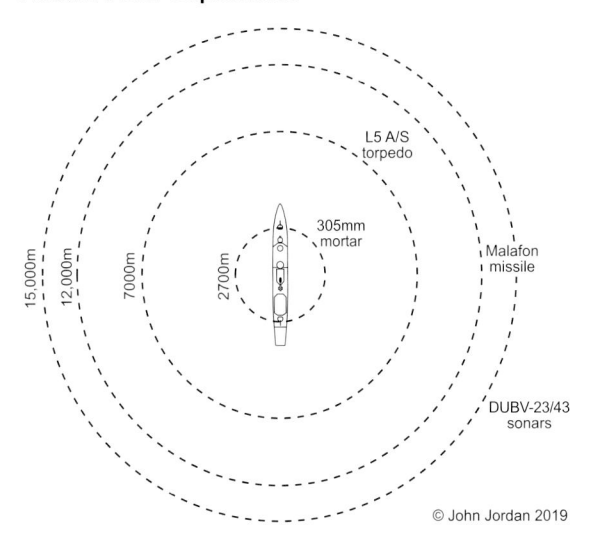

Aconit departs Portsmouth on 22 February 1977 after her weekend visit.

Deployment of Malafon A/S Missile

Aconit: Malafon Hangar & Torpedo Room

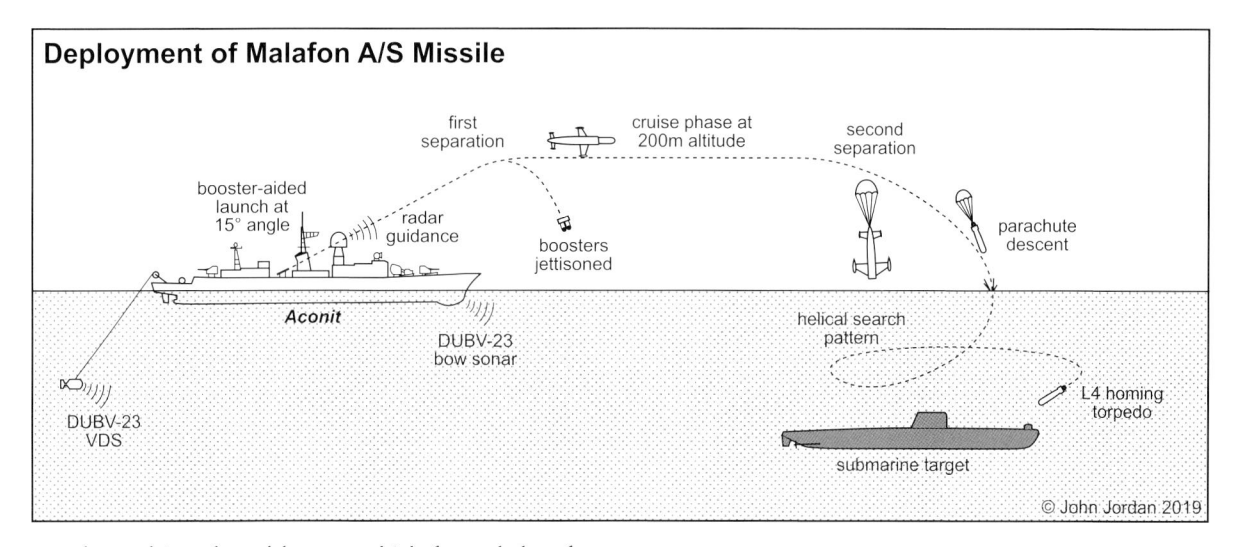

were located in a broad hangar which formed the after superstructure (see drawing). The missile was stowed without its wings and tail fins, the booster fins being folded. The wings and fins were attached on an angled ramp at the forward end of the hangar before the missile was run out onto the launcher, which was powered by hydraulic motors and operated by the launcher commander (*chef de rampe*). Malafon was removed from *Aconit* in 1997; it was to have been replaced by the Franco-Italian Milas with the MU-90 torpedo, but France pulled out of the programme in April 1998 for budgetary reasons.

The other major antisubmarine weapon system was the ship-launched L5 homing torpedo, eight of which were embarked on cradles in the after part of the Malafon hangar. The cradles were angled at 30 degrees forward of

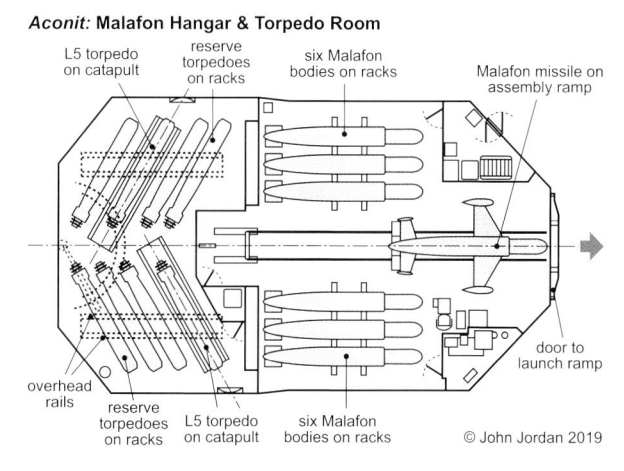

The bow section of *Aconit*. From left to right: the forward 100mm Mle 1968 dual-purpose gun, the 305mm A/S mortar, and the DRBC 32C fire control director.

Aconit: Operations Room

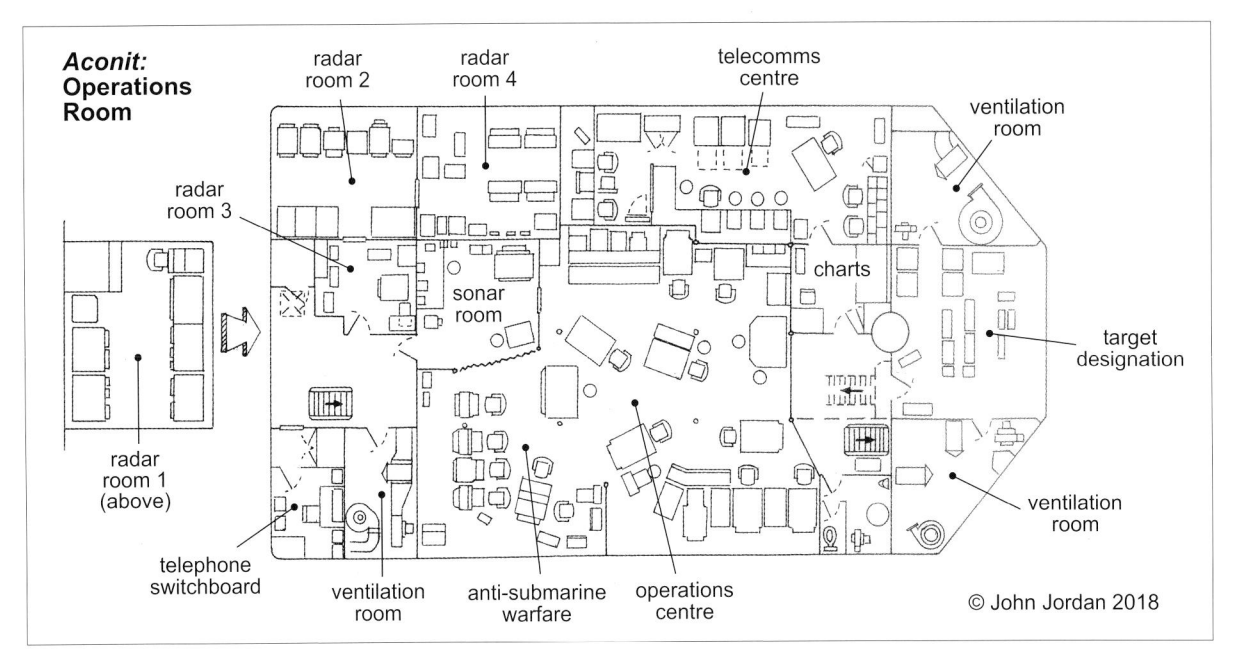

[Diagram labels: radar room 2, radar room 4, telecomms centre, ventilation room, radar room 3, sonar room, charts, target designation, radar room 1 (above), ventilation room, telephone switchboard, ventilation room, anti-submarine warfare, operations centre, © John Jordan 2018]

the beam, and the torpedoes were handled by a system of overhead rails (see drawing). There were two Type KD 59C torpedo catapults, one to port and the other to starboard, which launched the torpedoes over the side of the ship using compressed air. The L5 Mle 4 torpedo had a length of 4.4m and a diameter of 533mm, and weighed 935kg (warhead as in the L4); it was therefore a comparatively 'heavyweight' model compared with the short-range 12.75in (324mm) A/S torpedoes fitted in contemporary RN and US Navy escorts. It had a maximum range of 7,000m at 35 knots, had an acoustic homing head and could descend to a depth of 500m. Initial guidance was provided by a Type D2A torpedo fire control system (DLT).

The Malafon launcher amidships; the hinged door for the missile hangar is in the lowered position and the nose of a missile can be seen projecting above it. Atop the hangar are the two Syllex chaff dispensers (with a Mk I seagull between them!). The two blackened 'stove-pipes' above the port-side Syllex launcher are the exhausts for the diesel alternators.

Electronics

The primary sensors fitted in *Aconit* were the paired DUBV 23 and DUBV 43 sonars; these could be used independently or in combination. The DUBV 23 bow sonar operated between 4.9kHz and 5.4kHz, which was at the high end of the low-frequency spectrum. The French wanted to achieve reliable direct-path detection at ranges of 10–15km with a view to providing target data for the Malafon missile; long-range detection using the bottom bounce or convergence modes, which were key features of the US SQS-26, was of less interest. Accuracy was claimed to be 150m (±1%) in range and one degree in bearing. The DUBV 43 variable depth sonar, which featured a large 'fish' housing a similar array to the DUBV 23, was towed at 4–24 knots at a depth of 300m, and was useful for looking beneath the thermal layers that were particularly prevalent in the Mediterranean. The requirement for the sonars was drawn up in 1963–64 and the prototypes trialled on the Type 47 EE *D'Estrées* following her A/S conversion.

The DRBV 13 target designation radar was specially developed for *Aconit*. It was housed within a large, distinctive radome atop the bridge. Performance proved to be disappointing, and it was replaced in 1985 by the DRBV 15, a frequency-agile S-Band radar with pulse compression. The DBRV 13 was complemented by the standard DRBV 22A air/surface surveillance radar, which had a maximum 280km range; it was mounted on the short lattice mainmast.

There was an ARBR 15 ESM outfit, complemented shortly after *Aconit*'s completion by two Syllex chaff launchers – a French version of the British Knebworth Corvus.

The sensor and weapon outfits were coordinated by a SENIT 3 tactical data system,[2] the consoles for which were grouped together in the Operations Centre (*Central Opérations* – see drawing). SENIT 3 permitted the presentation of raw or synthetic data, while Link 11 allowed VHF or UHF exchanges of data with other units. The SENIT 3 installation in *Aconit* comprised a central module plus modules dedicated to Antisubmarine Warfare (ASM), Air and Surface Warfare, Electronic Warfare (GE) and Communications.

Service Career

On completion of her trials, *Aconit* was initially based at Toulon, serving there until 1975. She spent the remainder of her career at Brest, and was stricken on 28 February 1997. During her time in service she underwent the following modifications:

Jan 1974	Pennant number F 703 > D 609.
Jul–Oct 1983	305mm mortar disembarked; replaced by two Exocet ramps 1984–85, each for four missiles. ARBR 16 ESM fitted.
1985	DRBV 13 replaced by DRBV 15 inside same radome. DRBN 32 navigation radar replaced by Decca 1126.
1991/92	100mm Mle 68 replaced by CADAM; DRBC 32B FC radar replaced by DRBC 32D. Syracuse and Inmarsat SATCOM embarked. ETBF DSBV 62C towed array fitted. 2 x 20mm, 2 x 12.7mm.

Endnotes:

[1] This was a new designation for the French postwar navy. The destroyers and escorts built in the 1950s had been designated 'escorts' (*escorteurs*), the contemporary guided-missile destroyers *Suffren* and *Duquesne* 'frigates' (*frégates*), like their US counterparts of the *Leahy* and *Belknap* classes. The three larger anti-submarine ships (F 67 type) which followed *Aconit* were likewise classified as *frégates*, but the term *corvette* was again employed initially for the smaller A/S ships of the *Georges Leygues* class (C 70 type).

[2] SENIT stands for *Système d'Exploitation Navale des Informations Tactiques.*

The stern, showing the heavy winches used to deploy the massive body of the DUBV 43 low-frequency variable depth sonar – partially visible above the stern.

WARSHIP NOTES

This section comprises a number of short articles and notes, generally highlighting little known aspects of warship history.

SHIP MEDALLIONS OF THE *REGIA MARINA*
Enrico Cernuschi recounts a little-known story of tradition and female fashion.

The Italian Navy introduced ships' badges incorporating heraldic crests in 1951, when the first former US Navy destroyers and DEs were tranferred under the Mutual Defense Assistance Program (MDAP) and the Italian sailors discovered what was, for them, a novelty. Prior to that time, during the tenure of the *Regia Marina* (1860–1946), the unofficial rule was that the choice of a coat of arms rested with the current commander of the ship. The badge could thus change over the years until the vessel was paid off; even so, custom and practice generally dictated that the first (or the most striking) badge would become associated with that ship. Given that the selection of an emblem was dictated by the dominant style of the time, it was nevertheless possible for two different warships with the same name commissioned, say, in the 19th and the 20th century respectively, to have different badges.

Heraldry was not decisive, even if the name of the ship carried a certain weight among possible options. The motto of the ship may have been a privilege of the first commander, but once approved by the Navy Minister (who had the last word) it became permanent for that particular vessel. Given this relative freedom, which allowed the Navy to proceed in harmony with the mood of the times, ships' badges in the Italian (and French) navies materialised as medallions for more than 80 years. Old habits died hard, and although crests were adopted by the *Marina Militare* during the 1950s, medallions survived until the late 1960s.

The medallions had a diameter of about 2.5 centimetres (one inch) and were made of silver by mints or by specialised medal manufactures such as Picchiani e Barlacchi (Florence), Castelli–Pialorsi (Brescia), the Lorioli Brothers and their traditional competitor Stefano Johnson (Milan), Alberti (Bergamo) or Corradini (Bologna). Among the most illustrious engravers were Romano Romanelli, a Florentine sculptor and former *Regia Marina* officer during the Great War, the famous jewellers of the Janesich family from Trieste, masters of Art Deco with offices in Bond Street, Paris and Monte Carlo, Sergio Vatteroni and Gaetano Orsolini. The very few solid silver versions of these medallions (paid for by the Ministry, with a contribution from the CO of the ship if he was changing the original badge) were made by the best jewellers and were presented as gifts, during official visits, to the '*vere signore*' (the most honourable ladies), together with a bouquet of flowers tied by a blue ribbon with the name of the ship and a small commemorative certificate of parchment. The lady in question might be a baroness, the female member of a very wealthy or illustrious family, the wife of the commanding officer or, if he was unmarried, his mother or sister. Medallions of sterling silver were used for female visitors during less prestigious events; for non-commissioned officers and sailors' cooperatives the medallions were of light alloy, and these are by far the most commonly available for collectors today. In the late 19th century gold medallions were fashionable among commanding officers' wives to mark their elevated status, but they were abandoned after the wife of the new Italian naval attaché in Washington DC was arrested by American customs officials and charged with smuggling gold coins into the USA.

Although the Italian Navy has a rich collection of these medallions, their recording and classification has proved difficult. Fortunately one of the best sets of medallions was donated in 1970 to the Italian Navy's Museo Tecnico Navale of Spezia by Luigia Frattoni, the widow of Signor Giuseppe Guidin. Kept by the Director of the Museum in the personal safe behind his desk, this collection was shown to the author in 2017 – the first time since the original inventory had been drawn up – by the former Director Captain Giosuè Allegrini. Signor Guidin was an expert on the medallions, and his research revealed previously-unknown details about the history of the Italian Navy.

Medallions were commissioned not only to celebrate individual warships, but also actions and significant events. The Italian occupation of Saseno Island, which controlled the Bay of Valona, took place in 1914, and received little publicity at the time as Rome was still technically neutral. The first significant break with Austra-Hungary had occurred in late July 1914, when King Vittorio Emanuele III refused to join Vienna's aggression against Serbia, citing the essentially defensive nature of the alliance with Germany and Austria-Hungary signed by the House of Savoy in 1882. On 31 July 1914 the last Austro-Hungarian warship present at Durazzo, the cruiser *Szigetvár*, faced with an Italian squadron despatched from Taranto, left Albanian waters, which had been internationally patrolled since 1913. The last link with Vienna having been thus severed, the newly-appointed monarch of Albania, Prince Wilhelm of Wied, resisted Austrian and German requests for a declaration of war against Serbia – he was promised Kosovo as a reward – and was forced into exile on the Italian navy yacht *Misurata*, returning to his home in Germany via Venice and Switzerland.

WARSHIP NOTES

The medallion commemorating the landing on Saseno Island. (All images are courtesy of the Museo Tecnico Navale di La Spezia)

The Albanian refusal to join the attack on Serbia led, in turn, to Vienna's refusal on 4 August 1914 to despatch the Austrian battle fleet (or at the very least a pre-dreadnought squadron) to operate off the coasts of Albania, Greece – where the pro-French Prime Minister Venizelos was abroad on vacation, leaving power in the hands of his pro-German rival, foreign minister Georgios Streit supported by the Hellenic King and the Army – and Turkey, thereby completing a coastline route from the Black Sea to Trieste and consolidating the diplomatic success achieved by the small German Mediterranean squadron with the Sultan. When confronted with these ambitious plans, the Austro-Hungarian Navy's response, which was supported by the old Emperor Franz Joseph, was always the same: the first task of the KuK *Kriegsmarine* was to defend the Empire's shores from the threat of an Italian landing in Dalmatia or in Istria. When on 13 August Paris and London reluctantly declared war on Austria-Hungary, any further idea of a break-out beyond the Adriatic Sea was abandoned.

After this silent strategic success, the Italian Navy needed a second pillar to seal the gates to the Adriatic, and this became particularly urgent due to the approaching autumn gales. It was therefore decided to land a *Regia Marina* force on Saseno and to set up a number of 120mm (4.7in) gun emplacements. This was a covert operation, and even after the conclusion of the Great War the date was difficult to determine. The medallion found at La Spezia (Fig 1) finally establishes that date as 29 October 1914.

Figure 2, which shows a large bronze disk rather than a medallion, was made by the Genoese jeweller Chiappe to celebrate the launch of the battleship *Impero* on 15 November 1939. The reverse of the disk records the month and the year but not the day, as Chiappe was well aware of the potential for delay in the launch of a hull of more than 10,000 tonnes due to the weather at that time of the year, and was concerned that he might have to melt down the original batch.

The history of *Impero* is a curious one. According to *Janes's* 1940, the battleship suffered serious damage during her launch. However, no mention of this is made in the Italian Navy's archive file for *Impero*. On 24 May 1940 Mussolini informed the Navy Minister and CoS Admiral Domenico Cavagnari – three days before the other chiefs of staff – about the impending Italian declaration of war against France and Britain. As *Impero* was currently an almost empty, large and vulnerable hull with no watertight compartments or armour, which could be sunk by a single bomb in the deep water of Genoa by a raid coming from nearby French airfields, it was decided to transfer the battleship to Trieste in the Adriatic. An unfavourable weather forecast, together with the imminent declaration of war, caused this operation to be halted on 8 June 1940, when the hull had been towed to the large but ill-equipped harbour of Brindisi. It was here that *Impero* had her machinery partially completed by herculean efforts, and on 22 January 1942 she sailed first to Venice, then to Trieste, where she finally arrived on 15 November 1942. In the meantime, weapons, propellers, spares and other items destined for *Impero* were used for the rapid repair of her sisters *Littorio* and *Vittorio Veneto*, while replacements were ordered from the various factories involved. Priority following the Italian declaration of war was accorded to the construction of destroyers, escort vessels and submarines, and when the Armistice was announced on 8 September 1943 *Impero* was still a long way from completion, which was scheduled for the autumn of 1945. Had she been completed before the Armistice she would have been

Impero (= 'Empire'), the fourth (and uncompleted) battleship of the *Littorio* class. The 'XVIII EF' on the reverse denotes not the day of launch, but the 18th year of the Fascist Era (*Era Fascista*), which began on 28 October 1922 and ended in 1945.

The protected cruiser *Etna*, named after the volcanic mountain. The engraved image is of the *Trinacria* (incorporating the head of Medusa), the ancient symbol of Sicily.

Irrequieto ('Exuberant'), one of a class of six destroyers completed just before the Great War.

The torpedo boat *Partenope*, completed in 1938. Partenope was one of the Sirens of classical mythology and the symbol of modern-day Naples, where the ship was built.

hamstrung by the same fuel and personnel shortages that led to the decommissioning of the older battleships *Cesare*, *Doria* and *Duilio* in late 1942.

Other medallions deserve to be recorded both because of the quality of the artwork and as a testimony of changing times and trends. Designs illustrated here include the Art Nouveau style adopted for the medallion of the protected cruiser *Etna* (Fig 3), the classical designs for the Great War destroyer *Irrequieto* (Fig 4) and the torpedo boat *Partenope* (*Spica* class – Fig 5), or the one inspired by the poet Gabriele D'Annunzio for the converted seaplane carrier *Giuseppe Miraglia* (Fig 6). The medallion for the cruiser *Bande Nere* has, appropriately, an image in the Renaissance style of one of the *condottieri* (the captains of bands of mercenaries during the Middle Ages – Fig 7), while that of the heavy cruiser *Zara* (Fig 8) was entirely conventional, featuring one of the standard Republic of Venice themes. More modern was the Futurist medallion of the submarine *Ametista* (Fig 9), which was enamelled. The square form of the medallion for the battleship *Vittorio Veneto* (Fig 10) represented a break with tradition; on the reverse is the 'sailor's prayer' written by the poet Antonio Fogazzaro in 1901 and which, every day at sunset, is read on board the Italian warships at sea with the crew deployed in front of the ensign. The Republic of Venice is again the inspiration for the medallion of the destroyer *Sebenico*, seized from the

The seaplane carrier *Giuseppe Miraglia*, converted from a liner during the 1920s (see *Warship 2020*).

WARSHIP NOTES

The 'scout' *Giovanni dalle Bande Nere*. It depicts Ioannes Medices (subsequently nicknamed 'Giovanni of the Black Band'), one of the leaders of the mercenary bands prominent in Italy during the 14th and 15th centuries, the *condottieri*.

The unusual 'Futurist' design adopted for the medallion of the submarine *Ametista*.

The cruiser *Zara*, named after the town in Dalmatia following its cession to Italy after the First World War. The inscription on the reverse reads 'TENACEMENTE' ('Tenaciously').

The Art Deco design of the medallion of the battleship *Vittorio Veneto*; the reverse features the 'sailor's prayer'.

201

The medallion of the destroyer Sebenico, *the former Yugoslav* Beograd.

The Roman tag of the battle fleet based at Ravenna.

The original medallion of the French contre-torpilleur Panthère, *later the Italian* FR 22.

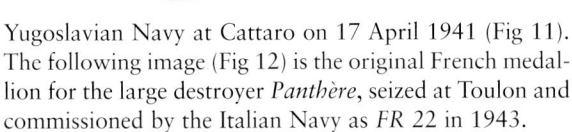

The commemorative medallion devised for the journal Rivista Marittima. *The coat of arms is that of the* Marina Militare, *not that of the* Regia Marina, *which was nevertheless similar and had in its centre the crest of the House of Savoy: a white cross on a red shield dating from the crusades.*

Yugoslavian Navy at Cattaro on 17 April 1941 (Fig 11). The following image (Fig 12) is the original French medallion for the large destroyer *Panthère*, seized at Toulon and commissioned by the Italian Navy as *FR 22* in 1943.

The first and the final items in the Guidin collection deserve a mention. The oldest medallion is an ancient Roman one of the Adriatic Sea; it bears the inscription *Classis Praetoria Ravennatis Pia Vindex* (Fig 13), one of the two battle fleets created by the Emperor Augustus around the year 27 AD. This was not intended for presentation to illustrious visitors, but simply as a military tag (*nihil sub sole novum*) to identify the sailors of the Eastern Mediterranean fleet from their colleagues of the *Classis Misenensis* based at Miseno in the Gulf of Naples, which was tasked with patrolling the Western Mediterranean.

The other, probably the last medallion to be minted, was created by Admiral Aldo Cocchia, editor of *Rivista Marittima*, the Italian Navy Staff monthly since 1868 (Fig 14), for the regular contributors to the magazine. Admiral Cocchia was awarded a gold medal (the highest Italian award for gallantry), and conceived the commemorative medallion in 1955 in conjunction with the then-Chief of Staff Admiral Corso Pecori Giraldi, a former classmate.

During the interwar period Italian naval wives had the habit of wearing a charm bracelet combining the (silver) medallions of their husbands' commands at formal events. Following a worldwide revolution in dress in the 1960s, this fashion was abandoned. It has now made a come-back, and stylish ladies from old Italian naval families have reinstated the bracelets that formerly belonged to their mothers, grandmothers and ancestors as an element of distinction on social occasions. It appears that everything is subject to change except salt water and human beings.

THE ROYAL NAVY AND NATIONAL NAMES

Kenneth Fraser continues his series on the naming of ships with a note on the use of 'national' names for RN warships.

In many navies, warships of the largest type have been named after the country itself (*eg Deutschland, Italia*), and nations organised on a federal basis have commonly used the names of the federal states for classes of ships (US and German battleships). The United Kingdom, a state composed of several nations endowed with alternative names in literature and ruling over a host of overseas territories during the imperial era, presents a more complex picture, and the adoption of these names at a particular moment in time has often been governed by political expediency. Given the Navy's long history, most of those featured here originated in the days of sail. Also, when choosing geographical names the Admiralty was generally of the view that the more important warships should take the names of the more important places, and *vice versa*: thus the First World War would see armoured cruisers named after counties, light cruisers after large towns and minesweepers after smaller ones, while London, as the imperial capital, merited a battleship.

Britannia was used for a continuous series of first-rates from the 17th to the 19th centuries. The first was launched in 1682, and it is probably not coincidental that this occurred in the reign of Charles II, for that monarch had in 1667 introduced the figure of Britannia onto the coinage, with a corresponding inscription. The model for the figure was one of his mistresses, Frances Stewart, Duchess of Richmond, although the design derived from a Roman original. The fourth *Britannia* ended her days as a training ship at Dartmouth, and her successor (formerly the 121-gun *Prince of Wales*) adopted that name, but lost it when replaced by the buildings of the Royal Naval College in 1905. This allowed a *King Edward VII*-class battleship to be so named, but following her loss in 1918 the name was out of use for a generation. The name of the Royal Yacht *Britannia* of 1953 may have derived from that of George V's famous sailing yacht (1893–1936).

Although the name *Albion* has for hundreds of years been in literary use only, we know from a classical source that it had been the earlier name of the island of Britain. It too was used in the 18th and 19th centuries for a series of ships of the line (though never first-rates), but was chosen for a battleship of 1898 and an aircraft carrier of 1947. The present landing ship *Albion* derives her name from the last of these, as her sister ship *Bulwark* is also named for a sister of the older *Albion*, and both of these had been converted to helicopter carriers.

But what about the individual nations of the UK? Significantly, there has never been an HMS *Anglia*, possibly because of the traditional habit of confusing England with Britain.[1] *Hibernia* was first proposed for a 74-gun laid down in 1761, but she was renamed *Prince of Wales* before her launch, in honour of the birth of the future George IV. We may take the later usage of that name as a tribute to the Royal Family rather than to Wales. However, the 110-gun *Hibernia* of 1804 had been laid down in 1792, at which time there was a pressing political reason to make a gesture to Ireland, where nationalism reached a high pitch in the 1780s and 1790s. This vessel survived for a hundred years, latterly as a base ship at Malta, the same name being taken by her successor, formerly the *Achilles* of 1863, which was again renamed in 1904 to free her name for another *King Edward VII*-class ship, which served until 1921.

In 1914 the Navy acquired three battleships building for foreign powers and had to find names for them quickly. The choice of *Erin* (never used previously) for the former Turkish *Reshadieh* must have been prompted, again, by a wish to make a gesture to Ireland, where the Home Rule crisis had become acute. Following the independence of the Irish Free State in 1922, neither of these names could be repeated; Northern Ireland, on the other hand, was remembered in 1942 by the name of the destroyer *Ulster*, used only once before, in 1917, possibly to commemorate the numerous casualties of the 36th Ulster Division at the Battle of the Somme.

It is tempting to suppose that the building of the first *Hibernia* caused an influential person at the Admiralty (conceivably the Scots-born Admiral Middleton?) to realise that Scotland had been neglected. It is certain that in 1794 the 120-gun *Caledonia* was ordered, though not launched until 1808. She was followed by an ironclad which served from 1862 to 1886, but since that time the name has been used only for stationary training ships and a shore establishment. In the course of his disagreement with Winston Churchill, then First Lord of the Admiralty, over the names of forthcoming battleships, King George V suggested in August 1913 that *Caledonia* should be considered. Churchill did not take the royal hint, even when the purchases of 1914 offered unforeseen opportunities; and thus, when war broke out, Ireland had given her name to two battleships and Scotland to none. The First Lord would have been aware that there were then 84 Irish Nationalists in Parliament but not a single Scottish Nationalist. However, the poetic variant *Caledon* was used for a light cruiser in 1916; the Admiralty may have taken the view that the name *Caledonia* had too high a status for a comparatively small warship. Scotland's alternative poetic name, *Scotia*, has never been used for a warship, only for a shore establishment, but we may note the destroyer *Scotsman* of 1918 (a time when there had been complaints that not enough warships were being given Scottish names). That name was used again for a submarine in 1944, but never since.

Wales, not having been a kingdom in its own right as Scotland and Ireland had been, has had less prominence in British official symbolism, and it is possibly for this reason that the name *Cambria* has been attached only briefly to an RNVR training ship (the former frigate *Derg* in 1954–59) and to the succeeding shore establishment. However, there have been several ships named *Cambrian*, none of them very large: the first a fifth-rate of 1797 and the latest a destroyer of 1943.[2]

Jersey, Guernsey and the Isle of Man are regarded by their citizens as nations, and their names have featured several times on the Navy List; the first two, however, reached it from contrasting political directions. The government of the Commonwealth named most of its warships after Parliamentary victories in the Civil War, and the capture of Jersey, which had held out for the Royalists until 1651, gave rise to one. Since then there have been five more *Jerseys*, of which the latest was the Offshore Patrol Vessel (OPV) of 1976. Another of those 'victory' names had been the 22-gun *Basing*, which however was renamed *Guernsey* under the restored monarchy in 1660; Pepys tells us that Charles II and his brother decided on the new names during their voyage back to England. *Guernsey* would have been chosen because the island had been the last British territory to fall to the republicans, in 1651. There have been five more *Guernseys*, the latest also an OPV in 1977.

The Isle of Man had to wait much longer for a warship name; this was the seaplane carrier *Manxman*, a passenger steamer commissioned into the Navy in 1916 but retaining her mercantile name. She was followed by the minelayer *Manxman* of 1940 which, significantly, had a sister ship *Welshman*: the one name must have suggested the other (which had not been used before).

In the days of the British Empire, a large number of warships were named for Britain's possessions, of which two cases are particularly noteworthy. The names of the great self-governing Dominions were, in general, kept for capital ships. The *King Edward VII* class presents a rare example of a whole class of battleships with related names: apart from the lead ship, each was named for a part of the Empire, but sometimes not in an obvious manner. There were then only two nations yet recognised as Dominions, Canada and Australia, but Australia had taken the title of Commonwealth. Thus the battleship *Dominion* was implicitly named for Canada and the *Commonwealth* for Australia; neither name had ever been used before. To these were added *Britannia* and *Hibernia* (see above), together with *New Zealand* (soon renamed *Zealandia* to make way for the battle-cruiser), *Hindustan* for India and *Africa* (where the Union of South Africa had yet to be established).

The other significant case is that of the battleship *Canada*, renamed in 1914 when the Chilean *Almirante Latorre* was purchased. In that year the Government still entertained faint hopes that their Canadian counterparts might agree to finance three dreadnoughts to join *Australia*, *New Zealand*, and *Malaya*, which had been funded by the respective countries. However, if the name was meant to encourage the deed, it failed to do so. By the time of the Second World War, the major self-governing Dominions had a higher constitutional status, but the larger colonies had cruisers named after them, the smaller ones only frigates. However, the Admiralty was not entirely consistent; if their plans had been fulfilled, one of the largest ships in the fleet, an aircraft carrier of 46,000 tons, would have been named for one of the smallest colonies, *Malta*; the explanation lies in the island's distinguished war record.

The majority of the names mentioned here, enjoying high prestige, have been attached to ships of great importance: first-rates, battleships, and so on. But as the Royal Navy can now afford so few of their modern equivalents, it may be a long time before they can be used again.

Sources:

JJ Colledge & Ben Warlow, *Ships of the Royal Navy*, Chatham Publishing (London, 2006).

Rif Winfield, *British Warships in the Age of Sail 1714–1792*, Seaforth Publishing, (Barnsley, 2007).

Rif Winfield, *British Warships in the Age of Sail 1793–1817*, Chatham Publishing (London, 2006).

David Lyon & Rif Winfield, *The Sail & Steam Navy List: All the Ships of the Royal Navy 1815–1889*, Chatham Publishing (London, 2004).

Aidan Dodson, 'The Incredible Hulks: The *Fisgard* Training Establishment and its Ships', *Warship 2015*, 29–43.

Randolph Churchill, *Winston Spencer Churchill Volume II Companion Part 3, 1911–1914*, Heinemann (London, 1969).

FJ Dittmar & JJ Colledge, *British Warships 1914–1919*, Ian Allan (London, 1972).

HT Lenton & JJ Colledge, *Warships of World War II*, Ian Allan (London, 1964).

Jane's Fighting Ships 1920, 1962–63, 1986–87.

Endnotes:

[1] We also have to reckon with the fact that during the Great War the Admiralty commissioned several merchant ships with national names some of which had been identical to those of warships. Specifically, the London and North-Western Railway's fast Irish mail steamers *Anglia*, *Cambria*, and *Scotia* kept their names, whereas the *Hibernia* had to change hers. She was appropriately renamed *Tara* after the capital of the ancient Irish high kings. P & A Campbell's Bristol Channel paddle steamers *Britannia* and *Cambria* (again) served as paddle minesweepers and were renamed *Britain* (the only example of this name on the Navy List) and *Cambridge*: that name must have been suggested by its similarity to the original. (These two vessels later served as auxiliary AA ships during the Second World War under the names of *Skiddaw* and *Plinlimmon*; the authorities were then giving the names of mountains to any such ships that required a new name). Eventually, the Caledonian Railway's Clyde paddle steamer *Caledonia* was also called up, but her name did not need to be changed.

[2] There appears to be more sensitivity nowadays to the smaller home nations. The first three Type 26 frigates are to be named *Glasgow*, *Cardiff*, and *Belfast*.

A's & A's

STALIN'S SUPER-BATTLESHIPS: THE *SOVETSKII SOIUZ* CLASS (*WARSHIP 2021*)

Author Stephen McLaughlin has written in with two corrections to his article. The first concerns the photograph on page 26, which claims to show a barbette of *Sovetskii Soiuz* under construction. Russian colleague Sergei Vinogradov has pointed out, in a letter dated 12 December 2020, that the barbette in question belongs to the dreadnought *Sevastopol* and that the photo dates from 1913! The second error appears on the same page, where it is stated that the 'reserve command post' beneath the after main battery director had 20mm protection; in fact it was 220mm (Vasil'ev, 68), which makes more sense. In the author's defence, this mistake originates in the published armour diagrams which show the thickness as 20mm, which is clearly a typo for 220mm.

POSTWAR RADAR DEVELOPMENT IN THE ROYAL NAVY (*WARSHIP 2022*)

Further to the article on pages 98-114 of this year's annual, author Peter Marland notes that, given the recent take-over and absorption of the former *Collingwood* collection by the National Museum of the Royal Navy (NMRN), it is the latter that should now be approached for access to the collection and display items.

NAVAL BOOKS OF THE YEAR

William H Garzke Jr, Robert O Dulin Jr and William Jurens with James Cameron
Battleship *Bismarck*: a Design and Operational History
Seaforth Publishing, Barnsley 2019; hardback, 610 pages, 160 photos (60 in colour), 66 figures, diagrams and maps; price £55.00.
ISBN 978-1-5267-5974-0

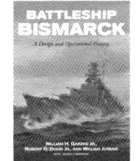

According to an old saying, a camel is a horse designed by committee. This majestic volume, crediting three co-authors and a prominent film-maker as its creators, is in effect the product of such a committee. The first two co-authors started researching 1939–1945 battleships at the height of the Cold War, over sixty years ago; their well-known books on Axis and Allied battleships of the Second World War were published in the early 1980s, with Garzke as lead writer and Dulin assisting with research and analysis. The books made full use of information made available at that time by the various navies and archives. However, much remained outside the public domain and, even if known to the authors, not for publication. In the ensuing forty years, much new archival material has been released, many more survivors have narrated their reminiscences, and most particularly the wrecks of *Hood* and *Bismarck* have been located and surveyed. Gartze and Dulin are again in partnership, with Jurens providing technical data and drawings, while James Cameron, who assisted with this volume, provided much the best visual presentation of the wreck to date in a 2001–02 diving expedition. They have woven all currently known information into this expanded and updated account.

The book begins with the origins of *Bismarck*. The post-war naval programmes of the United States, Britain and Japan led to the era of naval conferences and treaties, a 'battleship holiday', standard displacements, 35,000-ton capital ships and 16in gun calibres. Weimar Germany, restricted by the Versailles Treaty to 10,000-ton 'replacement' capital ships with 11in (28cm) maximum gun calibre, decided against heavily protected coast defence ships in favour of three relatively fast and moderately protected *Panzerschiffe* (armoured ships) for commerce raiding – the so-called 'pocket battleships' that could outrun conventional battleships and outfight conventional cruisers. The *Reichsmarine*'s move towards full-sized battleships began in 1934 when the fourth and fifth armoured ships, *Gneisenau* and *Scharnhorst*, were enlarged to fast battleships, well-protected but armed with the relatively light 28cm guns of the *Panzerschiffe*. When, following Hitler's abrogation of the Versailles Treaty in March 1935, Germany opted for full-sized battleships her naval architects lacked the full array of naval design information available in Britain, the USA and Japan. Thus while the British *King George V* class had the latest side armour, deck protection and anti-torpedo measures, Hermann Burkhardt, the lead designer of *Bismarck* and *Tirpitz*, had to work upwards from the First World War *Baden* and *Bayern*.

Chapters on the career of *Bismarck* focus on the fraught journey up the Kattegat in the proximity of Swedish warships and the closely-watched stay in Bergen. The Royal Navy was kept abreast of events and all available ships were called into the chase that led to the Battle of the Denmark Strait. The battle damage to the two battleships and the assumed sequence of events leading to the loss of *Hood* are treated in great detail. Further chapters on German command decisions after disengagement in the Denmark Strait, the torpedo attack by Swordfish from *Victorious* and the subsequent shaking off of the pursuers lead to eight chapters – some 120 pages – on the *Bismarck*'s last fight. Analysis of the final duel is particularly thorough, with tables, Polar plots [*ie* showing relative positions from the points of view of individual ships] and diagrams to show the effect of major hits. The authors conclude that torpedoes from *Dorsetshire* combined with German scuttling charges administered the *coup de grâce*.

Overall, the book is a very complete and technically precise account, with excellent photos, diagrams, maps and drawings (though it is *very* large). The 'camel committee' moments are mostly in the text, which is occasionally obscure and sometimes sufficiently badly expressed as to be ambiguous or incomprehensible; it could have been better checked for repetitions and inconsistencies. Important errors are few; in 1941 *Repulse* and *Renown* had a 9in (not 6in) main waterline belt, while Kurt von Schleicher, not Wilhelm Groener, was German defence minister in November 1932. Despite these flaws *Battleship Bismarck* will undoubtedly become a classic, and is a 'must' for the bookshelf.

Ian Sturton

Frank Jastrzembski
Admiral Albert Hastings Markham: A Victorian Tale of Triumph, Tragedy & Exploration
Pen & Sword Maritime, Barnsley 2019; 197 pages, 41 B&W photographs, nine maps, two appendices, bibliography, index; price £19.99.
ISBN 978-1-5267-2592-9

Andrew Gordon's seminal work *The Rules of the Game* includes a detailed analysis of the tragic collision between HMS *Victoria* and HMS *Camperdown* in 1893, which led to the loss of 358 lives. A central tenet of Gordon's book is that the Royal Navy of the Victorian age was too hidebound by regulations that caused many of its officers to lose the ability to take independent decisions at crucial moments. It is therefore ironic that Admiral Markham, as commanding officer of *Camperdown*, was criticised for his lack of initiative in not countermanding orders before HMS *Victoria* was fatally rammed by his ship. This incident aside, Frank Jastrzembski's work is a catalogue of Albert Markham's bold and extraordinarily adventurous life during which, on countless occasions, he had to rely on his own decision-making in a range of life or death situations.

The *Victoria* tragedy coincided with Markham's last sea-going appointment, in the prestigious and coveted role of second-in-command of the Mediterranean Fleet. The events that followed tarnished his reputation and haunted him for the rest of his life. The other incident that caused controversy over his actions and incurred the 'displeasure' of the Admiralty Board occurred shortly after the start of his long naval career.

Markham's first two major overseas appointments were to the Far East. When he was just 14 years old he was posted to China, and later to the Australian Station. On both occasions he witnessed and participated in the harsh realities of colonial rule. While on the Australian Station, Markham was required to deal independently with a particularly unpleasant form of slavery called 'blackbirding' that involved the unscrupulous exploitation of inhabitants of some of the most remote South Sea Islands. The decisions he took in order to try to root out the slavers, keep the peace and, most controversially, to teach the indigenous population 'a lesson', was condemned in some quarters, not least by the Admiralty itself.

Much of this episode makes for uncomfortable reading, particularly at a time when Britain's colonial past is under intense scrutiny. Later, owing to the generosity of staffing afforded by the long period of *Pax Britannica*, Markham was able to escape the routine duties of a professional naval officer and indulge his all-consuming passion for adventure and exploration of the remotest parts of the globe and the Arctic regions in particular. In 1875 he was selected to join a 'chosen band' on an expedition to attempt to reach the North Pole. As Jastrzembski rightly points out, ground-breaking scientific and geographical advances aside, the true purpose of this well-funded expedition was 'to bring greater glory to England'. Ultimately, the expedition failed for a variety of reasons, including an outbreak of scurvy among the crews of HM Ships *Alert* and *Discovery*. The hand-picked volunteer crews endured unimaginable hardships, but Markham was undeterred and continued to find time and opportunity for expeditions across the North American continent and again in the Arctic.

Throughout the narrative, the author adopts an admirably neutral tone. He has a very easy, accessible style that in places makes Markham's extraordinary life read like a 'Boy's Own' adventure story. At the same time one gains a fascinating insight into the life of a Victorian gentleman naval officer, who lived through the apogee of the British Empire, loved his hunting and shooting and the opportunities he gained to explore, record and write about the scientific discoveries he made on his travels. However, reputation was everything among the upper classes of the time and it is unfortunate indeed that, in the ultimate reckoning, the name of Albert Markham has been associated with failure due to that fateful day in June 1893.

Jon Wise

Zvonimir Freivogel
Raiders: German Auxiliary Cruisers of World War Two

Despot Infinitus, Zagreb 2018; softback, 258 pages, more than 220 photographs, maps and diagrams; price €49.90.
ISBN 978-953-8218-21-7

Derived from a German book by the same author first published in 2003, *Raiders: German Auxiliary Cruisers of World War Two* aims to provide an overview of the operations of these enigmatic vessels using both text and photographs. The publisher has adopted a large (28cm x 23cm) format to allow the latter to be displayed to best effect.

Following the customary foreword and introduction setting out the author's aims, the opening pages briefly summarise mercantile raiders from the days of the corsairs through to the interwar years. An equally succinct summary of the German decision to adopt commerce disruption cruiser (*Handelsstörkreuzer* or HSK) warfare and the challenges it involved leads on to summaries of the HSKs' various cruises. Encompassing some 150 pages, these form the bulk of the book's content. The remaining pages include a brief assessment of the raiders' performance and the potential for such operations in the future, as well as appendices and tables covering equipment, brief technical specifications and lists of ships sunk or captured.

The cruise summaries provide a useful synopsis of raider activity, while occasional snippets of additional detail hint at the author's more in-depth knowledge of the subject matter. However, it is the illustrations that are the book's main feature. On the plus side, the photographic coverage is extensive and contains many interesting images, particularly those derived from German archives. However, the inclusion of photographs of peripheral relevance, such as of more conventional German warships used in operations against trade, or of sister ships to vessels intercepted by the various raiders, suggest something of a lack of focus. It is also disappointing that reproduction of images is of decidedly variable quality and that many captions are not more extensive. The overall impact of the visual material is often reduced by questionable choices with respect to layout and design.

In summary, the author has provided a useful introduction to German auxiliary cruiser operations supported by a heavy focus on visual content. However, both the limited level of detail and the various illustrative weaknesses prevent a more positive recommendation of what is an expensive book.

Conrad Waters

Paul Brown
Abandon Ship: The Real Story of the Sinkings in the Falklands War
Osprey Publishing, Oxford 2021; hardback, 320 pages, 31 B&W photographs; price £20.00.
ISBN 978-1-47284-643-3

The Falklands War was a wake-up call for the Royal Navy. Successive governments had cut the number of ships and aircraft, and the Navy had become focused on its NATO commitments centred around its North Atlantic anti-submarine role. The mobilisation of a significant portion of the fleet and merchant marine to conduct a war it never expected to fight, in remote and inhospitable waters, using technology never used in anger was not a decision taken lightly. From the outset there was no doubt that that there would be losses both in lives and equipment.

Many books have been written about the conflict over the past forty years. However, in *Abandon Ship* Paul Brown sets out to deliver the 'the true story behind the dramatic events' that led to the loss of seven major combatants. From the torpedoing of the Argentine cruiser *General Belgrano* the story is then relayed chronologically through the mounting losses of UK ships starting with the Type 42 destroyer HMS *Sheffield*, a victim of an air-launched Exocet missile, through the two Type 21 frigates *Ardent* and *Antelope*, another Type 42, HMS *Coventry*, SS *Atlantic Conveyor*, a merchant ship rapidly converted to aircraft ferry, and finally the landing ship RFA *Sir Galahad*.

The author has used the UK Freedom of Information Act to obtain the Board of Inquiry reports into the losses and in so doing is able to provide a new level of detail. This provides a gritty narrative with an inevitable focus on the human and material failings as the brutality of naval warfare was painfully re-learned by a new generation.

After a brief summary of the conflict, each individual loss is allocated its own chapter using a standard approach that includes a comprehensive background and full account of the sinking based on the detailed Board of Inquiry narrative. These give a fascinating insight into both material failings and also the impact of decision-making under exacting conditions. The Board's conclusions are then discussed, with additional analysis by the author. It is here that the book stumbles. The author's analysis often goes little further than the Board reports, otherwise relying predominantly on reflective personal opinions of senior survivors published some years later. Furthermore, there are a number of omissions and errors. While individually relatively minor, these discrepancies, together with a reliance on online sources, often lead to speculative assumptions and even a sweeping dismissal of the actions of those confronted for the first time by the realities of conflict. Lessons were learned even during the conflict itself, and the survival of HMS *Glamorgan* late on was as much to do with the preparedness of the ship's company as it was with the resistance of the ship to damage.

Abandon Ship is a gripping read for those interested in understanding the harsh realities of war and in particular its human aspects. However, the author has missed the opportunity to combine the new details gained from the Board of Inquiry reports with the wider picture that would have give a truly measured and balanced account in time for the 40th anniversary of the conflict.

Philip Russell

Stefan Draminski
Anatomy of the Ship: The Battleship *Scharnhorst*
Osprey Publishing, Oxford 2021; hardback, 336 pages, many B&W photographs, 600 line drawings and 400 colour 3D illustrations; price £40.00.
ISBN 978-1-4728-4023-3

In this follow-up to his book on *Bismarck*, published in September 2018, author Stefan Draminski focuses on her immediate predecessor *Scharnhorst*, which has generally received less extensive coverage. The book follows the format of recent 'Anatomies', combining line drawings with colour artwork generated by computer graphics modelling.

The first section provides a detailed technical history of the ship, and is illustrated with superb black-and-white images of the ship from the German archives plus a comprehensive set of data tables. The author's English is readily intelligible but not idiomatic, and would have benefited from editing by a native speaker with a sound grasp of the correct technical terminology. Draminski has an excellent command of German, and in some instances imprecisions result from a literal translation of the German term (for example *Zielgeber*, rendered as 'target giver'). However, none of this detracts from the quality of the information, and the author's grasp of the German terminology is a major plus. The tables detailing the guns and ammunition are particularly informative. Nor is the author uncritical of German engineering and design; defects included 'wetness', a discontinuity in the armoured deck to accommodate the boilers, unreliable propulsion machinery, and slow/heavy HA mountings and directors.

The main part of the book follows the traditional Anatomy format, with a drawing section divided into General arrangements, Hull structure, Superstructure, Rig, Armament, Fire control, Fittings, Aircraft and Boats. The line drawings and colour artwork show the ship at various key stages of her service life. They are simply stunning; Draminski is a wonderful draughtsman, equally comfortable with both of these genres, and the possibilities of combining the two are used to great effect. General views of the ship have the line drawings and colour artwork on different spreads; where they are run across the gutter there is a small divide that ensures no key features are lost.

For the 'detail' sections the 'flat' line drawings (profile, plan, face and rear) are juxtaposed with part views of the model turned to give a 3D perspective. In theory this is an excellent idea. However, in purely visual terms the fine B&W line drawings are often overwhelmed by the 'blocky' 3D colour perspectives. Also, the colour artwork is not always in the correct sequence (*eg* the cutaways of the forward magazines and shell rooms on page 103, which are not labelled, and the colour perspectives of the after superstructure from page 160 onwards).

However, these are minor blemishes in what is otherwise a hugely impressive book representing many hours of dedicated research on the part of the author. Production values are excellent, and at £40 the book represents very good value for money.

John Jordan

Celia Clark and Martin Marks
Barracks, Forts and Ramparts
Tricorn Books, Portsmouth 2020; large format softback, 450 pages, illustrated with many photographs, maps and drawings in colour and B&W; price £40.00.
ISBN 978-1-9128-2164-8

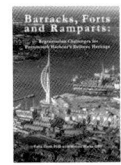

Although the subject matter of this book is arguably outside *Warship*'s remit, it is of interest because, as the subtitle 'Regeneration Challenges for Portsmouth Harbour's Defence Heritage' makes clear, it is about what happens to the built infrastructure in and around a major naval base when that establishment is subject to contraction over several decades. Ships do feature, mainly in terms of the Historic Dockyard's impressive collection of preserved vessels, but the book is mainly about buildings, their 'active' histories and what has happened to them since.

While up to 95 per cent of a redundant warship can be recycled today, its equivalent on land is more challenging to deal with. In addition to the well-known titular structures, this volume covers the Royal Naval Hospital Haslar, the Priddy's Hard ordnance depot, 'stone frigates' *Sultan*, *Dolphin*, *St Vincent* and *Daedalus*, the Royal Clarence victualling yard and the many buildings within the Historic Dockyard and the Naval Base itself. The underlying story is complex, embracing local and national politics, and interests that include the MoD, HM Treasury, English Heritage and assorted museums. All too often corporate and developers' concerns seem to have taken precedence over the wishes of local people, perhaps best exemplified by the sad tale of the redevelopment of the HMS *Vernon* site on Portsmouth's former Gunwharf. Here and elsewhere, listed buildings, some of considerable architectural or historical merit, were unfortunate victims of the exploitation of loopholes in existing legislation, conveniently 'fell down' or caught fire, or were simply demolished, to the detriment of future generations. Thankfully, in contrast, there are also many areas where the story is less gloomy, in particular with the ongoing conversion of the historic buildings of the dockyard itself to form the National Museum of the Royal Navy (Boathouse No 4 is a rather unlikely winner here).

As with any study of conservation, the ongoing nature of the subject means that aspects of the story have been overtaken by events. The recent controversial proposals for the former ordnance depot at Tipnor Point, and the damage caused by the sea to the area adjacent to Fort Blockhouse do not feature.

The book is clearly the product of a great deal of work on the part of Clark and Marks, and is very thoroughly referenced throughout. It is therefore disappointing that it has been so haphazardly put together: the layout is poor, the illustrations frequently too small and sometimes repeated, and the editing slipshod. The authors also suffer from the common failing of wanting to include everything: every reminiscence, every opinion (presumably in the interests of balance – though their own stance is pretty clear throughout). This, combined with the above problems, means that what could have been an interesting read is often very heavy going indeed.

Stephen Dent

Angus Konstam
Hunt the *Bismarck*: The Pursuit of Germany's Most Famous Battleship
Osprey Publishing, Oxford 2019; hardback, 336 pages, B&W plate section and maps; price £25.00.
ISBN 978-1-4728-3386-0

Angus Konstam's new book draws on extensive research – albeit using a number of dated secondary sources – and first-hand accounts (with a particular focus on that of the sole surviving German officer, Burkard von Müllenheim-Rechberg) to relate the epic story of *Bismarck*'s first and last sortie. The author claims a 'readable narrative … written by a naval historian who can cut through the technicality to retell this classic naval tale to a modern audience' (p 16).

Konstam is often prone to hyperbole. When *Hood* blows up, von Müllenheim-Rechberg 'felt every nerve being yanked out of his body' (p 9). The 'neat decks' of *Bismarck* would end up as 'a tangled mess of ripped steel' (p 21). And, channelling Disney and Coca Cola, the Blohm & Voss yard was 'a place where shipbuilding magic happened' (p 28).

Nor is the author a reliable narrator. *Scharnhorst* and her sister *Gneisenau* were not classified by the Kriegsmarine as 'battlecruisers'; they were initially *Panzerschiffe* (= 'armoured ships'), and subsequently *Schlactschiffe* (= 'battleships'). The second officer of a ship the size of *Bismarck* would not be known as the 'First Lieutenant' (p 18). The name of Vice Admiral Marschall is incorrectly spelled 'Marshall' (p 44) . The French port of St-Nazaire was not a 'naval base'. The CO of *Prinz Eugen*, Helmuth Brinkmann, is demoted from Vice Admiral to Captain on the same page (p 38). On page 157, a 'three-masted [sic] warship' was 'undoubtedly a British heavy cruiser'. It was *Repulse*, not *Renown*, that was despatched by Tovey to Halifax to refuel (p 231). Finally, *Bismarck*'s steering compartment was, according to the author, 'protected by a watertight bulkhead that was just 45mm thick' (p 265). Published plans of the ship show that the 45mm internal torpedo bulkhead extended only over the citadel; the steering compartment was protected by an armoured box with 80mm side walls and a 110mm roof.

There are also imprecisions of terminology that may result from an uncritical acceptance of the translation of von Müllenheim-Rechberg's memoirs. These include: 'Equipping Pier' (for 'fitting-out quay', p 30), 'electrical plant room' (p 187), 'front funnel' (p 188), and 'shot mats' (p 192) – the latter employed to cover the hole in *Bismarck's* bow resulting from the shell hit by *Prince of Wales*.

The book has its strengths. *Bismarck's* career prior to her sortie has tended to be neglected in other accounts, and the numerous maps are clear, useful and largely free of error. However, the author's lack of a sound technical understanding, which is highlighted by his implausible explanation for the loss of HMS *Hood*, means that the book will be of limited interest to readers of *Warship*.

John Jordan

Angus Britts
A Ceaseless Watch: Australia's Third-Party Naval Defense, 1919–1942

US Naval Institute Press, Annapolis 2021; hardback, 354 pages, B&W illustrations; price £49.95/$54.95.
ISBN 978-1-6824-7533-1

Following his previous book on the decline of Britain's naval power in the Far East, Angus Britts focuses here on Britain's failure to live up to her commitments to Australia and New Zealand in this new work, part of the series 'Studies in Naval History and Sea Power'.

During the days of Empire there was a fundamental difference of opinion between Australia and London regarding naval defence. Australians felt isolated and prey to attack and sought a strong naval presence. By comparison, the British Isles were the hub of a vast maritime Empire, with Australia just one of many territories to be defended. To the Admiralty's strategists, Australia was best defended by financial subsidies and a detached squadron, backed up by the might of Britain's battle fleet. Australians only acquired their 'own navy' in 1909, when the British invited the Dominions to fund 'fleet units', a policy enthusiastically embraced by Australia: in 1913 the battlecruiser *Australia* and the cruisers *Melbourne*, *Sydney* and *Encounter* (on loan from the RN) arrived at Sydney to a rapturous welcome.

The real strategic concern for Australia was the rise of Japanese power. By 1917 the focus of naval strategic thinking in London seemed to be shifting eastwards. However, economic and political constraints ruled out an Imperial Pacific Fleet, and Australia's naval defence remained tied to Britain's imperial strategy. Promises were made: a modern naval base was to be developed at Singapore and a battle fleet would be sent eastwards in time of tension. However, this new strategy was dependent on a rejuvenated and expanded Royal Navy, and Britain's 'stop-start' development of Singapore, the naval arms limitation treaties, debates over air power, economic constraints and the deteriorating situation in Europe all conspired to undermine it.

In 1937, the Imperial Conference finally determined that Europe, not the Far East, was the focus of Whitehall's strategic concerns. Despite this, Britain reaffirmed her commitment to developing Singapore and gave further reassurances to Australia that a battle fleet, or at least a 'flying squadron', would be sent eastwards when necessary. However, as Britts points out, this meant that Australia's naval security was dependent upon a stable political situation in Europe, something that by 1937 was increasingly improbable.

The outbreak of war in Europe in September 1939 saw Australia playing a full part, albeit in a defence strategy based on British priorities and subordinated to the assessments of Churchill and his Chiefs of Staff in their Whitehall bunker thousands of miles away. Following the outbreak of war with Japan and the failure of the joint defence initiatives with the Americans and Dutch (ABDA), Australia urged Whitehall to develop a coordinated strategy but was ignored. However, the military disasters of 1942 had increased the importance of Australia in American eyes, as Australia was the only Allied-held territory west of Hawaii and would become a key staging post for MacArthur's counter-offensive in the southwestern Pacific. Britain struggled to send forces to the region, traditional ties crumbled and Australia moved closer to the USA.

Britts has successfully illustrated the longstanding strategic tensions between Australia and successive British governments during the interwar period and the circumstances that impelled Australia into the American orbit. *A Ceaseless Watch* is a detailed and engrossing book which is highly recommended.

Andrew Field

John Roberts
Destroyer *Cossack*, detailed in the original builders' plans

Seaforth Publishing, Barnsley 2020; hardback, 128 pages, illustrated with over 230 full colour images from builders' plans, including gatefold; price £30.00.
ISBN 978-1-5276-7706-5

This is the eighth book in this series, based on the National Maritime Museum's collection of builders' drawings and, like some of the other volumes, concentrates on a particular vessel but then adds some plans of sister ships to illuminate a particular point missing from the name-ship's collection. The reproductions are mainly of tinted drawings, a skill now being revived out of necessity for 3D CAD drawings, but one which originally turned a basic construction drawing into a work of art. If there is a criticism of the format, it is that the reader really needs a magnifying glass to get the best out of some of the reproductions, particularly those of the sister ships, which are generally reproduced at a smaller scale.

Following an introduction covering the origins of the design and the career of HMS *Cossack*, the book is divided into nine sections, including the gatefold plans, covering specific aspects of the ship such as decks, armament and machinery. These are followed by a frame-by-frame description and explanation of the profile and section drawings, mirrored further on by a similar description of the deck drawings. The book concludes with a small section on later developments of the class, a bibliography and list of plans. There is no index, but it is probably rightly assumed that the reader knows enough of the ship to be able to find any specific feature. One further small criticism is that the deck drawings show the hammock positions but the explanation of the symbol doesn't appear until the very last mess deck shown, several pages on!

Author John Roberts points out that steam heating was only added to the surviving ships of the class from 1943 onwards. Winter in the North Atlantic could be pretty cold, almost as cold as the run to Murmansk and sufficiently cold as to reduce crew efficiency, so it is slightly surprising that it took so long. This prompts a memory of asbestos removal from HMS *Cavalier* prior to her restoration and voyage to Chatham. Every steam heating pipe throughout the ship had had its asbestos insulation removed, except that every pipe hanger displayed two raw edges where the asbestos had been exposed but left in place!

This is a fascinating book with a wealth of explanation, one which at an academic level well illustrates the work needed to achieve a workable layout which, in the opinion of the reviewer, was one of most difficult aspects of warship design.

W B Davies

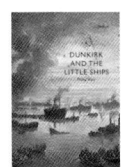

Philip Weir
Dunkirk and the Little Ships

Shire Publications, Bloomsbury, Oxford 2020; 112 pages, illustrated with 100 B&W and colour images; price £9.99.
ISBN 978-1-7844-2375-9

For such diminutive A5 titles, Shire books manage to pack in a great deal of information, and Philip Weir's *Dunkirk and the Little Ships* is no exception. The subject matter here needs little introduction, and Weir wastes no time in getting to the history. Importantly though, he gives the events at Dunkirk context by spending the first chapter describing the German *Blitzkrieg* through the Low Countries and the withdrawal to the Channel ports. This includes the story of the valuable work done by the small ships (predominantly destroyers) of the Royal Navy in evacuating Dutch ports and rescuing a number of incomplete Dutch warships.

This is followed by the story of the evacuation, both operationally and on the beach, weaving the story of several significant vessels into the narrative. However, the heart of the book is, as the title suggests, the vessels themselves, and the next 35 pages are dedicated to them. Weir skilfully includes a number of surviving historic vessels that, while not having direct connections to Operation 'Dynamo', are representative of vessels that were. Hence both the cruiser HMS *Caroline* and 'Q-ship' HMS *President*, which had both been converted to drill ships by the time of the Second World War, are included, not for their participation but because of their associations with vessels that did: in this case HMS *Calcutta* – *Caroline* was a sister ship, while *President* provided *Calcutta*'s crew for Dunkirk. Weir is to be applauded for making these connections, which are commonplace when looking at terrestrial heritage, but all too rare when looking at maritime equivalents.

Weir also covers the other evacuations from France, Operations 'Cycle' and 'Aerial', again eking out several maritime connections, before turning his attention to the origins of the Association of Dunkirk Little Ships and the regular pilgrimage of vessels across the Channel. There is a list of places to visit associated with the events in the book, and a surprisingly detailed index.

Shire books can only ever provide an overview of a subject, but even so this one contains more than enough information to satisfy anyone with an interest in Dunkirk. There is no bibliography, which would have been useful for those wishing to learn more, but perhaps that is asking quite a lot for a book of this size.

Stephen Fisher

Hugues Canuel
The Fall and Rise of French Sea Power: France's Quest for an Independent Naval Policy, 1940–1963

US Naval Institute Press, Annapolis 2021; hardback, 350 pages, small photo section, data tables, endnotes and extensive bibliography; price $54.95/£50.95.
ISBN 978-1-68247-616-1

This latest book in USNIP's series Studies in Naval History and Sea Power explores the renewal of French naval power from the fall of France in 1940 through the political turmoil of the period 1940–45 and the first two decades of the Cold War. The author is a defence attaché in Japan, and the present work has been expanded into book form from a PhD thesis written under the aegis of the Canadian Royal Military College.

This is a most impressive piece of research and covers ground which has previously received only piecemeal coverage in both English- and French-language histories. Beginning with the Armistice of June 1940, by which time the French Navy had reached the peak of its interwar development, Canuel traces the British aggressions of 1940, the foundation of the Free French Naval Forces (FNFL) under the spiky Admiral Muselier, the chaotic interplay and political manoeuvring between the Free French, Darlan and Giraud following the 'Torch' landings of November 1942, and the ongoing antagonisms between the FNFL and the forces that had remained loyal to Vichy following the Armistice but which had now opted to rejoin the war against Hitler's Germany.

The main body of the magnificent French fleet that had been the subject of envy of Churchill (and arguably of Hitler) had been scuttled at Toulon during that same November 1942, and the admirals of the FNFL and of the *Forces maritimes d'Afrique* (FMA) aspired to rebuilding a major fleet commensurate with France's status as an independent colonial power with an empire which stretched from the Caribbean to the South Pacific. The composition of the new fleet would take into account the latest technical developments, and would be centred on carrier task forces able to project power around the world.

However, neither the British nor the Americans were willing to grant French wishes. Britain, although supportive of De Gaulle and the Free French, was prepared to allow the FNFL only a subordinate role under RN command. Transfers comprised A/S corvettes of the 'Flower' class and frigates of the 'River' class, together with a handful of small submarines. The Americans were less keen on De Gaulle, preferring to negotiate with representatives of the 'legitimate' French government, and actively opposed the revival of the former French empire. They were prepared to refit some of the more modern French units based in North and West Africa with up-to-date AA weaponry, but transfers of newly-built ships were again limited to destroyer escorts (DEs) and subchasers. The role of the FMA would therefore likewise be antisubmarine warfare and coastal patrol, the only concession being the use of the modern cruisers for fire support during the landings in Provence.

These tensions continued post-war. France's aspirations to rebuild the fleet were initially thwarted by the devastation wrought on French industry and infrastructure, and by a crippling lack of funding. The advent of NATO and the Mutual Defense Assistance Program (MDAP) brought financial relief, but the USA continued to insist on a naval force structure subordinated to Allied missions. However, the loans of the former British carrier *Colossus* (as *Arromanches*) and the US light carriers *Langley* and *Belleau Wood* (to bolster the French position in Indochina) enabled the French to expand their experience of carrier air operations, and under CNS Admiral Nomy (himself a former naval aviator) ambitious plans were drawn up for three carrier task forces, to be used both to support NATO missions and for independent national purposes ('*Union française*'). This progamme was developing nicely when De Gaulle, now President, decided that France's future independent defence strategy would be based on a nuclear triad of aircraft, land-based missiles, and SSBNs. The huge cost of the missile submarines and their associated development and infrastructure effectively deleted the third carrier task force from the roster.

Canuel's book is not without its flaws. The table of French Naval Strength 1 September 1939 (page 5) has a number of serious (and puzzling) errors, and there are numerous smaller factual errors scattered throughout the book. However, the account of the political history of the period is commendable, and the author's judgements are admirable and precise. This is an essential read for anyone interested in the wartime troubles and the postwar development of the French Navy.

John Jordan

Philippe Caresse
The Battleships of the *Iowa* Class: A Design and Operational History
US Naval Institute Press, Annapolis 2019; hardback, 534 pages, heavily illustrated; price £75.00.
ISBN 978-1-5911-4598-1

First published in two French-language volumes as *Les Cuirassés de la classe Iowa* in 2015, this book aims to provide a technical and historical account of the US Navy's ultimate battleship class. Running to well over 500 pages, it is a large and heavy book that contains a substantial amount of illustrative material, much in colour. Needless to say, it also carries a hefty price tag to match.

Following a brief foreword by David Way, curator of *Iowa*, and an equally short preface by the author, the book is divided into two principal parts. Part 1, Genesis and Technology, runs to a little over 200 pages and is divided into twelve separate chapters. In addition to describing the origins of the class, these encompass a range of technical features such as armament, propulsion and shipboard equipment as well as broader subjects such as lists of commanding officers and battle honours. There is brief reference to the fates of the incomplete *Illinois* and *Kentucky*. Given that a large part of these chapters is taken up with illustrative material, technical discussion is relatively limited, and any reader wanting a detailed understanding of the design would be better served by some of the many other books previously published on these ships.

Much of the remainder of the book, around 300 pages, comprises the four chapters that form Part 2. Each of these is dedicated to describing the career of one of the four *Iowas* that were ultimately completed. In general terms, these chapters contain much more information than the technical chapters and provide a reasonably comprehensive account of the key events in the lengthy US Navy service of the class. As is the case for Part 1, a key element of these descriptions is the extensive selection of photographs that accompany the written text. The book concludes with a short note on sources – although references to specific archival documents are omitted – plus the usual acknowledgements and index.

The book's illustrations are both a major strength and a weakness. On the positive side, the extent and quality of photographic coverage is impressive, encompassing a wide range of external and internal views from construction through to the four battleships' current status as museum ships. Disappointingly, some images are incorrectly captioned, while more could have been done to indicate the approximate timeframe of the photographs that have not been precisely dated.

A much greater limitation is the restricted selection of technical drawings, which are often reproduced at too small a scale for the labelling to be legible (see the general arrangement diagrams that seemingly represent *Missouri* and *New Jersey* at the time of their final reactivation). While these weaknesses are partly counterbalanced by some nice external renderings, this aspect of the treatment of the ships' external appearance is more than a little haphazard. One of a number of curiosities is the inclusion of a table of British Admiralty camouflage colours alongside the description of US Navy camouflage schemes adopted for the class.

The Battleships of the Iowa Class provides a well-illustrated operational history of these famous ships but is much less successful in dealing with their design aspects. The book's limitations are a disappointment given its eye-watering price tag.

Conrad Waters

Gérard Garier & Alain Croce
Les Cuirassés Echantillons Tome 01: *Brennus, Carnot, Charles Martel*

Editions Lela Presse, Le Vigen 2020; large format hardback, 260 pages, many photographs, maps, plans and drawings; price €55.90.
ISBN 9-782374-680231

This is the first of two French-language monographs on the six battleships of the notorious *Flotte d'échantillons* ('Fleet of Samples') completed during the mid-1890s. Built to competing designs drawn up by the preeminent French naval architects of the day, they were nevertheless similar in respect of their armament, designed speed and thickness of protection. This volume covers the earlier *Brennus* together with the first two ships of the 1891 programme, *Carnot* and *Charles Martel*. Following a brief Preface and a lengthier Introduction, the authors allocate a separate section to each ship detailing technical characteristics, followed by accounts of the service history of each of the ships; this inevitably results in a degree of repetition.

There are many data tables, plans and photographs. As the authors have pointed out, these three ships are the least well served when it comes to 'as-fitted' plans, which have survived only for the later ships to be covered in Volume 2. Despite this, they have accumulated a veritable treasure trove of plans and drawings, and the latter have been worked up into a series of excellent colour profiles by Gérard Garier to show the ships at various stages of their careers. The photographs are generally postcards from the period and are not always of the highest quality, but the collection is admirably comprehensive, covering the different periods of the service life of the ships and presenting the opportunity for informative captions.

The design of the book is 'busy' and a little garish to English eyes, with extensive use of colour for headings and tables, and the bizarre typeface used for the Foreword and the Preface (small, 'scratchy' and italic on a blue-squared backing) is not an easy read. A more serious problem is that, in their desire to include every surviving plan and drawing they can find for these ships, the authors have often opted to reproduce material that is inappropriate and confusing. This is particularly true for *Carnot*. For example, on page 95 it is stated (correctly) that the main guns were Modèle 1887, but the plans reproduced on the opposite and following pages are of the Mle 1893–1896 and 1893–1896M series, guns that had a longer firing chamber better suited to the new slow-burning powders, and which were not in service with the French Navy until the early 1900s. This is also true of the plans of the 274.4mm turret on page 97: the only installation of the Mle 1893–96 turret was on the coast defence battleship *Henri IV* laid down in 1897.

There are other errors that suggest that the authors are not entirely comfortable with the technical aspects of these ships. In the Introduction it is stated that *Brennus* featured 'cemented' armour; however, Augustus Harvey did not demonstrate his face-hardening process until 1890–91, and although there were experiments with cemented plates for the main turrets of *Bouvet*, the first French battleship to have a cemented belt was the much later *Iéna*. The authors also seem unable to distinguish between the 65mm gun Mle 1891, the newly-designed QF model fitted in these ships, and the Mle 1881, which was a field gun used by the landing party; two of the latter were generally embarked and secured to (but not fired from) the Battery Deck. Finally, it is stated for both *Carnot* (page 107) and *Charles Martel* (page 140) that the main and auxiliary machinery was fuelled by coal and wood(!); the small quantity of wood on these ships was used in the galleys.

There is a lot of excellent and previously unpublished material in this first volume on these unjustly neglected ships. However, the data is unreliable and there are frequent confusions and conflicts in the authors' text and illustration that will need to be resolved by the serious reader.

John Jordan

Alessio Patalano & James A Russell (Eds)
Maritime Strategy and Naval Innovation: Technology, Bureaucracy, and the Problem of Change in the Age of Competition

US Naval Institute Press, Annapolis 2021; softback, 307 pages, 11 tables/diagrams, endnotes, index; price $75.00/£54.59.
ISBN 978-1-6824-7525-6

It is now more than thirty years since the 'binary' world of the Cold War passed into history. Since then there have been what the editors describe as 'profound shifts in the maritime climate', not the least of which has seen the rise of China and its aggressively outward-looking stance in the Indo-Pacific region. Additionally, technological innovation threatens to render obsolete many of the naval weapons platforms which Western navies, and in particular the USA, have taken for granted over many years.

Editors Alessio Patalano and James Russell 'bookend' this erudite collection of essays by thirteen distinguished writers and lecturers with their own individual contributions. The book is divided into four sections that address successively 'strategy and innovation' from a historical and a contemporary viewpoint, then the organisational and bureaucratic factors affecting naval innovation, with an analysis of how this might impact on the future of war at sea.

As one might expect from a Naval Institute Press publication, the book deals with its subject very much from the perspective of the US Navy as the most powerful navy in the world, with China as the main aspirational contender to that title. Britain and Germany occupied similar positions prior to the Great War of 1914–18, and the 'historical' section looks at maritime strategy issues from the perspectives of both countries. The potency of submarines as weapons of war and the increasing importance of naval intelligence in the age of wireless communication are considered as examples of the innovations of the day.

Three contributors address some of the 'grand strategy' issues facing the USN and its allies in the second decade of the 21st Century. The efforts of the United States to stay ahead of rivals China and Russia, the changing face of US naval strategy over the past 75 years, and the fundamental question 'What are navies for?' are all considered within the context of affordability. There is the ever-present danger that innovation can outrun the ability of nations to build the ships to carry and deliver the new weapons, let alone keep pace in a rapidly evolving geopolitical world. For example, advances in the deployment of unmanned aircraft and the new threat posed by land- or submarine-launched hypersonic missiles have, quite suddenly it seems, called into question the future of one of the bedrocks of US naval strategy, the carrier battle group.

It comes as something of a relief to find that one chapter in this book is devoted to what the author calls the 'human capital'. He argues that despite the ability of quantum computing to enable AI to deliver complex, technological solutions ever more quickly and efficiently, 'the most critical facets in combat will continue to be human'. So we are not quite obsolete yet!

Jon Wise

JJ Colledge, Ben Warlow and Steve Bush
Ships of the Royal Navy: The Complete Record of All Fighting Ships of the Royal Navy from the 15th Century to the Present
Seaforth Publishing, Barnsley 2020; 493 pages; price £25.00.
ISBN 978-1-5267-9327-0

The late Jim Colledge published the first volume, covering major warships, of the first edition of what was then subtitled '*an historical index*' in 1969, with a second volume covering Navy-built trawlers, drifters, tugs and requisitioned vessels appearing the following year. A supplement was issued in 1986, and an updated two-volume set published during 1987–89. Further updated versions of the first volume (now not called such, and with the present subtitle) by Ben Warlow were issued in 2003 and 2006, supplemented by the latter author's work on static ships and shore establishments in 1992 (revised edition 2000). Now, Steve Bush has joined Lt-Cdr Warlow in producing this further revision of the 'fighting ships' volume.

As with previous editions, the Introduction begins with a statement of the 'Scope of the Book' – which remains even now incorrect. It states that in addition to ships of the British Royal Navy 'it is hoped that most of the ships which have served at any time in the British Commonwealth navies are included', but in fact the listing is restricted to ships of the Dominion navies (*ie* those that have been formally His/Her Majesty's Ships). Thus, while the Australian, Canadian and New Zealand navies are covered down to the present day, units belonging to *republics* of the Commonwealth of Nations are omitted: for example, the Indian *Delhi* (ex-*Achilles*), acquired in 1948 before India became a republic in 1950, is listed, but not the 1991-launched current bearer of the name.

In contrast to previous editions, Royal Fleet Auxiliaries are included in the book, on the basis of their increasing use in front-line roles – a most valuable enhancement. Trawlers, tugs etc, continue to be excluded except 'as a continuation of a ship name (but see Preface)', The parenthetic remark leads to a degree of head-scratching, as there is no mention of the topic in the 'Preface to the 2020 Edition' (the only one included). However, consultation of the 2003 edition's Preface indicates that the intention appears to be to refer the reader to Volume 2 (last re-issued in 1989).

Although the Introduction appears not to have been fully checked and revised when the new edition was undertaken, its 'Details of ship "classes" mentioned in the text' section has added the Types 26 and 31e frigates, Batch 2 'River'-class OPVs, and *Dreadnought*-class SLBMs. The value of this section is questionable, as it is very far from comprehensive, and the data, apart from (in some cases) armament, is repeated in individual ship entries, except for the very latest additions. Taken together, any future revision might wish to revisit the whole Introduction, and perhaps the way in which it relates to the main ship listing.

The core of the book remains this alphabetic listing of ships, giving a vessel's type, principal characteristics, builder, date of launch, and fate, including re-namings, re-classifications and transfers. While all relevant new vessels, built, ordered or projected since the last edition appear to have been added, the revision of existing entries has not been comprehensive. A random check has yielded, *inter alia*, *Carrick* (ex-*City of Adelaide*), listed as 'for BU 2007', but which actually arrived in Adelaide in 2014 for preservation.

The latter problem is, of course, but a very minor issue in a book that is indispensable to any naval historian, and whose updating is extremely welcome – especially the addition of the RFAs. Nevertheless, such omissions and the issues with the Introduction noted above are to be regretted, at the same time as one warmly welcomes this new edition.

Aidan Dodson

Norman Friedman (Editor)
British Naval Weapons of World War Two. The John Lambert Collection, Volume III: Coastal forces Weapons
Seaforth Publishing, Barnsley 2020; hardback, 208 pages, illustrated with approximately 130 drawings and photographs in B&W; price £40.00.
ISBN 978-1-5267-7710-2

When John Lambert died in 2016, the world lost a superior naval historian and illustrator. Lambert's drawings of small military craft and naval armament can be found in many books on the subject, and his technical drawings of every detail of fixtures and fittings were unsurpassed. It is wonderful then to see so many of his drawings published posthumously, many for the first time, in this multiple-volume series by Seaforth.

The book is divided into two parts: an extended introduction by Norman Friedman, and Lambert's drawings. Perhaps oddly for a book about weapons, Friedman spends most of his 56-page introduction discussing the various types of boat commissioned into Coastal Forces, and only the final eight pages are dedicated to their armament. The overviews of each class do at least include their typical armament, and the text draws heavily on the Admiralty's procurement policy and build programmes. It is well illustrated, and Friedman refers to the specific drawing numbers where relevant, making it easy to match this first part of the book with the second. However, the overreliance on build programmes as a source leaves a gap in vessel service histories and a few errors have crept in. Friedman mischaracterises the CMB of the First World War as a concept that never saw action, when in fact the boats took part in numerous operations and sank at least one German destroyer, not forgetting their successes in the Baltic in 1919. The account of the evolution of the various classes likewise contains errors.

Lambert's drawings were often to different scales and sometimes on multiple sheets and there is no attempt to scale them here – a 60ft boat may appear the same length on one page as a 72ft boat on the next. This does mean that some of the larger drawings have had to be compressed in order to fit on a 29cm x 24cm page. However, all captions and notes are legible and each drawing has its own scale bar. Where a drawing straddles a double page, the drawing has been split into two to avoid some of it disappearing into the gutter. This may not please everyone, but is probably the best way to reproduce Lambert's work in a book.

In total there are 58 original drawings (some of multiple sheets) reproduced over 143 pages. The first nineteen pages are individual vessel drawings: some exterior views, some interior and some technical, showing framing and machinery. After this, the drawings show weapons: torpedoes and their firing equipment, guns and their mountings, depth charges, Holman projectors, mines, radar, cranes and small arms. Some of the drawings show little-known components, such as the power supply to gun turrets, or magazine feeds.

The book itself is a pleasingly weighty hardback, with a matt white dustcover and bold, bright title colours. It will undoubtedly become the standard reference for anyone with an interest in naval weapons or Coastal Forces.

Stephen Fisher

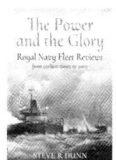

Steve R Dunn
The Power and the Glory: Royal Navy Fleet Reviews from the Earliest Times to 2005
Seaforth Publishing, Barnsley 2021; hardback, 320 pages, illustrated with numerous photographs, paintings and plans in colour and B&W; price £29.99.
ISBN 978-5267-6902-2.

Fleet Reviews have demonstrated national might, commemorated success, celebrated royal events, and in 2005 marked the bicentenary of what is, arguably, Britain's most famous naval battle. Although costly to conduct, and something of a distraction for an armed force that is normally out at sea in peace as well as war, they retain their value for any navy anxious to make a statement.

The author discusses the purpose and conduct of the Royal Navy's reviews in the context of the long history of Britain and the sea. The restored Stuarts used fleet reviews to bolster their European status, and to reaffirm the link between fleet and throne. George III made ample use of them to mark his 'Britishness', notably to celebrate the 'Glorious First of June' in 1794. His son the Prince Regent presided over a victory celebration in April 1814, consciously created by the Government to celebrate Britain's role in the defeat of Napoleon. The intention was to impress the Tsar of Russia, the King of Prussia and the Chief Ministers of Austria in order to reinforce British diplomacy in Europe; the strategy proved to be effective. The St George's Day Review of 1856 used the fleet assembled for an attack on St Petersburg to stress the central role of naval power in the defeat of Russia in the Crimean War. Once again foreign dignitaries were the main audience, and once again the impact more than met expectations.

In the late 19th century new traditions were created: 'Jubilee' Reviews to mark the 50th and 60th anniversaries of Queen Victoria's reign, then a 'Coronation' Review to mark the accession to the throne of each new monarch, starting with Edward VII. These reached a zenith in 1911, when the review coincided with the critical point of the Anglo-German arms race, displaying an immense modern fleet to the visiting German 'battle cruiser' *Von der Tann*, along with ships from Britain's ally Japan, and Entente partners France and Russia.

After 1918 the size of the Navy declined, as Britain entered into arms limitation treaties and lost the economic power to sustain a dominant fleet. Coverage of the last major event to date, the 2005 Trafalgar bicentenary, leads the author into a discussion of decline and sea-blindness, with the review as the litmus test of national direction and consequence. The Cold War locked Britain into a European defence pact, and the focus on land and land-based air forces created a strategic posture that long outlived the collapse of the Soviet Union and is only now being unravelled with naval deployments to the Pacific and final withdrawal from two long-term continental commitments: Germany and Afghanistan. Going forward the UK's national strategy will become more maritime, the combined action of all armed forces and Government agencies unified by a focus on the critical role of the sea in national life and security. Perhaps the next Fleet Review will mark that shift…

This book provides a comprehensive account of Royal Navy fleet reviews and boasts a particularly complete bibliography. The excellent illustrations are worthy of a chapter of their own. Highly recommended.

Andrew Lambert

David Hobbs
Taranto and Naval Warfare in the Mediterranean, 1940–1945

Seaforth Publishing, Barnsley 2020; hardback, 440 pages, numerous B&W photographs, five appendices, endnotes, bibliography, index; price £35.00.
ISBN: 978-1-9383-6

David Hobbs, captioning a fine photograph of three sub-lieutenants and two aircraft handlers posed against a backdrop of HMS *Indomitable's* island, writes: 'it is dedicated as a tribute to them all; those that returned and those that will never grow old'. Those words neatly sum up the essence of this quietly passionate account of the Fleet Air Arm's actions in the Mediterranean during the Second World War.

The emphasis in the book title on the night-time assault on the Italian Fleet at Taranto, ground-breaking though it was, is in fact misleading. The author covers the entire Mediterranean Campaign: in addition to Taranto, it takes in the action off Cape Matapan, the Malta convoys, the carrier 'club runs', the shore-based operations on land and at sea, and the amphibious assaults. They are all afforded an equal importance. It is indeed 'a tribute to them all'.

The victories, the defeats, the achievements and the draining attritional losses, both human and materiel, must all be viewed from the position from which the Fleet Air Arm (FAA) was working. In almost all respects, it was one of disadvantage. The aircraft were mostly slow, outdated and sometimes inadequately adapted for the role. The carriers from which they operated were often too small and old while the newer ones suffered from design faults which might have been rectified had there been greater financial investment in procurement during the 1930s. Hobbs also points the finger at the wartime Ministry of Aircraft Production for failing to recognise the FAA's critical need for new aircraft during 1940–41 capable of undertaking a wide variety of missions in the hostile environment of the 'narrow seas'. Moreover, during the early parts of the campaign particularly, there were simply too few of them.

Time and again it would seem the FAA was forced to make do and improvise with what it had during those desperate times when the Axis forces threatened to dominate the skies, use the seas with impunity and, on land, threaten to reach ever further eastwards through the Levant to the prized oilfields and beyond. Moreover, the air crews and their support had to 'learn on the job' not only to operate from the confined and yet highly demanding environment of an aircraft carrier but also, as the situation required, from improvised bases ashore.

David Hobbs is scrupulous in his approach, apportioning equal importance to each phase in this long conflict, detailing squadrons, flights, names of crew members and, importantly, individual losses of life. He is unsparing of reputations: Admirals Cunningham and Somerville among the 'top brass', for instance, come under scrutiny regarding their sometime lack of appreciation and understanding of the importance of naval air warfare in a rapidly changing world. By contrast others, such as Admiral Boyd, are praised for their far-sightedness. The author uses his intimate knowledge and professional expertise to great effect: his annotations of the great number of photographs frequently point to details that the ordinary reader might miss and which add a further, fascinating dimension.

Strategists have questioned the significance of the Mediterranean conflict in terms of the outcome of the Second World War. Be that as it may, what is certainly true is that the mantle of naval power projection moved irrevocably from the battleship to the aircraft carrier during the course of this campaign, and the achievement and cost of its transference is set out very ably in this authoritative and well-written account.

Jon Wise

Hans Lengerer & Lars Ahlberg
Capital Ships of the Japanese Navy 1868–1945, Vol III: Battleship *Tosa* Demolition Tests to the Modified *Yamato* Class

Despot Infinitus, Zagreb 2019; hardback, 494 pages, illustrated with 244 line drawings, 15 maps, and 141 photographs (21 in colour); price €89.90.
ISBN 978-953-8218-57-6

This is a notably expensive book, and one that does not appear to be easily available. There is something to be said for having read Volumes I and II before embarking on the present volume, as it is in many ways the culmination of the authors' research set out in the two earlier books and in numerous papers and articles. Volume III has a simple contents list setting out each of the seven parts and their contributory chapters or reports, twelve full chapters and three reports. Unfortunately there is no index, the book concluding with only a very select bibliography.

The principal criticism is that the text appears to have been translated from the authors' native languages, and while this is a considerable improvement on on-line translation, the meaning of some sentences is unclear.

There are also some repetitive misuses of language, such as compartments being 'immersed' when flooding is meant, while the illustration of the cruiser *Yoshino* on pages 416 and 417 has had the split halves of the photo reversed. These criticisms aside, the authors' research is impressive in its depth and spread. Although the cover claims the book to be an 'outline history' it is much more than that; there are, for example, detailed accounts of Admiral Hiraga Yuzuru's ideas on design.

Much of the information about the *Tosa* trials, in which an unfinished WWI-era battleship hull was subjected to live explosions to test the designed protection system, comes from original IJN documents, as do the design evolutions that arose from these tests, forming Part I of the book. The resultant design considerations for the 'super-battleships' of the *Yamato* class are presented in Part II with extensive tables and plans, many of the latter in Japanese or with German translations. The ships' operational employment, loss and the subsequent report on the sinking of *Musashi* form Part III. (*Yamato*'s loss was apparently too late in the war to justify making resources available for an evaluation.) Part IV is dedicated to the planned follow-on to the *Yamato* class.

There then follows a section of colour photographs of the large-scale model of *Yamato* at Kure Maritime Museum (Part V), while Part VI covers the propellants used by the IJN and their design and manufacture, rangefinders, optical equipment in general, and rocket launchers. Part VII is an addendum to Volume I.

This is an immensely immersive volume from two widely-published acknowledged experts, and merits careful reading.

W B Davies

John Henshaw
V & W Destroyers:
A Developmental History
Seaforth Publishing, Barnsley 2020;
159 pages, 89 photographs, 48 drawings; price £25.00.
ISBN 978-1-5267-7482-8

This is a rather odd book. While the V&W-class destroyers of the late First World War do indeed make up the largest single part, the first 54 pages are taken up with a fairly detailed account (including drawings) of British destroyers since HMS *Havoc* of 1895. The author justifies this, quite reasonably, as providing background to the V&Ws themselves, but there is little real attempt to make the links that would make such an approach meaningful. There is also a lack of any significant detail of German destroyer development that contributed to the genesis of the V&Ws: all we have is a mention of 'reports of German advances [that] produced the usual reactive response'.

It also seems clear that the author has not consulted any of the key primary sources, most importantly the relevant Ships' Covers. Rather, he simply quotes previous authors who *have* done such work, using the format 'according to [author]' in preference to footnote references. Where sources disagree, Henshaw simply records the conflicting claims without attempting to resolve them. The writing is often discursive, with many first-person asides (not all apposite), and frequent use of language that might be appropriate to a public lecture but not a book that purports to be a serious study.

On the other hand, the book contains a large number of excellent drawings, which provide a good overview of the differences between the various batches of the class and also detail changes in appearance over time. These include some interesting depictions of one of their 'cousins', the *Scott*-class flotilla leader *Stuart*, including the first drawing the reviewer has seen of her configuration as a destroyer transport. There are also sketches of some of the unrealised projects for Second World War conversions of V&Ws, although it is not wholly clear upon what these are based. Drawings of individual fittings employed in the class, in particular weapons and sensors, accompany descriptions of their use in service, and there are sketches of a selection of Second World War camouflage schemes.

The author's extensive 'prehistory' section means that we also have a good range of drawings of earlier British destroyers (plus one German – actually *S115*, although not captioned as such), as well as the V&Ws' 'descendants' of the British A–I classes. However, when considering the wider influence of the V&Ws the author displays a poor understanding of foreign destroyer developments; for example, he seems unaware of the Italian design origins of the Soviet Project 7 and Project 7U vessels.

Photographic coverage is extensive, and covers most aspects of the history of the V&W class. However, many images have been taken from on-line sources (with inadequate credits) and some have been printed far larger than justified by their low resolution. Accordingly, while the book has value for its drawings and its handy digest of the vital statistics of the class, the text has very significant flaws that detract from the volume's overall value.

Aidan Dodson

Jim Crossley
Churchill's Admiral in Two World Wars:
Admiral of the Fleet Lord Keyes of Zeebrugge & Dover
Pen and Sword Maritime, Barnsley 2020; 197 pages,
7 B&W illustrations, 5 maps; price £25.00/$49.95.
ISBN 978-1-52674-839-3

Cecil Aspinall-Oglander's 1951 biography of Keyes is long out of print, and this book appears to be aimed at a new generation that has little knowledge of this fascinating individual. It is written in an easily readable style and moves from 19th century anti-piracy patrols through the Boxer Rebellion to the development of submarines, Gallipoli, the Dover Patrol, the Mediterranean Fleet and the birth of Combined Operations.

After his brief description of the Gallipoli campaign, Crossley observes that 'Keyes, Churchill and their allies seem to have lost that grip on hard reality which is essential to sound strategic thinking'. This is a valid appraisal, but further analysis of Keyes' own role would have added greater depth. Despite this, Crossley does succeed in stimulating interest in the admiral that will hopefully encourage many readers to study his strengths and weaknesses more deeply. He balances descriptions of Keyes' love of action with comments about the impracticability of some of his theories. An outstanding junior officer during the Boxer Rebellion, was he a good submarine force commander in the North Sea or chief of staff in the Gallipoli Campaign? There is much to ponder in these topics and in his relationship with the Belgian Royal Family in two World Wars.

Crossley explains that many details of the attack on Zeebrugge were inherited from Admiral Bacon, and questions whether Keyes' frequent presence in small ships in dangerous waters off a hostile coast was the best way of exercising command. His role in establishing how the Fleet Air Arm was to be administered in discussions with his brother-in-law, Lord Trenchard, during 1924 is mentioned but hardly given the degree of analysis such an important topic merits. Keyes never ceased to yearn for a major command in the Second World War, thinking himself to be expert on naval air and amphibious matters. He was not, and the degree to which he was out of touch was revealed in his visit to the US Navy in the Pacific during 1944, although he was received kindly. Crossley concludes with some justice that 'it is impossible not to regard his career in the Second World war as pathetic' and believes that his time as an MP was 'woeful', meriting no serious examination.

Minor errors include references to submarine 'telescopes' rather than periscopes and a photo caption that refers to its subject as 'Keynes'. The five maps at the beginning of the book are very basic, and lack both orientation and any indication of scale; they make no real contribution to the subsequent text. Overall, however, this book is a good read and is recommended as an introduction to an admiral who saw remarkable and unprecedented changes during his long career.

David Hobbs

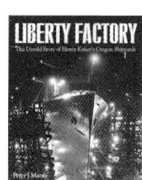

Peter J Marsh
Liberty Factory: The Untold Story of Henry Kaiser's Oregon Shipyards
Seaforth Publishing, Barnsley 2021; hardback, 256 pages, illustrated with 3 maps and 219 B&W photographs; price £35.00.
ISBN 978-1-5267-8305-9

Much has been written on the mass-produced 'Liberty' ships of the Second World War, with Peter Elphick's 500 page *Liberty* particularly prominent. Peter Marsh's new book has a narrower focus: Kaiser's Oregon yards. The only other book that covers some of this is Fred West's *Kaiser Carriers*, currently unavailable other than in an almost unreadable electronic version.

Liberty Factory is set out in three parts totalling twenty chapters, bookended by an introduction and a tailpiece, plus lists of sources and suggested further reading. Part 1 deals with the historical background, the setting up of emergency shipyards, Henry Kaiser's involvement, and the origins and engineering of the ships and their building. Included are not just the 'Liberty' ships, but the later 'Victory' class, the Kaiser-built escort carriers and various smaller craft.

Part 2 deals specifically with the Kaiser yards themselves and their performance, and includes a lot of social detail, while Part 3 covers the other yards and war industries in the area, many of which supported Kaiser's yards. These include aspects of the human side of the story with a few more links to the RN and UK side of things.

While much of Part 1 may be familiar, it is the depth of coverage on the social side that is really new, in particular the notes from the late Lawrence Barber, then marine editor at *The Oregonian* newspaper, who tried to cover every aspect of the yards and their people. Initially this reflects a natural conservatism, with the city of Portland being felt to be the most inhospitable place in the US at the time; more detailed human interest stories record how it later blossomed with social changes that came to the people of the area as the population grew in numbers and diversified. Also detailed at some length in Part 2 is Henry Kaiser's own social engineering, providing free medical care, schools, housing and *crèches* for the children of the many mothers who came to work at the yards.

If you feel that there was a lot unsaid in previous books on the 'Liberty' ships, then this is the book for you, highly recommended.

W B Davies

Hans Joachim Koerver
The Kaiser's U-boat Assault on America: Germany's Great War Gamble in the First World War
Pen & Sword Military, Barnsley 2020; 355 pages, 96 images; price £25.00/$34.95
ISBN 978-1-5267-7386-9

The title of this book is misleading: rather than an account of the German submarine campaign in US waters during April–November 1918, which the reader might have expected, it is actually a study of US–German relations from 1914 to 1917 through the lens of submarine warfare. Nevertheless, it tells this story well, and highlights the chaotic way in which the First World War was managed in Germany.

Koerver argues that 'semi-autonomous and competing institutions and single-minded interest groups paralysed each other in their attempts to pursue selfish interests and opposing strategies'. As a result, the German Chancellor's efforts to maintain good relations with the USA by restricting submarine operations to 'cruiser warfare' – where ships were stopped and inspected before a decision was made to sink them, with the safety of crews guaranteed – were frequently frustrated. Submarines 'accidentally' torpedoed ships (both enemy and neutral) without warning or were simply kept in port, rather than carrying out the 'cruiser warfare' that key members of the naval hierarchy despised. For the latter, an 'unrestricted' campaign was the only way ahead, despite the risk of provoking the USA into declaring war on Germany.

The levels of duplicity revealed of Tirpitz himself and others, including Scheer, is remarkable, and their actions progressively undermined Germany's standing in Washington. Thus, as time went by, German actions and failures to implement agreements on sparing US ships (not to mention the Zimmermann Telegram inciting Mexico to attack the USA as Germany's ally) pushed the anti-war US President Wilson progressively into such a corner that the sinking of the American SS *Vigilancia* on 16 March 1917 left him with no option other than to ask Congress to declare war.

The story is told through a wide range of published and unpublished contemporary documents (including the diaries of protagonists) and data extracted by the author. The latter underline that there was actually little military advantage in moving from submarine 'cruiser warfare' to the 'unrestricted' variety. Indeed, it is argued that had 'cruiser warfare' been maintained only a few months longer, not only would shipping losses have pushed the UK just as close to the tipping point as did the 'unrestricted' campaign, but that the effect would also have been reinforced by the exhaustion of the UK's foreign currency and gold reserves. Faced with such a situation, the UK might well have been forced to sue for peace.

The text is supported by a variety of images, in particular maps, graphs and tables. Of the photographs, one captioned 'Allied Warship with "Dazzle" camouflage' (p 187) is actually the French cruiser *Gloire*, pictured in 1944. There are also a number of typographical and factual errors: Winston Churchill was First Lord of the Admiralty, not First *Sea* Lord, while Tirpitz was State (not 'Assistant') Secretary of the Reich Navy Office. The English is not always idiomatic, but the book is recommended as an important contribution to the political background of the naval war 1914–18.

Aidan Dodson

Aaron S Hamilton
Total Undersea War: The Evolutionary Role of the Snorkel in Dönitz's U-Boat Fleet, 1944–1945
Seaforth Publishing, Barnsley 2020; hardback, 416 pages, 44 photographs, nine drawings and six charts, five appendices, bibliography and index; price £35.
ISBN 978-1-5267-7880-2

Following an initial summary of the U-boat war prior to the introduction of the snorkel, Aaron Hamilton provides an informative and well-illustrated account of the development of the two main types of folding snorkel: the Type I with ball float and a troublesome flanged joint connecting the snorkel inlet tube to the boat's interior; and the Type II with ring float (a hollow sliding cylinder). Both were intended to be installed in existing boats by dockyards and bases. It was unfortunate that the less reliable Type I snorkels were fitted to the long-range Type IX U-boats, which were exposed to the rigours of long patrols to North America and were responsible for a number of fatal failures. The more numerous Type VIIs were given both Types I and II. While snorkelling with the diesels running, crews could be subjected to large pressure drops when the float valves submerged, while poisonous carbon monoxide was more likely to be released into the boat's interior, as was explosive oxyhydrogen from venting batteries. A snorkel had a low radar return but the Germans developed two coatings, one of which used rubber loaded with iron powder (see *Warship 2021* Gallery) to reduce returns further. A rubber coating for the whole boat, codename 'Alberich', also reduced sonar reflections, but its application was very laborious and only thirteen boats were fitted.

At the end of the war, Germany was building the high-speed, Walter-inspired Types XXI and XXIII Electroboats. However, they were designed before it was realised that even they would need snorkels to survive. The decision to adopt extensible snorkels led to vibration problems that were a major cause of delays, especially to the Type XXI which made only one inconclusive operational patrol at the very end of the war. If any of the Walter-turbine-powered Type XXVIs had been completed, they would have had folding snorkels.

Hamilton's second part addresses the operations and the evolution of tactics that followed the introduction of the snorkel. After its first ineffective use against Operation 'Neptune' and the withdrawal to Norwegian bases, the snorkel-equipped U-boats conducted two inshore campaigns: by Type VIIs around the British Isles; and by Type IXs on the east coasts of Canada and the US. Based on his review of all the surviving *Kriegstagbücher*, Hamilton describes a selection of patrols in British waters and every patrol to North America. From the text and appendices C and E, it appears that the latter accounted for only about one sixth of the total Allied losses to U-boats with snorkels, but it is regrettable that no overall analysis is provided of the total losses suffered by the opposing sides in these campaigns.

Admiral Dönitz coined the phrase 'Total Undersea War' for this final phase of the U-boat war, mainly, it seems, to encourage his crews and retain Hitler's support. The reality was that the snorkel created a stalemate. The U-boats were less vulnerable while snorkelling to their inshore patrol areas: and, once there, they were able to spend most of their time on the bottom, where they were hard to detect. But this so limited their mobility that they seldom found targets. They made hardly any wireless transmissions by which they could be located or from which Ultra intelligence could be obtained, but U-boat command lost the means to coordinate their activities. This reality is not emphasised by Hamilton, though he does acknowledge that the snorkel boats had little impact on the outcome of the war. Despite some omissions, his book is a valuable reference, notably to the technology and the tactical employment of the snorkel.

John Brooks

WARSHIP GALLERY

The scrapping of HMS *Agincourt*, *New Zealand* and *Princess Royal* at Rosyth, 1923–25

Aidan Dodson provides a commentary on a selection of photographs documenting the breaking-up of three iconic British warships.

The photographs published here come from an album that belonged to Wallace Cowan, Chairman of the short-lived Rosyth Shipbreaking Company, which broke up the battleship *Agincourt* (1913) and the battlecruisers *Princess Royal* and *New Zealand* (both 1911) between 1923 and 1925. The album was presented – presumably because of its inclusion of images of the last-named ship – to the Royal New Zealand Navy by Cowan's daughter in 1968, and is now held by Archives New Zealand, which the author wishes to thank for permission to publish.

The three ships were originally sold to the Exeter-based (but Edinburgh-originating) firm of electrical engineers J&W Purves on 22 January 1923: *New Zealand* for £21,000 and the other two for £25,000 each. The contract was, however, immediately transferred to the newly-established Rosyth Shipbreaking. As the sales were explicitly to meet the UK's obligations under the Washington Naval Treaty of the previous year, the contracts for demolition required work to be completed by '18 months from date of ratification of [the Washington] Treaty' (*ie* 17 January 1925).

The ships were taken over 'as lying' against the south wall of the main basin at Rosyth Dockyard on 28 January. Work began with *New Zealand*, and once she was well-advanced the demolition of *Agincourt* was begun, dismantling continuing until she was shorn of all superstructure except her disarmed turrets. Attention then shifted to *Princess Royal*, so that by March 1923 *New Zealand* was ready for transfer to the beaching ground outside the northwest dockyard wall; the other two had now been stripped to upper deck level, other than the remains of their heavy gun mountings. As soon as *New Zealand* had been moved out of the basin and around to the beaching ground, her place against the dockyard wall was taken by *Princess Royal*, whose stripping was then given priority, with the result that once *New Zealand* had been reduced to the last few scraps, *Princess Royal* was able to take her place on the beaching ground at the end of September 1924.

However, as the January 1925 deadline approached, *Agincourt*, although now well cut down, was judged not yet sufficiently mutilated to meet contractual commitments.[1] Discussions with the Admiralty concluded with an agreement that cutting the hull in two on the beaching ground would discharge the company's obligations, but bad weather prevented the hulk from passing through the

Princess Royal, *Agincourt* and *New Zealand* berthed along the south wall of the Main Basin at Rosyth Dockyard in late 1923. They had been handed over to the shipbreakers on 25 January that year. The battlecruisers had been in the Rosyth Reserve, while *Agincourt* had been beginning conversion to a mobile fleet base. *New Zealand*'s demolition is well advanced, with *Agincourt* shorn of all superstructure apart from barbettes and turrets, but *Princess Royal* appears superficially intact. (All photographs courtesy Archives New Zealand)

WARSHIP 2022

View from the top of the 30-ton crane (the base of which is visible in the previous image), showing further detail of the state of the three ships in late 1923.

Dockyard locks on the only day prior to the deadline that would allow its beaching on the highest Spring Tide. As the next such tide was after the deadline, it was ultimately agreed that the hulk would be taken into one of the Dockyard's dry docks and cut in half there. This proved successful and within 48 hours of entering the dock both halves had been floated out, to be taken to the beaching ground for final breaking up on the next convenient tide.

Once *Agincourt* was finally cut up, Rosyth Shipbreaking's only further purchase was the ex-German floating dock *Kiel VII*, which arrived in the Forth on 23 September 1926. In November, however, it was one of the company's remaining assets sold to the Alloa Shipbreaking Co for £41,000. Rosyth went into liquidation the following month, Cowan joining the Alloa board. Alloa Shipbreaking later became Metal Industries, one of the mainstays of British shipbreaking down to 1980.[2]

Notes
[1] The Washington Treaty required that breaking up 'shall always involve the destruction or removal of all machinery, boilers and armour, and all deck, side and bottom plating'.
[2] See I Buxton, *Metal Industries: Shipbreaking at Rosyth and Charlestown*, World Ship Society (Kendal, 1992); *Shipbreaking at Faslane*, World Ship Society (Windsor, 2020).

Although late in starting, work on stripping *Princess Royal* was, nevertheless, well advanced by 21 Mar 1924, with funnels and turrets all gone, and just the breech ends of the 13.5in guns left in their cradles. *New Zealand* had been towed away earlier that day, and the battlecruiser now occupies her former berth, where dismantling could be expedited. The ships laid up ahead of *Princess Royal* are 1916-launched Admiralty 'Leaf'-class oilers, with *Pearleaf* on the right; the ship directly inboard of her is probably *Appleleaf*.

Agincourt, viewed from the bridgework of *Princess Royal* in late 1923. All funnels and superstructures have been removed from this part of the ship, as have all guns except for the severed pair of 12in weapons in the superimposed turret aft. The vessel at top left is the storeship *Impérieuse*, originally the ironclad *Audacious* (1869), which would be broken up at Inverkeithing in 1927.

View from the top of the crane down into *New Zealand*'s after boiler room and engine room, showing exposed boiler tubes and turbine blades.

WARSHIP GALLERY

The guns of 'Y' turret of *Princess Royal* are cut into slices for removal; the wire-wound construction of the guns is clearly visible. The battlecruiser *Tiger* is visible under refit in the background, adjacent to the Dockyard's 100-ton crane; scaffolding is visible around her derrick mast, which is in the process of being heightened and a starfish installed.

Agincourt's hull is cut through between the former locations of her midships turrets. Note the gas cylinders bottom left for the oxy-acetylene cutting-torches.

WARSHIP 2022

Left: *Agincourt*, well cut down but hitherto still insufficiently mutilated to meet the terms of the Washington Treaty, lies in one of the three Rosyth dry docks after being cut in two.
Below: *Agincourt*'s bow section arrives at the beaching ground.

The after part of *Agincourt* after being floated out of dock. The ship visible on the left is the light cruiser *Centaur*, under refit.